Integrating Advancements in Education, and Society for Achieving Sustainability: Research and Evidence-Based Strategies from the Developing world

About the Conference

This book is the collection of selected articles that appeared at the First International Analytics Conference 2023 held in Hyderabad in virtual mode on February 2nd the 3rd 2023. In an era defined by the twin imperatives of knowledge and sustainability, this compelling volume explores the powerful synergy between advancements in education and the transformation of society towards a more sustainable future. Delve into the heart of progress as educators, innovators, and change-makers come together to catalyze positive change.

Within these pages, you'll witness the transformative potential of education as it equips individuals with the skills, knowledge, and perspectives necessary to address the multifaceted challenges of sustainability. Discover groundbreaking pedagogical approaches, innovative technologies, and visionary programs that are revolutionizing learning and inspiring the next generation of leaders.

Simultaneously, the book sheds light on the interconnected relationship between education and society, illustrating how an informed and engaged citizenry is driving sustainable practices, community resilience, and global change. From eco-conscious urban planning to social entrepreneurship, it showcases the societal initiatives that emerge when education and sustainability combine forces.

'Integrating Advancements in Education and Society for Achieving Sustainability' is a compass for those navigating the path towards a brighter, more sustainable future. Join us in this thought-provoking exploration of how education and society are jointly shaping the way we think, act, and work together to achieve a harmonious balance with the world around us.

Integrating Advancements in Education, and Society for Achieving Sustainability: Research and Evidence-Based Strategies from the Developing world

Edited by

Dimitrios A. Karras

Srinesh Thakur

Sai Kiran Oruganti

Routledge
Taylor & Francis Group
New York London

First published 2024
by Routledge
4 Park Square, Milton Park, Abingdon, Oxon OX14 4RN

and by Routledge
605 Third Avenue, New York, NY 10158

Routledge is an imprint of the Taylor & Francis Group, an informa business

British Library Cataloguing-in-Publication Data
A catalogue record for this book is available from the British Library

Library of Congress Cataloging-in-Publication Data
A catalog record has been requested for this book

ISBN: 978-1-032-70841-6 (pbk)
ISBN: 978-1-032-70846-1 (ebk)

DOI: 10.4324/9781032708461

Typeset in Times LT Std
by Aditiinfosystems
Printed and bound in India

Contents

List of Figures

Integrating Advancements in Education, and Society for Achieving
Sustainability – Dimitrios A. Karras et al. (eds)
© 2024 Taylor & Francis Group, London, ISBN 978-1-032-70841-6

List of Tables

Integrating Advancements in Education, and Society for Achieving
Sustainability – Dimitrios A. Karras et al. (eds)
© *2024 Taylor & Francis Group, London, ISBN 978-1-032-70841-6*

Editor's Biography

Proceedings of the International Analytics Conference (IAC 2023)

Edited by

Dimitrios A. Karras received his Diploma and M.Sc. Degree in Electrical and Electronic Engineering from the National Technical University of Athens (NTUA), Greece in 1985 and the Ph. Degree in Electrical Engineering, from the NTUA, Greece in 1995, with honours. From 1990 and up to 2004 he collaborated as visiting professor and researcher with several universities and research institutes in Greece. Since 2004 he has been with the Sterea Hellas Institute of Technology, Automation Dept., Greece as associate professor in Digital Systems and Signal Processing, till 12/2018, as well as with the Hellenic Open University, Dept. Informatics as a visiting professor in Communication Systems (the latter since 2002 and up to 2010). Since 1/2019 is Associate Prof. in Digital Systems and Intelligent Systems, Signal Processing , in National & Kapodistrian University of Athens, Greece, School of Science, Dept. General. He is, also, adjunct professor with GLA University. Mathura, India and BIHER, BHARATH univ. India as well as with EPOKA and CIT universities Tirana. Moreover, he is with AICO EDV-Beratung GmbH as senior researcher as well as Director of Research and Documentation at ADIafrica N. G. O. He has published more than 80 research refereed journal papers in various areas of pattern recognition, image/signal processing and neural networks as well as in bioinformatics and more than 185 research papers in International refereed scientific Conferences. His research interests span the fields of pattern recognition and neural networks, image and signal processing, image and signal systems, biomedical systems, communications, networking and security. He has served as program committee member in many international conferences, as well as program chair and general chair in several international workshops and conferences in the fields of signal, image, communication and automation systems. He is, also, former editor in chief (2008-2016) of the International Journal in Signal and Imaging Systems Engineering (IJSISE), academic editor in the TWSJ, ISRN Communications and the Applied Mathematics Hindawi journals as well as associate editor in various scientific journals. He has been cited in more than 2560 research papers, his H/G-indices are 20/52 (Google Scholar) and his Erdos number is 5. His RG score is 32.78.

Srinesh Singh Thakur is a distinguished hardware engineer with a profound expertise in FPGA and embedded systems. With a notable tenure at the Laser Research Institute of the esteemed IIS Fraunhofer in Germany, Srinesh contributed significantly to cutting-edge research in the field. Holding a Masters in Electrical Engineering from Darmstadt University of Applied Sciences, Germany, he honed his skills in the heart of technological innovation. Currently, Srinesh wears multiple hats, demonstrating his entrepreneurial prowess. He stands as the founder of Anvita Electronics Pvt. Ltd., a company at the forefront of technological advancements. Likewise, as the founder of Atya Technologies Pvt. Ltd., he continues to push boundaries in the ever-evolving tech landscape. Additionally, Srinesh serves as a co-founder of the SPAST Foundation in Hyderabad, India, demonstrating his commitment to social causes and community development. With a rich tapestry of experience and a visionary approach, Srinesh Singh Thakur is a driving force in the realms of technology and philanthropy, leaving an indelible mark on both the industry and society at large.

Sai Kiran Oruganti, a seasoned academic and researcher, held the position of Assistant Professor at the Indian Institute of Technology Tirupati from 2016 to 2017. His Ph.D. work made significant strides in the domain of Zenneck Wave Power transmission across shielded metal zones, resulting in the filing of 16 patents. Following his tenure in India, Dr. Oruganti played a key role in establishing ZN Ocean Technologies in South Korea. Between 2019 and 2022, he continued his academic journey as a Full Professor at Jiangxi University of Science and Technology in China, under the Foreign Talents Program, where he continued to specialize in Zenneck Wave type Wireless Power Transfer. He also co-founded the SPAST Foundation and currently holds the position of CEO at the Technology Innovation Hub of the Indian Institute of Technology Patna.

*Integrating Advancements in Education, and Society for Achieving
Sustainability – Dimitrios A. Karras et al. (eds)
© 2024 Taylor & Francis Group, London, ISBN 978-1-032-70841-6*

Foreword

In a world that continues to undergo profound changes and challenges, the synergy between education and societal progress stands as a beacon of hope and possibility. The volume you hold in your hands, "Integrating Advancements in Education and Society for Achieving Sustainability," is a testament to the transformative power of knowledge and innovation when harnessed for a sustainable future. As we embark on this journey, we explore the pivotal relationship between education and the sustainability of our societies. In these pages, you'll find a wealth of insights, solutions, and inspirations that underscore the indomitable spirit of human endeavor. Education is not merely a process but a profound force, nurturing individuals who are not only equipped with knowledge but also empowered with the drive and vision to shape a more sustainable world.

The chapters within this book illuminate the innovative pedagogical approaches, the imaginative use of technology, and the pioneering programs that are transforming education. They demonstrate how education is evolving to meet the dynamic challenges of our times and, in turn, nurturing the seeds of sustainability among learners of all ages. Education, as we will see, is not confined to the classroom but permeates every facet of society. This book celebrates the interplay between education and society, highlighting the initiatives and movements that have emerged as a result. From eco-conscious urban planning to community-driven sustainability projects, we witness how society is adapting and evolving in response to the principles of sustainability.

The stories shared within these pages are not just accounts of progress; they are narratives of change, of vision, of determination. They are stories of educators and students, of community leaders and innovators, all contributing their part in building a sustainable future. These stories exemplify the notion that a brighter and more harmonious world is possible when we synergize the power of education and societal progress. As you read through these chapters, I encourage you to contemplate the profound impact that education can have on society's trajectory. Education has the potential to nurture not only the leaders of tomorrow but also the changemakers of today. It is a transformative force that holds the key to a more sustainable, equitable, and hopeful world.

So, as you embark on this enlightening journey into the integration of education and societal advancement, remember that the challenges we face are opportunities for change, and that each one of us has a role to play in the pursuit of sustainability. I am excited to be your guide on this voyage, a journey into the transformative power of knowledge and innovation, and I invite you to explore the remarkable stories of educators, learners, and society at large who are actively contributing to a more sustainable world. Welcome to a world where education and society collaborate, where knowledge is a catalyst for change, and where the pursuit of sustainability is not just a noble aspiration but a shared endeavor. Enjoy the journey.

Welcome to the future. Enjoy the journey.

Srinesh Singh Thakur

Preface

In our ever-evolving global journey, two foundational forces—education and societal progress—have consistently shaped the trajectory of our world. The volume you hold before you, "Integrating Advancements in Education and Society for Achieving Sustainability," is an exploration of the profound synergy between these two forces, and their collective influence on sustainability. This book serves as a comprehensive exploration of the dynamic interplay between education and societal progress in the context of sustainability. Education is not merely a conduit of knowledge; it is a powerful instrument for change, equipping individuals with the understanding, skills, and inspiration to tackle the multifaceted challenges of sustainability.

Within these pages, you will discover the ever-adapting landscape of education, where innovative pedagogical methods, technological advancements, and visionary programs are reshaping learning. These transformative approaches ensure that learners are not only informed but also equipped with the critical skills and perspectives necessary for a sustainable future. Simultaneously, the book underscores the intricate relationship between education and society. It is within the context of society that education finds its purpose and relevance. You will encounter stories of societal initiatives and movements that have been sparked by an informed, educated citizenry. From grassroots community projects to large-scale environmental endeavors, it becomes evident how education serves as a catalyst for societal progress and sustainability.

The stories presented in these pages are not merely accounts of achievements; they are tales of inspiration, aspiration, and transformation. They emphasize that we are not passive bystanders in the face of change but active contributors to the creation of a sustainable world. These chapters symbolize the hope, determination, and relentless pursuit of a brighter, more harmonious world that is well within our reach. As you navigate through this volume, we encourage you to reflect on the profound implications of the integration of education and societal advancement. These stories will not only showcase remarkable achievements but will also inspire you to recognize your own potential as an agent of change and an advocate for sustainability.

This book is dedicated to those who envision a sustainable future and who understand that education is a potent catalyst for change. It is a tribute to educators, students, visionaries, and innovators who are leading the way toward a world where knowledge is not a resource, but a dynamic force for progress. Welcome to a world where education and societal advancement are seamlessly intertwined, where knowledge is the spark that illuminates the path to sustainability. We invite you to join us on this journey into a future where hope and possibility converge to create a brighter, more sustainable world.

Dimitrios A Karras, Sai Kiran Oruganti

Integrating Advancements in Education, and Society for Achieving
Sustainability – Dimitrios A. Karras et al. (eds)
© 2024 Taylor & Francis Group, London, ISBN 978-1-032-70841-6

Acknowledgements

Chairs

Nita Sukdeo
University of Johannesburg, South Africa

Dilrabo Bakhronova
Uzbek State University of World Languages, Tashkent, Uzbekistan

Organizing & Publishing Commitee

Dimitrios A Karras, PhD
University of Athens (NKUA), Greece

Sudenshna Ray,PhD
RNTU, AISECT University Bhopal, India

*Integrating Advancements in Education, and Society for Achieving
Sustainability – Dimitrios A. Karras et al. (eds)*
© 2024 Taylor & Francis Group, London, ISBN 978-1-032-70841-6

1

Adopting Sustainability in Resolving Legal Issues in E-commerce

N. Garg*
Department of Law, Bharati Vidyapeeth (Deemed to be University)
Institute of Management and Research, New Delhi, India

B. Sharma[1]
Associate Professor, Birla School of Law,
Birla Global University, Bhubaneswar, Odisha, India

G. Gupta[2]
Department of Law, Bharati Vidyapeeth (Deemed to be University)
Institute of Management and Research, New Delhi, India

Abstract: Evolution of internet has led to remarkable changes in the online business all over the world. E-Commerce transactions are not only confined to the consumers and retailers but they also comprise of economic, social, and environmental aspects pertaining to overall development in e-commerce in a sustainable manner. E-commerce provides for sustainability by protecting the environment and natural resources which is considered indispensable part of e-commerce. A new regime is being established pertaining to the consumption of goods and services as economy is becoming more inclined towards the sustainable consumption of the resources. The Global E-Commerce suffers from various challenges and opportunities and there need to be a policy redrafting promoting sustainable development, integrating e-Commerce, strategy inclusion of international organizations to embark upon the benefit of e-commerce and it is strengthening particularly in the developing countries.

Keywords: E-Commerce, Sustainable development, Environment concerns, Legal issues, Indian legislations

If your business is not on the internet then your business will be out of business.

-Bill Gates

1. Introduction

E-Commerce denotes a form of business whereby the buying and selling of goods

*Corresponding author: neha.shrutika.jain1@gmail
[1]g09gupta@gmail.com, [2]sharmabhavana44@gmail.com

DOI: 10.4324/9781032708461-1

and services takes place in an online medium without using the traditional paper documentation or monetary system. E-Commerce is a technological advancement that involves the secured market system involving instant payments verification with authenticated payment systems. It provides for a global reach without stepping out from the home. E-Commerce transactions are associated with robust legal frameworks and the business flourishes under the umbrella of restrictive rules and regulations providing for a harmonized environment.

But these regulations vary from state to state and due to the expansive approach of E-Commerce difficulty arises pertaining to certain laws such as tax, sale and purchase invoices, procedural and substantive regulations. E-commerce provides for convenient shopping with a variety of products, brands and services involving shopping through debit cards, credit cards, net banking, UPI and even cash on delivery. (Svetlana Revinova. (2021) SHS Web Conf. Vol. 114.) The success of e-commerce involves efficient mail delivery, infrastructure pertaining to telecommunications, secured payment systems, a conducive cross border online business environment and receptive legislative enactments as per the telecom developments.

2. Legal Issues in E-Commerce

Jurisdiction issues in e-commerce

The Cyberspace is not confined to a particular territory as it does not have a specific boundary. The issues raised are due to lack of harmonisation laws pertaining to Cyber space. The websites operating in a particular territory can be active websites or fully interactive websites that work for commercial gains in a particular territory; in such cases the jurisdiction shall lie of the court where the website is interacting with the customers for commercial gain. There are fully passive websites also where the information circulated is just for viewing by the prospective customers and there is no commercial gain out of it. The principal applied over here is minimum contract theory and effects principle whereby if a particular website affects the population, then it would have the jurisdiction.

Issues pertaining to Intellectual Property Rights

IPR laws lay strong foundation in physical legal enactment but they face diverse issues in online medium. The use or deceptive similarity of domain names is the major issue but there is no provision specifically in IT act pertaining to domain names and this concern is the subject matter of trademarks that provides for infringement of identical, deceptively similar or well-known marks. The digital copyright infringement involves the transfer of copyright protected data like research papers, website design, pirated videos, and MP3 for free download without the consent or authorization of the owners. Online copyright issues involve deep linking, framing, piracy of software's, circulation of unauthorized videos or audio files or circulation of movies even before their release. The issue of ownership of website's design is also a major concern particularly when the companies outsource such services from third party.

Security or data protection in e-commerce

As the E-commerce transactions have become voluminous, the data breach has emerged as a major concern. (ASEAN Framework Agreement. (2015) The concerns of identity theft and data impersonation are increasing in case of E-commerce, as the buyers and the sellers cannot see each and there is only existence of an online contract between them. The personal information, preferences and financial information of a person is the pre requisite for any online transaction and such information is misused on online platform by extracting credit card, debit card details, net banking, and bank account details via phishing, smishing or unsolicited phone calls.

Regulation of Content

The content on the e-commerce pertaining to the exchange or distribution of any information that belongs to the third party requires a proper regulation of content through legislative framework. There are certain contents that may raise objections in different legislations or may be considered as obscene. In determining whether the content depicted on an e-commerce website is lascivious or appeals to the prurient interest, the court (General of Doordarshan & Ors. vs Anand Patwardhan & Anr.) would take into consideration factors such as - (a) whether the work taken as a whole appeal to the prurient interest; (b) whether the work is patently offensive; (c) whether the work taken as a whole, lacks serious literary, artistic, political, or scientific value.

Taxation of E-Commerce

It consists of taxation pertaining to the income earned by NRI by sale of goods, services, database, income generated by the business and the other issue pertain to the establishment of permanent websites, software services and supply of technical services. The policies for imposing tax on e-commerce does not confer with the international standard policies and as a result of which there is significant increase in litigation pertaining to the withholding of taxes as per the income of the tax payer. Certain issues pertaining to the income derived, permanent establishments, points of sale and purchase have been left unaddressed or are not dealt adequately. Sometimes it is very difficult to determine the location of the buyer or the seller and this results in loss of tax revenue.

3. Legal Framework of E-Commerce in India

(a) Information Technology Act, 2000- The IT Act, 2000 is the legislation that has provided authenticity to E-Commerce transactions and Legal sanctity of e-contracts. Section 84A promotes e-commerce and e-governance and protection of data by making use of secure electronic means. Section 79 provides for protection of the intermediaries if they have no control on user or cannot modify or control the origination of data.

(b) Consumer protection Act, 2019-The Consumer Protection E-Commerce Rules (2020) provide for sale, purchase, marketing, and services in online medium. These rules apply to the E-commerce entities and not on individual activities in which a person is engaged in personal capacity.

Foreign Exchange Management Rules, 2019-It defines the E-Commerce entities as the Companies that are incorporated as per the Companies Act, 2013.

Legal Metrology Act, 2009-Also includes the e-commerce entities as Companies incorporated under the Companies Act, 2013 and it must comply with all the legal Metrology rules pertaining to the packaging of products and it lays down that the

(c) Indian Contract Act, 1872-Section 10A of the Act provides for the validity of E-Contract. It provides legal recognition to the offer, acceptance, communication, and revocation of contract in online medium provided that the contract is entered with free consent and it is for a lawful consideration.

(d) Competition Act, 2002-It governs the online agreements, abuse of dominant position, predatory offers in online sites, neutrality of platform and anti-competitive agreements in online medium.

4. E-Commerce and Sustainable Development Goals

A new sustainable development framework called "Transforming our world: the 2030 Agenda for Sustainable Development" consist 169 tasks and 17 goals in this plan. (Chaudhary Sanjay. (2017)) The purpose of this initiative is to enhance human prospects and quality of life. Implementations of sustainable development objectives may have a direct or indirect impact on e-commerce, and the concept of e-commerce might have had an impact on some of those goals as well (Singh P. (2011).).

Goal 8 is related to promoting sustained, inclusive, and sustainable economic growth, full and productive employment, and decent work for all. Goal 8 aims to maintain per capita economic growth in accordance with national circumstances, in particular, at least 7% annual GDP growth in the least developed countries. It also aims to increase economic productivity through diversification, technological advancement, and innovation, including by putting an emphasis on high-value-added and labour-intensive sectors. This goal also aims to gradually improve global resource efficiency in consumption and production through 2030. It also encourages the formalisation and growth of micro, small, and medium - sized enterprises, including through access to finance.

Innovation and technical progress, which support equitable and sustainable industrialization and encourage innovation, are provided by Goal 9 of the SDGs. They are crucial in the introduction and promotion of new technology, in facilitating global trade, and in enabling resource efficiency. Goal 9 places a focus on developing high-quality, dependable, sustainable, and resilient infrastructure, which would include regional and trans-border infrastructure, to sustain economic growth and social well-being, to enhance inclusive and sustainable industrialization & to increase small-scale industrial and other enterprises' access to financial services.

The Sustainable Development Goals cannot be attained without attaining goal 10, which is about decreasing inequities and making sure no one is left behind. This includes empowering and promoting social, economic, and political inclusion of all people, regardless of age, sex, disability, race, ethnicity, origin, religion, or economic or other status. By 2030, it aims to gradually achieve and sustain income growth for the bottom 40% of the population at a rate higher than the national average.

Experts claim that e-commerce has made the biggest contribution to the achievement of goals 12 (Ensure sustainable consumption and production patterns) and 13 (Take urgent action to address climate change and its repercussions), which are interconnected. (Hoffman A. J. (2018)

Strong global alliances and cooperation are provided by SDG 17. (Gorgani, Reza. (2014) This goal states that for a development agenda to be successful, inclusive partnerships are needed at the global, regional, national, and local levels. E-commerce can support Goal 17 by bolstering the methods of implementation and reviving the Global Partnership for Sustainable Development.

E-commerce has a beneficial effect on the global and national sustainable development. The impact of this is particularly seen in the labour sector, where the number of Internet businesses is outpacing job growth.

5. Conclusion

The E-Commerce industry is growing at a fast pace providing its acceptability by many people but it has raised several concerns in the legal systems of the countries due to jurisdiction issues, tax applicability, data privacy and cross-border consumer protection. In order to ensure the secure working of E-Commerce it is very essential to analyze the issues, prospective risks and challenges as the industry is very dynamic due innovations in technology; therefore, to adopt a sustainable development in the E-Commerce it is very essential that the legal regime should be made flexible to adapt to the changes and at the same time ensuring accountability pertaining to the activities concerned with E-Commerce.

REFERENCES

1. Svetlana Revinova. (2021). SHS Web Conf. Vol. 114.
2. ASEAN Framework Agreement. (2015) Article 5; ASEAN Economic Community Blueprint 2025.
3. Reference Director General, Directorate General of Doordarshan & Ors. vs Anand Patwardhan & Anr. (Appeal (Civil) 613 of 2005 of Supreme Court).
4. Chaudhary Sanjay. (2017). Effect of E-Commerce on Organization Sustainability. IOSR Journal of Business and Management. 19. 15–24.
5. Singh P. (2011). Environmental Impacts of E-Commerce. International Conference on Environment Science and Engineering. Vol. 8. 202–207.
6. Hoffman A. J. (2018). The Next Phase of Business Sustainability. Stanford Social Innovation Review, 16(2), 35–39. https://doi.org/10.48558/1C0C-0N15
7. Gorgani, Reza. (2014). Electronic Commerce as a Sustainable Business. European Journal of Sustainable Development. 3(3). 141–148.

Integrating Advancements in Education, and Society for Achieving Sustainability – Dimitrios A. Karras et al. (eds)
© 2024 Taylor & Francis Group, London, ISBN 978-1-032-70841-6

2 Infertility Quality of Life Questionnaire—Construction and Validation

Deviga Subramani*

Assistant professor, Department of Psychology,
CHRIST (Deemed to be University), Bangalore, India

Maya Rathnasabapathy

Associate professor, Department of Psychology,
School of Social Sciences and Languages, VIT University Chennai Campus, India

Abstract: Background: Quality of life denotes the well-being of an individual in a given context. Measuring quality of life has become a vital practice in the health sector to ensure the overall well-being of the patients.

Objectives: The study investigates the multidimensionality of quality of life among infertile women and developing scales to measure the quality of life in infertile women.

Method and results: The researcher used the qualitative and quantitative method to understand the quality of life of infertile women. The researcher identified eight primary constructs. They are Emotion, Belief, Behaviour, Familial, Marital, Religiosity, Social Self and Treatment that influence the quality of life of infertile women using Exploratory Factor Analysis (N=150).

Conclusion: The Infertility Quality of Life questionnaire developed is a highly reliable valid questionnaire that can be used as a diagnostic or screening questionnaire.

Keywords: Infertility, Psychological, Infertility QoL, Exploratory factor analysis, Confirmatory factor analysis

1. Introduction

Infertility or "the inability to produce a live birth after adequate sexual exposure without the use of contraception can affect both men and women. Globally, "Infertility affects up to 15% of couples who belong to the reproductive age". Infertility is a growing problem in India for the past two decades. Every year, approximately "60–80 million couples in the world are affected by infertility, out of which around 15–20 million (25%) are from India" (Singh et al. 2020). Worldwide, infertility is known as a medical condition.

*Corresponding Author: devigapsychologist@gmail.com

DOI: 10.4324/9781032708461-2

But it is influenced by social and emotional conditions. Inability to conceive becomes stressful. Women forgo their focus on goals of their life such as career, education, personal well-being and other life goals. It creates insecurity and sadness in the family. Women feel guilty and responsible for the state of childlessness. This affects their psychological well-being.

"Quality of Life" is a multidimensional concept. It involves the subjective judgment of both positive and negative aspects of life. According to World Health Organization, "quality of life is a concept used to describe development, growth, and well-being which reflects individuals' perceptions of their position in the community as well as their goals, expectations, standards, and priorities" (Bakhtiyar et al. 2019).

In the case of infertility, "Quality of Life" is used for psychological evaluation aiding in better psychosocial care and empirical research to explore the factors affecting the same. Chachamovich, Chachamovich, Ezer, Fleck, Knauth and Passos (2010), in his systematic review of the "quality of life", has reported that "quality of life" is significantly low in women diagnosed with infertility (Chachamovich et al. 2010). The factors affecting quality of life are broadly classified into individual and societal factors (Direkvand-Moghadam et al. 2014). Researchers have found that most women diagnosed with infertility experience high anxiety, stress and depression. While there are culture fair questionnaires such as Ferti-QoL available to screen or diagnose patients, there is a lack in the availability of tools (culture specific) to measure the same in India. The researcher aims to systematically construct and validate a tool to measure

infertility quality of life. The researcher also proposes that the infertility quality of life may consist of factors pertaining to emotional, belief, activities/behaviour, family, friends, religious/spiritual beliefs and society. The steps involved in developing, constructing and validating the tool is discussed in detail in the following sections.

2. Identifying the Factors of Infertility Quality of Life

The criteria for the sample were that men and women married for more than 2 years and are childless. The sample consisted of 20 patients. The researcher used semi-structured interview schedule to obtain information about patients' experiences and knowledge about infertility. The interview questions focused on the patients' experience of childlessness and factors influencing their psychological health such as the impact of infertility on emotional, cognitive, behavioural, social and relational aspects of their life. Interview duration was between 30 minutes and an hour long. The interviews were taped (with the consent of participant), transcribed and analysed using content analysis methods. From the results, the researcher identified significant emotion, beliefs, behaviour, coping strategies, support and discomfort caused by family, marriage, religious beliefs and perception of the of infertile women in social context, financial challenges, lack of trust in infertility specialists, high reliance on social media and internet for knowledge and information about infertility treatment are the major factors that influence quality living of an infertile women in treatment (Subramani & Ratnasabapathy, 2018).

3. Generation and Finalisation of Items

The study adopted two methods to identify the a priori domains. They are 1) review of existing literature and 2) in-depth interviews. The researcher used the inductive method to generate items using the responses from in-depth interviews. Following this, the researcher used reviews of existing scales to edit and add relevant items (deductive method) and generated 135 items. Then the researcher reduced 135 items to 95 on the basis of similarity, objectivity, ambiguity and repetition of items. The researcher wrote items using a five-point Likert scale.

4. Establishing Content Validity of Infertility Quality of Life

Face validity

The researcher requested a researcher and an educator (both women) for their feedback and critic on the 95 items of the questionnaire. The participants offered the following feedback "The questions are about the delay in pregnancy. Since it covers almost all aspects of a woman's life when she is struggling to conceive a child the questionnaire is very comprehensive. The questionnaire is a little lengthy and could be emotionally sensitive at times." Face validity helped the researcher reassure that the items are relevant and appropriate. The researcher proceeded with to content validity with caution that the questionnaire could be emotionally sensitive. We inferred that it might be advisable for a psychologist, an infertility specialist, social worker or a sociologist trained in interpersonal or counselling skills to administer the test. In this way, the researcher can ensure ethics and emotional protection of the participant.

Content validity

The researcher requested 11 experts requesting them to review and evaluate the items. The experts were Gynecologists, Nurses, Embryologist, Psychologists, Counsellors, Psychiatrists, Academicians from English language, and Social Work. The Experts' descriptive comments, Content Validity Ratio (CVR) and Content Validity Index (CVI) were used to review and modify the items of the scale. The Probability of chance occurrence and Inter-rater agreement for Relevance, Clarity and Necessity was calculated using Fleiss Kappa. The finding indicates that the inter-rater agreement between the experts' judgments is high and reliable. 15 items were eliminated and 4 items were modified using CVI for Relevance and Clarity and experts' comments (Subramani & Rathnasabapathy, 2021).

5. Cognitive Interview

Pre-testing questionnaires on the sample prove to be useful in adjusting or modifying tests to suit the need or objective to the tool. After the approval of the content validity of the questionnaire by the experts, the researcher requests the target participants to decide the representative property of the test. Cognitive interview is one such method where a sample of the population preferably between 2 and 5 are requested to verbalise their understanding of each item or statement in the questionnaire. The test was administered to a sub-set of the sample (N=5). The participants gave a written feedback for the overall questionnaire. The feedback was reassuring that the items are simple and easy to respond.

Testing the internal consistency of infertility quality of life

The infertility QoL was administered to a sample of 100 infertile women with a written consent. The Cronbach's Alpha coefficient is 0.92. The coefficient for the eight sub scales ranged from 0.56 to 0.77 which indicates acceptable internal consistency of data and acceptable validity for factor analysis.

Identifying the factors of infertility quality of life using exploratory factor analysis

The eight latent variables (Emotion, Belief, Behaviour, Familial, Marital, Social self, Religiosity and Treatment) were factor analysed individually using SPSS v25. Since the scale was developed with the above eight theoretical constructs, the researcher decided to explore the emerging construct under each eight major constructs. A of sample (N=150) was used to extract factors.

Verifying assumptions, extraction methods and factor rotation

In verifying assumptions prior to factor extraction and factor rotation, the researcher used 'Kaiser-Meyer-Olkin' and 'Bartlett's test of Sphericity' to measure sampling adequacy. The KMO test value is 0.572 and above, the value of the Bartlett's Test of Sphericity is significant at 0.00 levels. This indicates that the sample data is adequate for factor analysis. For the present study, the researcher used Principal Axis Factoring (Principal/Common Factor Analysis) with oblique rotation because it was expected that the factors of "Infertility Quality of Life" would be correlated. The factor analysis resulted in forming a total of 14 factors for the eight scales in total and the following are the factors for each scale. They are Emotional scale and Social self-perception scale with one factor namely 'Emotion' and 'Social Self', Belief scale, Behaviour scale, Familial scale, Marital scale, Religiosity scale and Treatment scale with two factors each namely, Self Belief, Self Tutelage, Behavioural Manifestation, Coping Behaviour, Family Stress, Family Support, Supportive Marital aspect, Counteractive Marital aspect, Religiosity – Negative, Religiosity – Positive, Treatment Attitude and Treatment Experience.

6. Reliability

"Cronbach's alpha was calculated to assess the internal consistency of the scale items, i.e., the degree to which the set of items in the scale co-vary, relative to their sum score" using a sample size of 60. The results indicate that only twelve factors had acceptable internal consistency. Two factors Self-Tutelage and Family support has reliability coefficient less than the required value of 0.40. The Cronbach's Alpha for the other twelve factors range from 0.5 to 0.89.

Table 2.1 KMO, scale items and factor loadings

Constructs	KMO	Bartlett's Test of Sphericity		
		Approx. Chi-Square	df	Sig.
Emotional Scale	.74	135.71	28	.000
Belief Scale	.61	291.62	45	.000
Behaviour Scale	.66	354.31	78	.000
Family Scale	.57	181.70	36	.000
Marital Scale	.69	476.41	78	.000
Religiosity Scale	.71	343.31	28	.000
Social Self Perception Scale	.75	264.38	21	.000
Treatment Scale	.60	386.79	78	.000

Emotion Scale

Items	Factor Loadings
I do not have peace of mind due to infertility	.55
I am scared that I will never be able to be a mother	.42
I feel guilty of not being able to give a child for my husband	.41
The sight of other kids with their parents upsets me	.53
I feel jealous when I see pregnant women	.21
I am sad because I am unable to conceive a child	.69
I am ashamed of infertility	.50

Belief Scale

Self Belief	Factor Loadings
I think I am pain for others because of my inability to give birth to a child	.74
I think I am unworthy because I am unable to become a parent	.49
I think I am cursed with infertility	.65
I worry about the consequences of others knowing about my infertility	.77

Self Tutelage	Factor Loadings
I want to experience parenthood	.88
I think it is my duty as a spouse to give a child to my partner	.47
I can recover quickly from the difficulties faced due to infertility	.51
I believe that I will become a parent one day	.48
Only when a child is born to me I will be saved from social ridicule	.50

Behaviour Scale	
Coping Behaviour	**Factor Loadings**
I visit temples very often to overcome infertility	.33
Rituals, poojas and prayers help me cope with infertility	.79
Meditation or exercises help me relax during infertility treatment	.73
I do not interact much with people due to infertility	.44
I try to find reason for my infertility	.33
Behaviour Manifestations	**Factor Loadings**
Crying to myself helps me relax when I am distressed because of infertility	.60
I do not sleep well because of infertility	.39
I avoid talking to people who hurt me with regard to infertility	.39
I eat healthy food to become fertile	.55
I have become ineffective in my work	.34
My primary longing for sex is to get pregnant	.52
Infertility hinders sexual satisfaction	.59
Familial Scale	
Family Stress	**Factor Loadings**
I get angry with those who ask me about pregnancy	.54
I get embarrassed when my relatives talk to me about my infertility	.77
I am unable to express openly to others what I experience due to infertility	-.49
My in-laws accuse me being a cause of infertility	.60
There is a lot of conflict in my family because of my infertility	.53
Family Support	**Factor Loadings**
I want to get pregnant to prove others wrong	-.69
My parents help me to overcome the issues involved in infertility	-.66
I readily follow all suggestions with respect to infertility	-.28
The sight of other kids with their parents upsets me	.53
I feel jealous when I see pregnant women	.21
I am sad because I am unable to conceive a child	.69
I am ashamed of infertility	.50
Marital Scale	
Counteractive Marital Facet	**Factor Loadings**
I blame my spouse for the pain I go through during infertility treatment	.30
Infertility treatment has caused strain in our marital relationship	.53
I feel insecure in my relationship with my partner as I am infertile	.69
I am scared of losing my husband in remarriage because I am infertile	.47
There is a lot of argument with my spouse due to infertility	.39

Supportive Marital Facet	Factor Loadings
My husband and I are supportive of each other in infertility treatment	.78
My spouse sacrifices a lot for me during infertility treatment	.61
My spouse and I satisfy each other sexually	.54
As a couple, we are happy irrespective of us not having a child	.21
I am protective of my spouse more during infertility treatment	.55
Fertility treatment has helped improve my relationship with my spouse	-.59
My relationship with my spouse has improved after trying fertility treatment	-.86

Social Self Scale

Social Self Perception	Factor Loadings
I avoid meeting people because I am infertile	.42
People avoid me because I am infertile	.41
I am ignored during rituals and ceremonies because I am infertile	.75
I am kept away from new born babies	.82
I am discriminated because I am infertile	.68
I am stigmatized as infertile	.57

Religiosity Scale

Religiosity - Negative	Factor Loadings
Certain religious believes limit me from certain fertility treatment	.38
I have lost faith in God after trying hard for a baby for all these years	.39
I believe my angry ancestors are punishing me by making me infertile	.58
Infertility is a punishment for my sins in my previous births	.80

Religiosity - Positive	Factor Loadings
I believe that the higher power will help me conceive	.71
I have entrusted myself to god with regard to fertility	.77
I will do my duty and God will take care of my fertility	.80
Faith in God is being an immense support in going through infertility	.72

Treatment Scale

Treatment Experience	Factor Loadings
I believe my use of contraceptives has caused infertility	.83
I believe my previous abortion causes infertility	.53
My interactions with the medical fertility staff is unpleasant	.44
Treatment negatively affects my mood	.47
I am frustrated with the procedures for infertility	.46
I develop an enormous amount of strain during treatment	.19

Treatment Attitude	Factor Loadings
I feel anxious during the ovulation period	.29
Obesity is a cause of my infertility	.58
I am open to many types of medical treatment	.40
I am scared to go through the treatment of infertility	.50
I am not able to conceive because all the sperms are coming out of the vagina	.32

Source: Aurhors

7. Discussion

Examining "quality of life" of patients enable health professionals understand patients' problems and facilitate their well-being. De Oliveira and colleagues (2015) said that measuring and enhancing "quality of life" is a way of treating mental health and physical health equally (de Oliveira et al. 2015).

The present study by gathering opinions of patients and reviewing relevant literature has found that there are eight major constructs that constitute "Infertility Quality of Life". They are Emotion, Belief, Behaviour, Family, Marital, Religiosity, Social Self and Treatment. These factors are in line with the findings of studies conducted by the WHO (1996), Rashidi (2008), Boivin (2011), Yaghmaei (2013) and Bakthiyar (2019).

Bakhtiyar (2019) confirmed in her study that infertility can possibly affect many aspects of women's "quality of life" such as physical health domain, mental health domain, social domain and the total "quality of life" significantly (Bakhtiyar et al. 2019). World Health Organization (1996) conceptualizes QoL as consisting of "physical", "psychological", "social" and "environmental" - and a "general factor".

"Fertility Quality of Life" questionnaire (Boivin et al. 2011) which comprises of four core factors and two treatment-related factors. The core factors are "Emotional", "Relational", "Mind/Body" and "Social" and "treatment-related" factors are "environment" and "tolerability". Yaghmaei and collegeues (2013) in Iran developed "Quality of Life in Infertile Couple Questionnaire" (QOLICQ) (Yaghmaei et al. 2013). The domains of the questionnaire are "physical", "psychological", "spiritual-religious", "economic", "affective", "sexual" and "social" (Yaghmaei et al. 2013).

Many researchers have used groups of indicators to define the domain and sub-domains of "quality of life". The domains of "quality of life" discussed in the systematic review of literature conducted by Brown and others in 2004 shows that all the indicators are relevant to adults but they can vary depending on the nature and need of the groups. Hence, there is no definitive, established or well-accepted theoretical framework of "quality of life".

In the present study, the researcher has based the study on social construction theory and biopsychosocial theory. According to Bliss (1999), social construction in application to infertility assumes that women with infertility interpret or understand their state of infertility based on the experiences of her environment and culture. Bliss (1999), also extends to explain application of biopsychosocial theory to infertility. The biopsychosocial theory is based on the family systems theory/approach. This approach is characterized by conscious attention to the medical condition which has implications on the personal and interpersonal life of a patient including her family. The eight major construct and 12 subconstructs identified in the study reflect the assumptions of social construction theory and biopsychosocial theory.

8. Conclusion

Quality of life is an important concept to be studied to understand the mental health of patients. This is true to infertility treatment. Infertility and its treatment affects women in various aspects. Infertility Quality of life questionnaire measures all the major aspects of a women's life affected by infertility. Infertility Quality of life questionnaire developed in the study would be an appropriate and accurate measure because it was developed using opinions of patients directly (Gill, 1994).

Research on psychosocial care, counselling or psycho-education requires a valid questionnaire to examine its effect on the "quality of life" infertile women. Due to the unavailability of appropriate and valid tools, psychologists, social scientists or infertility specialists in India are at disadvantage. By using a mixed-method approach, the researcher has carefully constructed and validated a multi-dimensional questionnaire measuring Infertility Quality of life specific to Indian women in infertility treatment. This questionnaire being culture-specific to India will prove to be useful for both clinical and research purposes.

REFERENCES

1. Bakhtiyar, K., Beiranvand, R., Ardalan, A., Changaee, F., Almasian, M., Badrizadeh, A., Bastami, F., & Ebrahimzadeh, F. (2019). An investigation of the effects of infertility on Women's quality of life: A case-control study. BMC Women's Health, 19(1), 1–9. https://doi.org/10.1186/s12905-019-0805-3

2. Boivin, J., Takefman, J., & Braverman, A. (2011). The fertility quality of life (FertiQoL) tool: Development and general psychometric properties. Human Reproduction, 26(8), 2084–2091. https://doi.org/10.1093/humrep/der171

3. Chachamovich, J. R., Chachamovich, E., Ezer, H., Fleck, M. P., Knauth, D. R., & Passos, E. P. (2010). Agreement on perceptions of quality of life in couples dealing with infertility. JOGNN - Journal of Obstetric, Gynecologic, and Neonatal Nursing, 39(5), 557–565. https://doi.org/10.1111/j.1552-6909.2010.01168.x

4. de Oliveira, M. F. M. F., Castro, R. C. M., Calou, C. G. P., de Oliveira, M. F. M. F., de Souza Aquino, P., da Silva, C. G. L., Meireles, C. B., Maia, L. C., Miná, V. A. L., Faé, B. N., Santos, M. F. A., dos Santos, M. D. S. V., Pereira, J. B., Lins, H. C. C., Neto, M. L. R., & Pinheiro, A. K. B. (2015). Ways to measuring quality of life in mental health. International Archives of Medicine, 8(1), 2013–2016. https://doi.org/10.3823/1690

5. Direkvand-Moghadam, A., Ali, D., & Azadeh, D. M. (2014). Effect of infertility on the quality of life, a cross-sectional study. Journal of Clinical and Diagnostic Research, 8(10), 13–15. https://doi.org/10.7860/JCDR/2014/8481.5063

6. Gill, T. M. (1994). A critical appraisal of the quality of quality-of-life measurements. JAMA: The Journal of the American Medical Association, 272(8), 619–626. https://doi.org/10.1001/jama.272.8.619

7. Singh, K., Shashi, K., Rajshee, K., Sinha, S., & Bharti, G. (2020). Assessment of depression, anxiety and stress among Indian infertile couples in a tertiary health care centre in Bihar. International Journal of Reproduction, Contraception, Obstetrics and Gynecology, 9(2), 659. https://doi.org/10.18203/2320-1770.ijrcog20200354

8. Subramani, D., & Rathnasabapathy, M. (2021). Content Validity of Infertility Quality of Life Questionnaire among women with infertility in treatment. Turkish Journal of Physiotherapy and Rehabilitation, 32(3), 2846–2854.

9. Subramani, D., & Ratnasabapathy, M. (2018). Understanding Psychological Health of Infertile Women Undergoing Assisted Reproductive Therapy: In-Depth Interviews with Infertile Women.

10. Yaghmaei, F., Mohammadi, S., & Majd, H. A. (2013). Developing and Measuring Psychometric Properties of "Quality of Life Questionnaire in Infertile Couples." International Journal of Community Based Nursing and Midwifery, 1(4), 238–245.

Integrating Advancements in Education, and Society for Achieving Sustainability – Dimitrios A. Karras et al. (eds)

Awareness and Agitation of Women Investors towards Derivative Market

S. S. Mageswari

Research Scholar, Department of Commerce, Avinashilingam Institute for Home Science and Higher Education for Women, Coimbatore, Tamil Nadu

P. Sasirekha*

Assistant Professor, Department of Commerce, Avinashilingam Institute for Home Science and Higher Education for Women, Coimbatore, Tamil Nadu, India

Abstract: Global finance has modified dramatically in current decades. Electronic processing, globalization, and deregulation have all converted markets, with most of the maximum crucial modifications related to derivatives. The by-product marketplace is newly begun in India and it isn't always recognized by every women investor, so there are numerous steps taken to create attention amongst women traders approximately the derivative segment. It is important for an investor to stay updated about these financial instruments if they are to stay in the game. Therefore, this study focuses on the "Awareness and Agitation of Women Investors towards Derivative Market", through which the women investor's investment behaviour is given much importance because much of the focus has been on the role played by small investors in the markets. The retail investors actually increase the price efficiency of the stock and their entry into the market also increases liquidity.

Keywords: Derivative markets, Women investors, Awareness, Agitation

1. Introduction

A derivative is a monetary tool that has a price decided with the aid of using the rate of the underlying asset. The term "Derivatives" means there is no independent price, *i.e.*, their prices are completely got from the price of the underlying asset. The underlying asset may be securities, commodities, bullion currency, livestock, etc. In different words, derivatives imply forward, futures, options, and different hybrid contracts, related for the motive of settlement success to the price of a designated actual or monetary asset or to an index of securities.

---

*Corresponding Author: Sasirekha_comm@avinuty.ac.in

DOI: 10.4324/9781032708461-3

2. Players in Derivative Market

The derivatives marketplace is much like different economic markets and has the subsequent three wide classes of participants:

Hedgers: These are traders who are having exposure to foresee the underlying asset which is tentative towards price risk. The hedgers make use of derivative markets basically for managing the price risk of portfolios and assets.

Speculators: These are persons who have a look at the future trends of the markets. They notice whether there is a chance for the price to rise or fall in the upcoming future and based on that accordingly they can buy or sell the options or futures and make benefit from the price movements of underlying assets in the future.

Arbitrages: They are another set of important participants in the derivative market. They position themselves in the financial market to earn profit without taking risks. Short and long positions are taken by them simultaneously to generate a riskless gain.

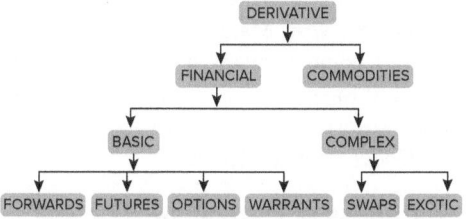

Fig. 3.1 Classification of derivatives [1]

3. Reviews for the Study

A. Sachin Jain studied the investor's awareness of derivative markets and the risk involved while trading. The study was based on primary data. The study concludes that investors very rarely trade in derivative markets and are yet to be aware of the risks while trading in derivatives. **Gunjan Tripathi** studied investors' perceptions of derivative trading. 100 investors' data were collected to understand the awareness and perception of derivatives. The study concludes that investor does not invest in derivatives on their own research due to a lack of knowledge instead they invest using their broker's advice. **Amala Sara John** studied that there is a high return with low investment in derivative markets.

Amalendu Bhunia et. al. studied the level of awareness of women investors in West Bengal towards various investment avenues. Primary data was used in the study. The study concludes that around 87.7 per cent of women are aware of the investment avenues. **Kamal Pant et. al.** studied diversity in gender which leads to an investment decision from 1145 respondents in Uttarakhand state. The study concludes that 81 per cent of investors invest in fixed deposits rather than investing in markets. **Dr. N. Selvaraj** studied the awareness and perception of traders on financial derivatives and concluded that stock brokerage should take steps to benefit the traders on a long-term basis.

4. Statement of the Problem

Now – a – days there are many retail investors especially women coming forward to invest in markets. Most retailers are aware and ready to invest in the stock market, but what about derivatives? "It is necessary to study the level of awareness among women in derivatives markets". Through this study, there will be a clear-cut vision of the derivatives, their role, and also the problems or risks involved while trading in derivatives.

Objectives

1. To study the awareness level of women investors in derivatives
2. The problems and difficulties of women investors in derivative markets

Methodology

Primary records were used. They were collected via a questionnaire from 100 women investors from Coimbatore city, Tamil Nadu with the aid of adopting a convenient sampling technique. The collected records were analyzed with the assistance of suitable tools specifically Descriptive Statistics and Garrett`s Ranking Technique. The study was based on the proposed model by N. Renuka (2019).

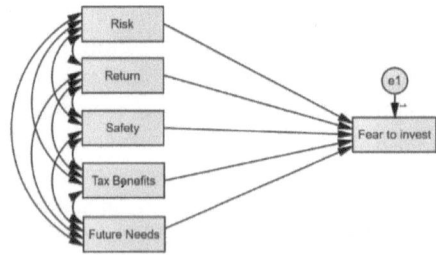

Fig. 3.2   Proposed model on agitation to invest in derivative markets

5. Results and Discussion

The summary of the major findings that emerged from the analysis was

Socio-Economic Profile of Women Investors

Socio-economic aspects such as age, educational qualification, and occupation provide insight to understand the level of awareness towards derivative of the women investors.

Table 3.1   Socio-economic profile N = 100

Variables		Number	(%)
Age	26 – 35 years	30	30.0
	36 – 45 years	52	**52.0**
	46 – 55 years	18	18.0
Marital Status	Single	19	19.0
	Married	69	**69.0**
	Separated	12	12.0
Educational Qualification	Up to school level	9	9.0
	Undergraduate	46	**46.0**
	Postgraduate	33	33.0
	Professional	12	12.0
Occupation	Government employed	29	29.0
	Private employed	30	**30.0**
	Self-employed	21	21.0
	Housewife	8	8.0
	Professional	12	12.0

Source: Primary data

Majority of the investors come under the age group of 36 – 45 years, through this it is clear that the young generation of women come forward to invest. Married women are more likely to come forward in investing when compared to single women. Most employed women investors are in the private sector and with Undergraduate degree.

Awareness of Derivative Instruments

The awareness of women investors towards derivatives is required. In this Table 3.2, the awareness of women investors in Derivatives Instruments is measured.

Table 3.2 Awareness of derivative instruments

ITEMS	EF	VF	MF	SF	NF	TOTAL	MEAN	RANK
Scores	5	4	3	2	1			
Futures	61	23	12	3	1	440	4.40	I
Forwards	53	25	11	6	5	415	4.15	II
Options	34	14	12	33	7	335	3.35	III
Commodity	22	18	19	38	3	318	3.18	IV
Swaps	12	24	30	22	12	302	3.02	V
Currency	19	17	15	45	4	299	2.99	VI
Interest rate	11	10	19	50	10	262	2.62	VII

Source: Computed data

The women investors' awareness level of derivative markets is calculated. Future contracts are ranked first which indicates that many women investors are aware. Though not many women investors are investing in derivative markets they are aware of the products in derivatives. Forwards are also one of the types of contracts under the derivative market which is known by the women investors and is ranked second and then comes options.

Problems while Trading in Derivative Markets

The problems and risks arising while trading in derivative markets is given in the form of a diagram.

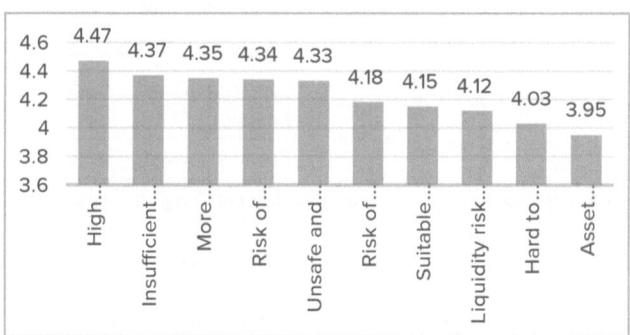

Fig. 3.3 Problems while trading in derivative markets

Source: Computed data

High brokerage fees are listed as the first problem by women investors because investments made by women will not be in bulk quantity. In that small amount kept for investment, a certain percentage must be given as brokerage fees since they have no knowledge or awareness regarding the derivative market which is the major problem while investing. The next is insufficient protection measures is there so investors must be very careful.

Reasons for not Investing in Derivatives

There are some reasons for women investors not investing in derivatives.

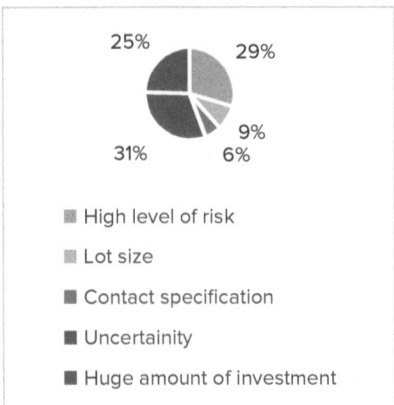

25%   29%

31%   9%
       6%

■ High level of risk

■ Lot size

■ Contact specification

■ Uncertainity

■ Huge amount of investment

Fig. 3.4   Reasons for not investing in derivatives

Source: Computed data

Figure 3.4, clearly explains that the reasons for not investing in derivatives are, the majority of the women investors feel that derivatives are uncertain, and secondly, they feel that there is a high level of risk while investing in derivatives and then the huge investment is required while investing in derivatives. Through this, it is made clear that uncertainty and risk factors are the most common reasons for women investors not to invest in derivative markets.

Model of Agitation to Invest in Derivative Markets

The SEM requires certain criteria to measure the model fit. Table 3.3 shows the derived value for the structural model under Derivative markets.

Table 3.3   Model of agitation to invest in derivative markets

Fit Indices		Model Value
Absolute Fit Measures		
Chi square	χ^2	50.20
	Degrees of freedom	12
	P value	.000
	χ^2/DF	3.256
RMSEA (Root Mean Square Error of Approximation)		0.071
GFI (Goodness of Fit Index)		1.677
Incremental Fit Measures		
AGFI Adjusted Goodness of Fit Index)		1.548
NFI (Normed Fit Index)		1.023
CFI (Comparative Fit Index)		1.320
RFI (Relative Fit Index)		1.002
IFI (Incremental Fit Index)		1.001
RMR (Root Mean Square Residual)		0.016
Parsimony Fit Measures		
PCFI (Parsimony Comparative of Fit Index)		1.190
PNFI (Parsimony Normed Fit Index)		1.233

Source: Computed data

The value of chi-square, RMSEA, NFI, and CFI is 50.20, 0.071, 1.023, and 1.320 respectively. The chi-square is significant, therefore could be observed as a perfect model and inferred that the model is suitable and accepted. It is discovered that the values are apt within the desired limit of SEM analysis.

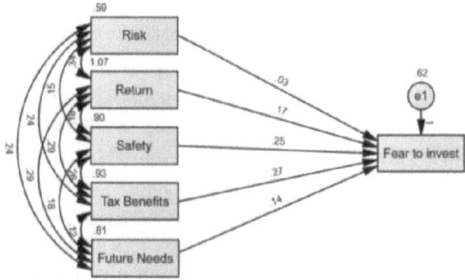

Fig. 3.5 Model of agitation to invest in derivative markets

Source: Computed data

6. Suggestions

The study was made to know the awareness and the problems of women investors while investing in derivative markets. Based on the findings, the following guidelines are listed:

- The regulating bodies should give awareness among women investors by educating them regarding trading in derivatives.

- The companies can conduct seminars, and live workshops and must provide and guide through which the women investors can come forward for investment.

- Though the majority of the investors are men the corporates must come forward in bringing the women investors especially the housewives for investment which would be very useful for them.

- Brokerage fees can be reduced for women investors because they feel that this is one biggest problem in derivative markets.

- Since the women investors feel that derivatives are for big investors the company must take initiatives to also bring in small investors in derivative markets.

7. Conclusion

In the present scenario, there is a boom in the stock market. Many new investors are entering the market. The majority of the investors are men compared to women investors and not many women investors are trading in the derivative market since they are not much aware of it. The study on the level of awareness of women investors towards derivative markets discloses that most women investors are aware of the derivative markets and their products. High risk, uncertainty, high brokerage fees, and a huge amount of investment are the main problem for women investors trading in derivatives. Most of the women come forward for investment but their investment in derivatives is still lagging. So required measures must be taken to bring small women investors into derivative markets.

REFERENCES

1. Gakhar, D. V. (2015). Indian Derivatives Market: A Study of Impact on Volatility and Investor Perception. *Available at SSRN 2659398.*
2. Mubeen, S., & Kumar, S. (2019). Awareness Of Woman Investor In Commodity Market In Bangalore. *Think India Journal, 22*(14), 13327–13338.
3. Pallavi, E. V. P. A. S., & Raju, T. K. (2014). An empirical analysis on perception of retail investors towards derivatives market with reference to Visakhapatnam district. *Indian Journal of Management Science, 4*(1), 54.
4. Raghavan, M. S., & Tomar, A. S. (2017). Derivatives market in India: An empirical analysis on perception of retail investors towards derivatives market with reference to Visakhapatnam district. *Journal of Advances and Scholarly Researches in Allied Education, 12*(2), 214–18.

5. Ravichandran, K. (2008). A study on Investors Preferences towards various investment avenues in Capital Market with special reference to Derivatives. *Journal of Contemporary Research in Management, 3*(3), 101–112.

6. Selvaraj, N. (2021). Traders Perception and Awareness on Financial Derivatives in Indian Stock Market. *International Journal of Business Management and Finance Research, 4*(1), 19–31.

7. Sharma, A. (2020). Effect of Demographic Factors in Investment Decisions of Individual Investors–A Case Study in Delhi NCR. In *International Conference of Advance Research & Innovation (ICARI).*

8. Tripathi, G. (2014). An Empirical Investigation of Investors Perception towards Derivative Trading. *Global Journal of Finance and Management, 6*(2), 99–104.

9. Umamaheswari, S., Anand, A., & Nithya, N. (2022, May). An empirical study on influential factor of investors' investment towards futures and options trading in India. In *AIP Conference Proceedings* (Vol. 2393, No. 1, p. 020052). AIP Publishing LLC.

10. Vohra, T., & Kaur, M. (2016). Awareness and Stock Market Participation of Women: A Comparative Study of Stock Investors and Non-Investors. *IUP Journal of Management Research, 15*(4).

*Integrating Advancements in Education, and Society for Achieving
Sustainability – Dimitrios A. Karras et al. (eds)*
© 2024 Taylor & Francis Group, London, ISBN 978-1-032-70841-6

Positive Psychological Capital Enhances Flourishing of Professors in Multicultural Work Environment in Universities of India

Ashraf Alam*, Atasi Mohanty

Rekhi Centre of Excellence for the Science of Happiness,
Indian Institute of Technology Kharagpur, India

Abstract: Multicultural university workspace comprises of professors and non-teaching staffs from different cultures, castes, and regional backgrounds. Diversity in culture, caste, religion, and gender is crucial for establishing a workplace that imbibes multiculturalism. Positive psychological capital (PsyCap) is important for optimal human functioning irrespective of inherent systemic biases. Four psychological constructs, namely hope, self-efficacy, resilience, and optimism, were found to best fit the inclusion requirements after drawing on positive psychology conceptions and past empirical research. The four constructs that constitute PsyCap were found to work together to generate a second-order core construct that had a greater association with performance and pleasure than any one of the four components taken alone. The research investigation found that positive PsyCap is crucial for a multiculturally diverse university workplace as they play an important role in developing the true potential of university professors, thus making their work more meaningful and productive.

Keywords: Psychological Capital, Professors, University, Higher Education, Multicultural Workplace, Hope, Optimism, Resilience, Self-Efficacy

1. Introduction

Positive psychological capital is the positive developmental state of an individual as characterized by high self-efficacy, optimism, hope, and resiliency. While every component has its own characteristics and interventions, the concept of PsyCap is greater than the sum of its parts. Based on appreciation and positive emotions, PsyCap is a core construct for wellbeing and thriving[1]. This research investigation focuses on studying analytically, the correlation between the importance of positive psychological capital and its impact on the flourishing of professors in the multiculturally diverse university workplace.

*Corresponding Author: ashraf_alam@kgpian.iitkgp.ac.in

DOI: 10.4324/9781032708461-4

2. Methods and Methodology

PsyCap index [2] was used to measure the psychological capital, and PMC Index was used to measure the performance and satisfaction of professors in a diverse university workspace. Both indices, *i.e.*, PsyCap Index and PMC Index were then parameterized and correlated with different components of PsyCap to come up with a conclusion. The researcher collected data through Google Forms and got a total of 30 responses from 30 professors. The professors were from 15 different universities belonging to 9 different states of India, such that seven were from Central, five from State, and three from Private Universities. We administered the Short Form of Positive Psychological capital questionnaire developed by Fred Luthans[2]. Using the 12-item Questionnaire, the Psychological Capital Index (PsyCap Index) of each professor was calculated. This 12-item questionnaire is divided into the four components of psychological capital, *i.e.*, Self-Efficacy (3 items), Hope (4 items), Resilience (3 items) & Optimism (2 items).

From this division, the researcher also calculated the component-wise index for each professor. The researcher added five questions related to the multicultural work environment, depending on whether it is a question of opinion or experience[3]. From those questions, like PsyCap Index, the researcher used Positivity towards Multi-Culture Index (PMC index). In the end, the researcher asked which component the professor believed to be the most important for good performance and satisfaction in a multicultural work environment. Thereafter, the relation between different indices was computed.

3. Findings

Here is the list of questions that were asked.

Questions on Multicultural Workplace:

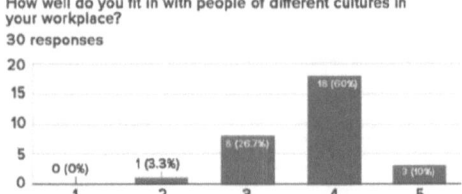

How well do you fit in with people of different cultures in your workplace?
30 responses

Fig. 4.1 Adaptability in multi-cultural workplace

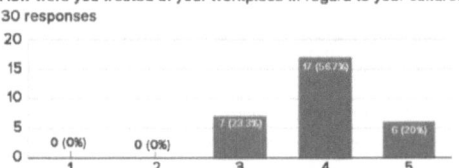

How were you treated at your workplace in regard to your culture?
30 responses

Fig. 4.2 Experience in multi-cultural Workplace

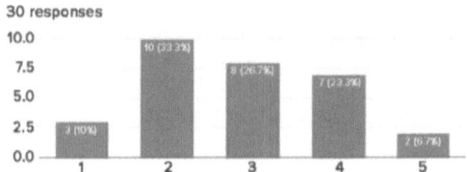

Your culture affects situations in your workplace.
30 responses

Fig. 4.3 Effect of one's own culture on their workplace

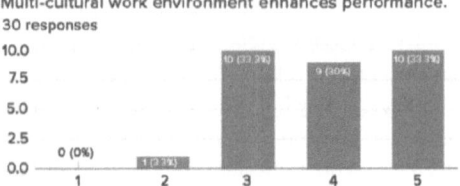

Multi-cultural work environment enhances performance.
30 responses

Fig. 4.4 Opinion on performance in a multicultural workplace

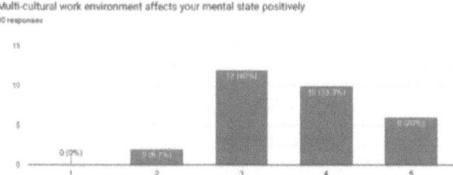

Fig. 4.5 Opinion on mental state in a multicultural workplace

Questions on Self-Efficacy:

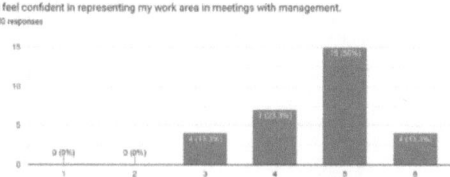

Fig. 4.6 Question 1 on self dfficacy

Fig. 4.7 Question 2 on self efficacy

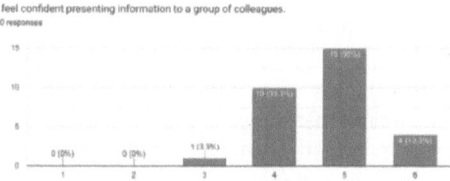

Fig. 4.8 Question 3 on self efficacy

Questions on Hope:

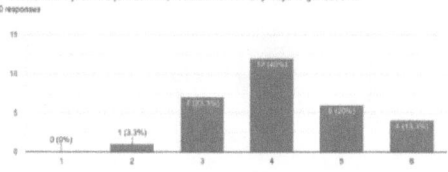

Fig. 4.9 Question 1 on hope

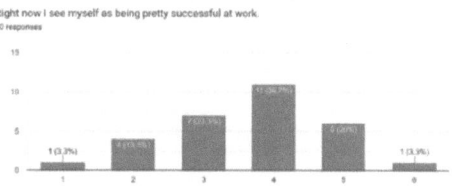

Fig. 4.10 Question 2 on hope

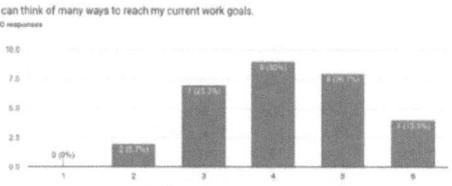

Fig. 4.11 Question 3 on hope

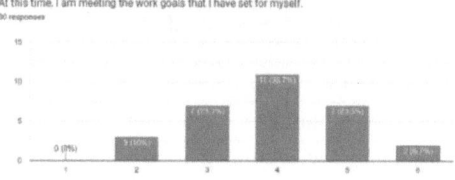

Fig. 4.12 Question 4 on hope

Questions on Resilience:

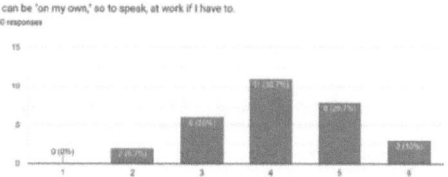

Fig. 4.13 Question 1 on resilience

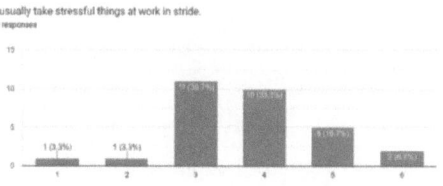

Fig. 4.14 Question 2 on resilience

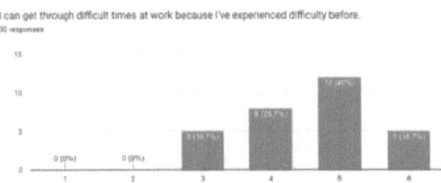

Fig. 4.15 Question 3 on resilience

Questions on Optimism:

Fig. 4.16 Question 1 on optimism

Data says that for most of the professors, optimism is the most important component, followed by resilience. Hope and self-efficacy, although important, are not something that comes to their mind when thinking about getting through the day in the workplace.

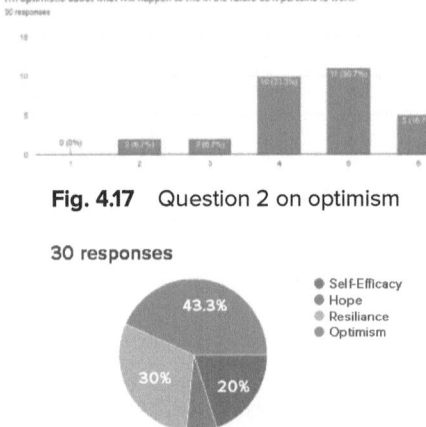

Fig. 4.17 Question 2 on optimism

Fig. 4.18 Hope is the most important component for satisfaction and performance

4. Data Analysis

We got the response data of all the questions in a .csv file. We then calculated each individual component of the PsyCap Index and PMC Index. Average indices for each component of the PsyCap index and PMC index for all professors were then calculated. The indices of professors based on the last question were then analyzed, i.e., the most important component in their opinion. The relation between various indices was then analyzed and plotted.

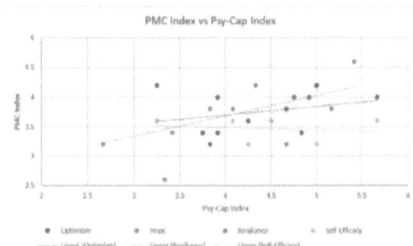

Fig. 4.19 Plot of PsyCap Index vs. PMC index

To observe the effect of the professor's individual opinion, the points were color-separated based on the professor's individual opinion on the most important component. Then trendlines were plotted for those different data sets. From what we see, we understand that the group of professors who believe in optimism and resilience have a positive correlation between their PsyCap and PMC indices. But those who believe in self-efficacy have almost zero correlation between the two indices.

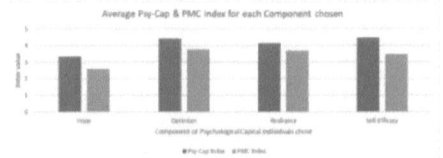

Fig. 4.20 Histogram of PsyCap & PCM Indices based on individual opinion

For PsyCap Index, professors who believe in self-efficacy have a slightly higher PsyCap index, followed by professors who believe in optimism. The trend is opposite in the case of PMC index, where Optimism takes the lead. On further observation, we can see that both indices follow almost the same trend when considering professors' individual opinions.

Correlation Map between various indices:

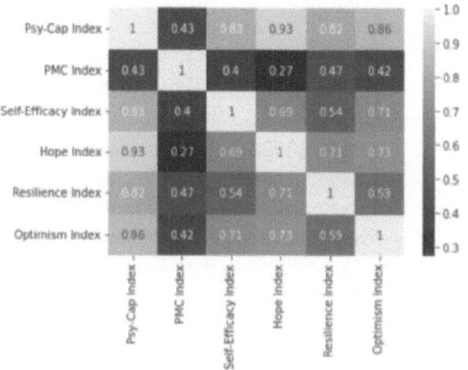

Fig. 4.21   Correlation map between various indices

Figure 4.21 shows the correlation map between various components and indices. We see that the PsyCap index is most closely related to hope, with a correlation value of 0.93, whereas PMC index is most closely related to resilience, with a correlation value of 0.47. The PsyCap index has correlation values of 0.86, 0.83, and 0.82, respectively, for optimism, self-efficacy, and resilience, and the correlation value between the PsyCap index and PMC index is 0.43. PMC index is most related closely to resilience with a correlation value of 0.47. It has correlation values of 0.42, 0.4, and 0.27 with optimism, self-efficacy, and hope, respectively.

Previously, from the histogram, we discussed that professors who believe in self-efficacy have the highest PsyCap index, slightly

above that of professors who believed in optimism. And in the case of PMC index, optimism was a bit higher than self-efficacy. Although we found that resilience is most closely related to PMC index, optimism is slightly higher than self-efficacy, which is the same as the trends from the histogram.

Table 4.1   Average values of components and indices.

Index	Average Value of Components and Indices
PsyCap Index	4.28
PMC Index	3.63 (out of 5)
Self-Efficacy	4.71
Hope	3.98
Resilience	4.16
Optimism	4.42

Table 4.2   Correlational values of components with average components and average indices.

Component Chosen	Average Component	Average PsyCap Index	Average PMC Index
Hope	3.25	3.67	3.1
Optimism	4.69	4.40	3.75
Resilience	4.15	4.12	3.68
Self-Efficacy	4.94	4.46	3.46

Professors who believed in optimism have the highest PMC index and have a decently good optimism value. Professors who chose resilience have an above-average PsyCap Index and PMC index while having a sufficiently good value for resilience.

5. Discussion and Conclusion

The correlation of PsyCap index with PMC index is quite high. Hope is the most related component to the PsyCap index, whereas it is resilience in PMC Index and vice versa. The correlation between the PsyCap Index and PMC Index, when done on the list of averages of components, is very high. From this, we may infer that professors who give importance to the same component of psychological capital have similar relationships with their PsyCap Index and PMC index. This investigation thus concludes that Psychological Capital has a good influence on the multicultural work environment. In cases where psychological capital is good, 'experience and positivity' towards multicultural work environment was also good. Each component of psychological capital has a similar effect on multicultural work performance/experience. It has also been observed that components like resilience and optimism have a very powerful impact, and thus enhance professors' thriving and flourishing capabilities in multicultural work environments.

REFERENCES

1. Alam, A. (2022). Positive Psychology Goes to School: Conceptualizing Students' Happiness in 21st Century Schools While 'Minding the Mind!' Are We There Yet? Evidence-Backed, School-Based Positive Psychology Interventions. *ECS Transactions*, *107*(1), 11199.
2. Luthans, F., Avolio, B. J., Avey, J. B., & Norman, S. M. (2007). Positive Psychological capital: measurement and relationship with performance and satisfaction. *Personnel Psychology*, 60, 541–572.
3. Alam, A. (2022). Investigating Sustainable Education and Positive Psychology Interventions in Schools Towards Achievement of Sustainable Happiness and Wellbeing for 21st Century Pedagogy and Curriculum. *ECS Transactions, 107*(1), 19481.

Note: All the figures and tables in this chapter were made the author.

Integrating Advancements in Education, and Society for Achieving Sustainability – Dimitrios A. Karras et al. (eds)
© 2024 Taylor & Francis Group, London, ISBN 978-1-032-70841-6

5

Role of Psychological Capital and Intragroup Conflict on Professors' Burnout and Quality of Service in Universities of India

Ashraf Alam*, Atasi Mohanty

Rekhi Centre of Excellence for the Science of Happiness,
Indian Institute of Technology Kharagpur, India

Abstract: In past research, intragroup conflict was found to have a detrimental impact on worker performance, including the quality of service provided by employees. However, depending on how the disagreement is managed, some authors have argued that these detrimental effects of intragroup conflict may not always occur. Contemporary research emphasizes the value of psychological capital as a predictor of employees' health and success. Thus, the current investigation examines the influence of psychological capital and intragroup conflicts on burnout and the quality of service of professors at the individual level. The data was gathered from thirty professors who are currently serving full-time in six different universities in India. The results demonstrated that psychological capital has a considerable influence on burnout and 'quality of service' of professors. Burnout and conflict among professors in intragroup interactions were found to be strongly related.

Keywords: Professors, Higher Education, Universities, India, Workplace Conflict, Psychological Capital, Performance, Burnout, Quality of Service

1. Introduction

Psychological Capital (PsyCap)

According to Fred Luthans et al. (2007), psychological capital is an individual's positive psychological condition of growth, which is characterized by possessing high levels of HERO, *i.e.*, hope, (self-)efficacy, resilience, and optimism [1].

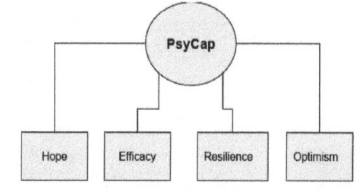

Fig. 5.1 Flow diagram showing the different components of psychological capital

*Corresponding author: ashraf_alam@kgpian.iitkgp.ac.in

DOI: 10.4324/9781032708461-5

Intragroup Conflict

The conflict that occurs within a group of people who share common interests, goals, or other defining characteristics is known as intragroup conflict [2]. Conflict within a group might take place on a small or large scale, for instance, at the workplace or among members of a particular group [3]. Even though it is frequently perceived as a problem, intragroup conflict can be a useful tool in some circumstances.

Intragroup conflict is categorized into three broad groups:

(a) Task-related conflict (TC), or disputes involving people's perceptions of disagreement regarding the details of their linked tasks, which typically involve divergent points of view, concepts, and opinions [4].

(b) Relationship conflict (RC) or disputes over perceptions of interpersonal incompatibility, frequently involving rumours and disagreements over personal ideas [5].

(c) Process conflict (PC) or disputes over how work will be completed, typically involving disagreement over protocols, rules, and processes [6].

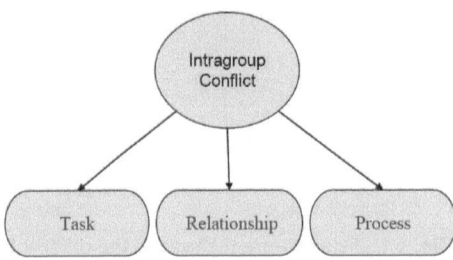

Fig. 5.2 Flow diagram showing the different types of intragroup conflict

2. Purpose of the Study

The objective of this study is to discover the role of psychological capital and intragroup conflict on professors' burnout and quality of service in universities of India.

3. Methods and Methodologies

Data were collected through online surveys that were circulated between May 2021 and March 2022, using google forms. All the professors who took part in this survey were full-time professors working in six Indian universities, namely Pondicherry University, Lucknow University, Jadavpur University, Anna University, Hyderabad Central University, and the University of Delhi. The purpose of the study and the requirements for participation were explained to the participants beforehand.

Measures

Psychological Capital (PsyCap)

To cut down on the number of questions and make the questionnaire succinct and to the point, we used the Psychological Capital Questionnaire (Luthans et al., 2007) in its 12-item condensed version [1]. This survey gauges psychological capital (PsyCap), which consists of four optimistic psychological states: efficacy (belief in one's capacity to carry out tasks successfully), hope (favorable assessments of one's capacity to achieve goals), resiliency (coping and recovery from adversity), and optimism (mental attitude to interpreting situations and events in a positive way).

A 6-point scale was used to score the 12 items on this questionnaire (Table 5.1).

Table 5.1 Table describing the questions of the survey for measuring the psychological capital

Psychological States	Statements
Hope	"If I were in a difficult situation at work, I could think of many ways to get out of it" "Nowadays, I try to achieve my goals with great energy" "For any problem, there are many ways to solve it"
Efficacy	"I feel confident when I'm looking for a solution to a long-term problem" "I feel confident in representing my work area in meetings with the organization's management" "I feel confident to contribute to discussions about the organization's strategy"
Resilience	"In one way or another, in general, I can manage work and its difficulties" "I feel that I can handle many things at the same time at work" "In general, I can easily step over the more stressful things at work"
Optimism	"When things are uncertain for me at work, I usually expect the best" "At work, I am optimistic about what will happen in the future" "At work, I always look on the positive side of things"

Intragroup Conflicts

To collect data for intragroup conflicts, an 8-item intragroup conflict scale was administered. All questions were to be answered based upon a 5-point Likert-type scale. The 8-item intragroup conflict scale that we administered is essentially composed of two types of intragroup conflict:

Task conflicts: These include disagreements over the way a particular task is performed, the type of results obtained from the task, or the contribution of the team members to the task [7]. These can lead to disharmony, dissatisfaction, and an unpleasant work environment and are crucial factors for considering intragroup conflict [8].

Relationship conflicts: These consider more interpersonal relationships of individuals with one another and certain problems that someone might have with another person regardless of the work. An excellent interpersonal relationship translates into lower conflicts within the group and helps in overall team building and morale.

Table 5.2 Table describing the questions of the survey for measuring the intragroup conflict

Measure	Questions
Relationship	"How much friction is there among the members of your work unit?" "How many personality conflicts are evident in your work unit?" "How much tension is there among members of your work unit?" "How much emotional conflict is there among members of your unit?"
Task	"How much conflict about your work is there in your work unit?" "How often do people in your work unit disagree about opinions regarding the work being done?" "To what extent are there differences in opinion in your work unit?" "How frequently are there conflicts about ideas in your work unit?"

Burnout

A state of physical or emotional tiredness that also includes a sense of diminished accomplishment and a loss of personal identity is known as job burnout, which is a sort of work-related stress. Job burnout drives down the efficiency at which an individual can complete tasks and be satisfied with their work life [9]. For the purpose of data collection, the Shirom-Melamed Burnout measure (also known as SMBM) was modified and administered, after proceeding with translation from Spanish to English. The metric for quantification was done by asking three questions to the professors regarding the physical, cognitive, and emotional aspects of burnout.

Table 5.3 Table describing the questions of the survey for measuring the burnout of the respondents

Measure	Questions
Physical Burnout	Do you feel physically drained during work hours?
Cognitive Burnout	Do you have trouble concentrating while at work?
Emotional Burnout	Do you have problems adjusting to co-workers' needs?

Quality of Service

Comparing how well a university meets the expectations of its professors may help measure the quality of its services. Students choose to study in a particular university to meet their individual specific needs. They have criteria and expectations for how a university will, consciously or subconsciously, provide services that fit their demands [10].

To analyze data for quality of service, we used a set of three questions, developed by Salanova *et. al.* (2005). These questions

were used to measure: service climate, empathy, and job performance. The metric for collecting the answers was on a 5-point rating scale ranging from one to five, with one being a sign of complete disagreement, whereas five suggesting complete agreement with the statement.

Table 5.4 Table describing the questions of the survey for measuring the quality of service

Measure	Question
Service Climate	Professors in my university have sufficient knowledge and the right kind of pedagogical skills to deliver superior quality lectures and perform outstanding research.
Empathy	Professors understand the specific needs of each student.
Performance	Professors deliver excellent lectures that are difficult to find in other universities.

4. Results and Analysis

Table 5.5 Descriptive statistics among the variables of the study (N = 30)

Variables	Mean	Standard Deviation
Age	40.73	8.60
Gender	0.70	0.47
PsyCap	4.08	0.75
Relationship Conflict	3.39	0.93
Task Conflict	3.63	2.41
Burnout	3.46	1.33
Quality of Service	3.46	0.99

Table 5.5 depicts that the data collected from the survey is mainly focused towards young professors, with the mean age being around 41 years. Most of the professors who participated in the study were males (70%).

The results for the psychological capital show that most professors have a high value for psychological capital of about 4 with minor deviations. A lower mean was found for relationship conflicts with also a greater spread across the values, as displayed by the higher standard deviation. Task conflicts varied greatly for individuals suggesting differences in ideas and opinions.

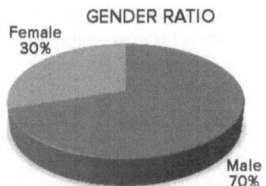

Fig. 5.3 Pie chart of the gender ratio of the subjects of the study (N = 30)

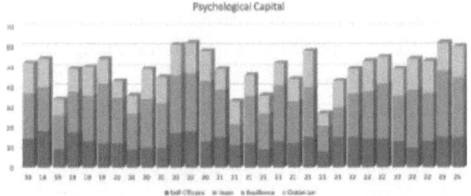

Fig. 5.4 Bar graph with various age group professors (x-axis represents ½ times the age) vs. their score of psychological capital (with subdivisions – self-efficacy, hope, resilience, and optimism) (N = 30)

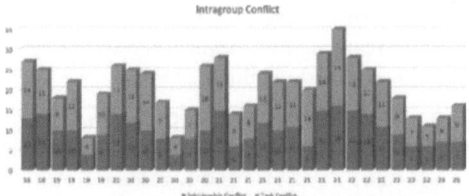

Fig. 5.5 Bar graph with various age group professors (x-axis represents ½ times the age) vs. their scores of intragroup conflicts (with subdivisions - relationship conflict and task conflict) (N = 30)

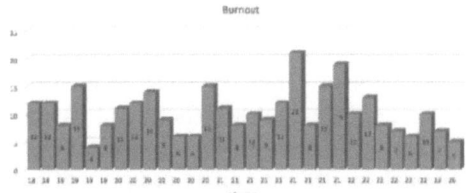

Fig. 5.6 Bar graph with various age group professors (x-axis represents ½ times their age) vs. their score of burnouts

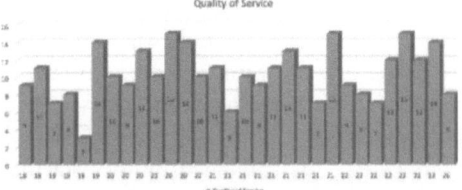

Fig. 5.7 Bar graph with various age group professors (x-axis represents ½ times their age) vs. their score of quality of service

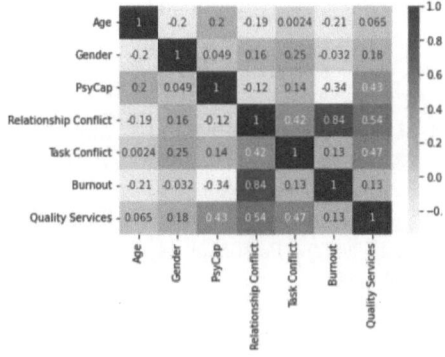

Fig. 5.8 Correlations between all the variables of the study (N = 30)

5. Findings and Conclusion

Psychological capital is associated with improved quality of service and reduced burnout. When the correlations chart is minutely observed, we see that psychological capital is having high correlation with the quality of service and a very low correlation with burnout. This explains that when professors have high psychological capital, they enjoy lesser burnout. Conflicts about personal issues or relationship conflicts are positively associated with burnout. Also, burnout has a high correlation with relationship conflict. So, when professors suffer burnout, one of the primary factors can be relationship conflict s/he has. It does not have a reasonable correlation with task conflict. Task conflict was not found to be associated with burnout. So, when professors suffer burnout, the reason need not necessarily be a task conflict. These findings may be explained by the fact that these disputes often involve lower levels of stress and anxiety than conflicts arising out of interpersonal relationships, which accounts for their non-significant link with burnout. Age and gender are insignificant factors when considering burnout and conflicts.

REFERENCES

1. Luthans, F., Avolio, B. J., Avey, J. B., & Norman, S. M. (2007). Positive psychological capital: Measurement and relationship with performance and satisfaction. Personnel psychology, 60(3), 541–572.

2. Bandura, A. (1994). Self-efficacy. Encyclopedia of human behavior, 4, 71–81.

3. Alam, A. (2022). Investigating Sustainable Education and Positive Psychology Interventions in Schools Towards Achievement of Sustainable Happiness and Wellbeing for 21st Century Pedagogy and Curriculum. ECS Transactions, 107(1), 19481.

4. Coutu, D. L. (2002). How Resilience Works. Harvard Business Review, May, 1 – 8.

5. Alam, A. (2022). Positive Psychology Goes to School: Conceptualizing Students' Happiness in 21st Century Schools While 'Minding the Mind!' Are We There Yet? Evidence-Backed, School-Based Positive Psychology Interventions. ECS Transactions, 107(1), 11199.

6. Fischer-Epe, M. (2016). Coaching (Vol. 5). Munich: Rowohlt.

7. Alam, A. (2022). Mapping a Sustainable Future Through Conceptualization of Transformative Learning Framework, Education for Sustainable Development, Critical Reflection, and Responsible Citizenship: An Exploration of Pedagogies for Twenty-First Century Learning. ECS Transactions, 107(1), 9827.

8. Luthans, F., & Youssef, C. M. (2007). Emerging Positive Organizational Behavior. Journal of Management, 6, 321–349.

9. Alam, A. (2022). Impact of University's Human Resources Practices on Professors' Occupational Performance: Empirical Evidence from India's Higher Education Sector. In Inclusive Businesses in Developing Economies (pp. 107–131). Palgrave Macmillan, Cham.

10. Reivich, K., & Shatté, A. (2002). The resilience factor: 7 essential skills for overcoming life's inevitable obstacles. New York: Broadway Books.

Note: All the figures and tables in this chapter were made the author.

Integrating Advancements in Education, and Society for Achieving Sustainability – Dimitrios A. Karras et al. (eds)
© 2024 Taylor & Francis Group, London, ISBN 978-1-032-70841-6

6

Application for Smartphones that Promotes Transgender Inclusion in Mainstream Society

Gnana Sanga Mithra S.*

Vinayaka Mission's Law School, Vinayaka Missions Research Foundation (DU)

Abstract: People whose gender identity differs from the gender assigned to them at birth are referred to as "transgender" under this general term. The term "transgender" was first used by American transgender activist Leslie Feinberg (1949–2014) to refer to all forms of gender nonconformity. For most transgender people living in Tamil Nadu, the exclusion is their main issue. Young trans people often end up in the care of deviant transgender people because they are unaware of the different welfare programs the government has launched. This research aims to analyze the awareness level of transgender individuals. A quantitative method was used to gather information from 56 transgender individuals residing in Chennai, Tamil Nadu. The outcome of the present study was the development of the smartphone application "Thirunar," which serves as a platform for raising awareness for young trans individuals.

Keywords: Transgender, Application, Welfare programs

1. Introduction

The World Health Organization states that "Gender refers to the socially constructed roles, attitudes, activities, and qualities that a given society believes suitable for men and women," while "Sex refers to the anatomical and physiological features that distinguish men and women." John F. Oliven, a professor at Columbia University, created the term "transgender" in his study "Sexual Hygiene and Pathology" (1965). People who identify as a gender other than the one that was assigned to them at birth are referred to as "transgender." The term "transgender" was used by American transgender activist Leslie Feinberg (1949–2014) to encompass all manifestations of gender nonconformity. Other transgender persons identify as non-binary or genderqueer, while some transgender people identify as either male or female.

2. Recognition and Acceptance

The landmark ruling by the Indian Supreme Court designated transgender people as the

*Corresponding author: sangamithra0212@gmail.com

DOI: 10.4324/9781032708461-6

third gender. Because transgender individuals are also Indian citizens, Justice K.S. Radhakrishnan, who presided over a two-judge supreme court panel, ruled on April 15, 2014, that they must be allowed equal opportunity to develop. Under constitutional protections, transgender people are entitled to fundamental rights. 2018 saw the decriminalization of homosexuality thanks to the judgment in the Navty Singh Johar v. Union of India case. The court was convinced that criminalizing sexual acts between consenting adults violated the equality guarantee of the Indian constitution. In his recent decision in the case of Arun Kumar and Sreeja vs. Inspector General of Registration and Others, issued by the Madurai bench of the Madras high court on April 22, 2019, Justice Swaminathan clearly distinguishes between sex and gender. The concept of gender—the idea that a person has the right to identify as a particular gender regardless of his or her sex—emerges when the two sexes of male and female do not apply to transgender people. According to Justice Swaminathan, transgender people are free to select their own gender. In accordance with Articles 14, 19, and 21 of the Indian Constitution, everyone living on Indian territory is entitled to equal protection under the law.

Transgender people were counted in the 2011 census as "other" after the Government of India approved the Technical Advisory Committee's recommendations. According to the Registrar General of India, as permitted by the Technical Advisory Committee, transgender people are given a specific code, "3," in the name of the "others" category, where "1" and "2" denote male and female, respectively. The 2011 census included data on transgender people's occupations, levels of literacy, and caste. Around 4.88 lakh transgender persons are thought to be living in India as a whole. In Tamil Nadu, there are about 22,364 transgender persons.

Tamil Nadu is a pioneer in providing comprehensive transgender-friendly policies. Inclusive Schemes for Transgender in the Indian States: The Tamil Nadu government established a Transgender Welfare Board in April 2008. The Social Welfare Minister serves as the board's president, and they are both joined by a transgender representative. Among the programmes are free housing and full scholarships for higher study. For their continued study, seats have been reserved at colleges and universities. Self-help groups are established and given loans with a 20% interest discount as alternative sources of income. The Welfare Board has provided transgender people with transgender identification cards. The state has announced separate ration cards for their free food subsidies.

A 1000-rupee pension plan for transgender people over the age of 40 was introduced. Government hospitals have made announcements about providing free sex reassignment surgery. The Tamil Nadu AIDS Initiative Voluntary Health Service established the "Manasu" (0091-44-25990505) helpline for members of the transgender community, and the Tamil Nadu Health Department's Principal Secretary launched it (TAI-VHS). The first LGBT helpline in India was established by Srishti Madurai, and it gradually grew to offer a 24-hour service with the tagline "Just having someone to talk to may save a life."

3. Methodology

Despite being recognized legally, individuals must still deal with problems like a lack of family support and unequal access to healthcare, education, and employment. Based on their gender transition, transgender people are split into two groups: transwomen (male to a female) and transmen (male to male) (female to male). Since transwomen disclose more frequently than transmen, who are a minority within the minority population, transwomen are more visible in society. Information was gathered from 56 transgender people in Chennai using the quantitative approach. based on the data received regarding the welfare measure's awareness. A smartphone application that contains numerous details regarding the plans announced by the Tamil Nadu state for inclusive development has been created.

4. Theoretical Framework- Goffman's Dramaturgical Perspective

Role Distinction: Following the transition process, transgender people distance themselves from their families while remaining connected to their peer groups. They become closer with other transgender people and join the transgender community in order to receive moral support.

Impression management: Transgender people present themselves to others in a way that makes them believe they are the gender they prefer, which is different from the gender they identify with at birth. They leave the house and assert their identification as the third gender if they are not welcomed by the family.

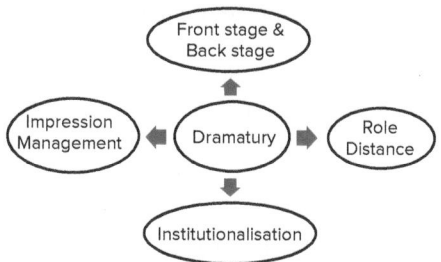

Fig. 6.1 Goffman's dramaturgical perspective

Institutionalization: People who identify as gender non-conforming leave their families and become a part of the community that embraces them for who they are. Within the group, there used to be a distinct hierarchy for transgender persons. These community members will assist the other members in improving their social standing.

Findings Nearly 37.50% of transgender women and 55.55 percent of transgender males fall within this age range. Nearly 23.08% of transwomen and 31.50% of transmen above the age of 31 fall into this category. Only 3.85% of transgender women and 7.40% of transgender men are older teenagers (16-20 years). 16.82% of transgender women and 5.55% of transgender men are older than 36.

Most respondents were between the ages of 26 and 30. In comparison to Transmen, transwomen come out earlier. The proportion of responders in their late 30s was lower. It foresees that many respondents who will emerge from the process will be engaged when they are older than 25.

The score of the relationship between gender and alienation within their community was significantly related (p=−291), where the p-value is lesser than 0.05. It is manifested that the gender transition of Transgender individuals influences the discrimination level within their community. The present research has evidenced that Transgender individuals do not have the proper support within the community. The acceptance level within the community is lower as the discrimination is higher among them. As a community, they have a different structural hierarchy through which everyone is related to the others.

Table 6.1 Spearman's correlation for measuring the relationship between gender and alienation within the community

Spearman's correlation		Gender	Isolation	
Spearman's rho	Current Gender	Correlation Coefficient	1.000	-.291**
		Sig. (2-tailed)	.	.000
		N	262	262
	Feel isolated	Correlation Coefficient	-.291**	1.000
		Sig. (2-tailed)	.000	.
		N	262	262

**. Correlation is significant at the 0.01 level (2-tailed).

Availability and Accessibility of Various Schemes Available for Transgender in Tamil Nadu.

For their overall growth, less than 3% of the respondents were successful in receiving educational subsidies. A little more than 15.27% of the transgender respondents reported receiving business grants from recognized self-help organizations. 11.07% of transgender people signed up for government-sponsored vocational training in fields like tailoring, soap making, and mask making. Only 3.05% of transgender people had obtained free Pattas from the government as part of the PMAY program. 29.77% of respondents who identified as transgender knew about the pension plan offered by the Tamil Nadu government. 4.81 percent of Transwomen people had gotten government-issued ration cards, and they were receiving food subsidies via the ration card.

5. Discussions

As a contribution of this research, "Thirunar" app has been developed which will be further channelized with the help of formal agencies. Through this app, gender non-conforming individuals can get contacted by NGOs headed by trans people who are working for the upliftment of Trans/gender non-conforming individuals where they will guide them in the proper way to get formal education and employment. As the world is moving toward digitalization, providing mobile phone applications might help young transgender to choose their future in the proper way.

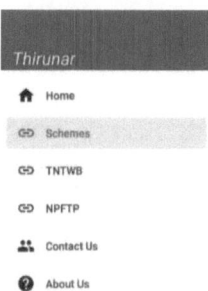

Fig. 6.2 Thirunar app home page

6. Conclusion

Thus, transgender individuals have been restricted from contributing to sociocultural, economic, and political activities. Lack of basic rights and recognition push them to do illegal activities which makes them deviant in society. Another aspect like a lack of awareness about the community, most of the young trans people who are in the initial stage of their transition get into contact with deviated individuals constrains their growth and development.

Social and economic aspects of transgender were identified with the level of family support, qualification, and income level. "After the 2014 judgment, their social identities have improved in society and they were also conscious of their rights and welfare measures." "The respondent feels that the lack of family support and low education chance out to be the major aspect to fall into deviant actions" (Gnana Sanga Mithra, Vijayalakshmi,2019). If parents understand the biological and psychological distinctions of transgender people, the pessimism towards them can be eliminated in society. In general, it is not easy to understand the state of mind of the transgender community. People need to

understand that feelings are diverse for every human being. The transgender community has a right to live as how they are and prompt their feelings without any distress. "Society needs to take care of social stigma towards the transgender community and give them a chance to stand equally and people should participate together in the developmental process of the community."

REFERENCES

1. Brumbaugh-Johnson, S.M., & Hull, K.E. (2019). Coming Out as Transgender: Navigating the Social Implications of a Transgender Identity. *Journal of Homosexuality*, 66, 1148–1177.

2. Carter SE, Ong ML, Simons RL, Gibbons FX, Lei MK, Beach SRH (2019),' Discrimination and suicidal Ideation among Transgender Veterans; The role of Social support and Connection, *LGBT Health*, 6.

3. Chakrapani, V., Vijin, P.P., Logie, C.H., Newman, P.A., Shunmugam, M., Sivasubramanian, M., & Samuel, M. (2017). Assessment of a "Transgender Identity Stigma" scale among trans women in India: Findings from exploratory and confirmatory factor analyses. International Journal of Transgenderism, 18, 271–281.

4. Dennis H. Li, Shruta Rawat, J. Michael Wilkerson (2017),' Harassment and Violence Among Men who have Sex with Men (MSM) and Hijras after Reinstatement of India's "Sodomy" Law', Sexuality Research and Social Policy, 14(3), 324–330.

5. Diamond, L.M., Pardo, S.T., & Butterworth, M.R. (2011). 'Transgender Experience and Identity.' *Handbook of Identity Theory and Research*, pp. 629–647.

6. Dutta, S., Khan, S.A., & Lorway, R. (2019). Following the divine: an ethnographic study of structural violence among transgender jogappas in South India. Culture, Health & Sexuality, 21, 1240–1256.

7. Gnana Sanga Mithra S and Vijayalakshmi V (2019), 'Changing Trends in Socio-economic condition of Transgender in Chennai city', International Journal of Engineering and Technology, 9(1).

8. S. Gnana Sanga Mithra & V. Vijayalaskmi (2022) A mathematical model for breaching the blockades-mainstreaming transwomen and transmen in Chennai by using block fuzzy cognitive, Journal of Interdisciplinary Mathematics, 25:3, 827–838, DOI: 10.1080/09720502.2021.2015100

9. Ristock, Janice (2005),' Relationship violence in Lesbian/ Gay/ Bisexual/ Transgender/ Queer [LGBT], *Violence against Women*, 367–375.

10. Roberts TK, Fantz CR (2014),' Barriers to Quality Health care for the Transgender population', *Clinical Biochemistry*, 47(10-11), pp. 983–987.

Note: All the figures and table in this chapter were made by the authors.

Integrating Advancements in Education, and Society for Achieving Sustainability – Dimitrios A. Karras et al. (eds)
© 2024 Taylor & Francis Group, London, ISBN 978-1-032-70841-6

Role of English in Enhancing Learning in Multilingual Classroom

Pritha Biswas, Divya R Krishnan, Umasankar M*

Christ Deemed to be University,
Bangalore, Karnataka

Abstract: Today, most educational institutions with English as their mode of instruction are a melting pot of multiple cultures and multiple languages. In such a heterogeneous milieu, English becomes the chosen language of communication. This study aims at understanding language anxiety and related emotional and psychological problems among first-year students from tier 3 cities and rural spaces in these multicultural and multilingual settings. A descriptive research design with an empirical approach will be adopted to address the objective of the study. A sample data size of 500 will be collected from the target population of first-year graduate students across disciplines from selected universities in Bangalore. A structured questionnaire will be used to capture the perception of students on the proposed research framework. The outcome of the study would help universities to understand the challenges of students in responding to an English-intensive environment and design an effective English pedagogical method.

Keywords: Multilinguistic, Multicultural, Language, Psychology, Pedagogy, Learning and Development

1. Introduction

The English language is the first global Lingua Franca and the commonly accepted language of scientific publications, entertainment, diplomacy, tourism, international business and trade and international telecommunications and media. Due to this, knowledge and a certain fluency in the language are desirable in a multinational and multicultural world to adequately comprehend, interact and engage in any field. In a country like India, English has long ceased to be a vestige of the colonial past and is now recognised as an official language used for administration and its related domains. However, the epicentres of engagements, usage and popularity of English have been concentrated in metropolitan and cosmopolitan cities. As

*Corresponding author: umasankar.m@christuniversity.in

DOI: 10.4324/9781032708461-7

a result, students from Tier III cities and rural spaces find communicating in English daunting due to a lack of exposure in their formative years of education. Therefore, when these learners enrol for higher education in institutions with multicultural and multilingual learning environments, they are unable to adequately communicate and comprehend with their peers and instructors. This causes anxiety, fear, and alienation while engaging with English, which has a debilitating effect on their confidence and learning. Our questionnaire is designed to elicit responses from first-year students of various universities to understand the nature of their language anxiety that proves to be a barrier to cohesiveness, involvement and cooperation in English intensive classrooms.

2. Literature Review

(Mazer et al. 2014), have focused their research on understanding the connection between the mode of communication employed by teachers in classrooms and its effects on the emotions felt and displayed by students. An analysis of the data collected by the authors depicts that when the teachers display chaotic and unclear means of communication for classroom subject content delivery, students are often left with negative emotions which by extension becomes a barrier to their effective learning. As an offshoot of these emotions, the students also become quite hesitant to approach the teachers when in doubt which further reduces their chances of having a meaningful outlook towards the teaching-learning process. The approach and findings in this article add to the premise taken by our research paper where we aim at understanding language anxiety and related emotional and psychological problems among first-year students from tier 3 cities and rural spaces in multicultural and multilingual settings. The first important variant of the learning space is the teacher-student communication and the resultant emotions invoked in the learner. As we set out to analyse language anxiety, an evocation of negative feelings, as understood from this research article can become a noteworthy factor.

(Qureshi, Javed, and Baig 2020), aims to locate psychological factors that influence speaking performativity in post-graduate students aspiring to become ELT and TESOL faculty. The results of the study indicate that psychological factors have a direct effect on anxieties among the students of the study. These factors mentioned in the article become pertinent to our research as we seek to recognise similar anxieties among the first-year students from tier 3 cities and the consequential effect on psychological and emotional encounters not only affecting their speaking skills but also their experience as learners.

(Pérez-García and Sánchez 2020) have studied the ability of Spanish EFL students to conceptualize emotion and express various emotions in English. The idea of emotions as an important linguistic category in language acquisition is investigated in the current research to acknowledge the anxiety, communication apprehension and emotional barrier towards the English language for students.

(Pickett and Fraser 2010) have put forth insights into the concept of the learning environment, the influence of learning environment on learner's cognitive and affective learning outcomes has been investigated along with perceptual differences in the learning environment for the tutor and learner. In terms of the current research work, students' mental barriers towards English usage can be seen as an important factor in the learning environment.

A review of the existing approaches to the current research area, a selection of which is given above, validates the relevance of the selected topic and given the student community and their barriers in terms of language and learning necessitates a study of the various factors that cause language anxiety among students of Tier III cities and explores the specific nature of this anxiety.

3. Aim of the Study

The current study is aimed at ascertaining the emotional encounters of students in a multilingual and multicultural environment thereby understanding why there is an emotional encounter and how it is affecting student engagement in the class. The study is also intended to identify the role of English Language Communication in overcoming cultural barriers. The study aims to understand what can be done further to promote healthy competitive culture among students in a multilingual and multicultural environment.

4. The Objective of the Study

The study is intended to understand the impact of emotional encounters of students on their classroom engagement in a multilingual and multicultural environment.

5. Methodology

An empirical research approach has been adopted in which this paper focuses solely on describing, emphasizing and measuring the phenomena in multicultural environments existing in the education sector. It is a theory-based method which is created by gathering data from undergraduate students studying in universities and analysing and presenting the collected data. The results will be based on a real-life experience that we come across by meeting and learning from the subjects and their responses to the survey. This can provide significant insights into why this problem exists and understand the need for change. We distributed questionnaires to the target population (Undergraduate students). A purposive sampling technique was adopted to select the respondent for the study. The study targeted for 250 responses and 300 Questionnaires were distributed, finally, 211 completed responses were received. Initially, a pilot study was conducted, and the questions and variables were adjusted accordingly to suit the study. The variables were tested and analysed using a hierarchical regression model.

6. Research Gap

Many papers in multilingual or multicultural studies have mainly focused on either the language aspect or the cultural aspect. But this paper concentrates on linking culture and language with the emotional and psychological aspects of students. This paper is intended to explore their emotional encounters rather than their performance. This paper covers only three emotions that are common and sudden reflections in the given situation namely Fear, Anxiety and Shyness.

7. Operational Model

An operational model was designed based on the factors identified from the existing literature and studying the current phenomenon in the multilingual and multicultural environment. The hypothesised model has three emotions influencing the students' classroom engagement concerning their environment and mediated by English Language Communication.

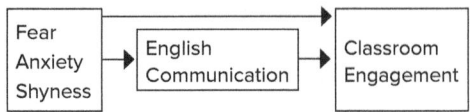

Fig. 7.1 Hypothesised model

8. Analysis

Table 7.1 Demographic profile of the respondent

Demographic Profile	Categories	Number of Respondents	% of Respondents
Gender	Male	103	48.8%
	Female	104	49.3%
	Prefer Not to say	4	1.9%
Types of School	State Board	33	15.6%
	Matriculation	4	1.9%
	CBSE	129	61.1%
	ICSE	41	19.4%
	Others	4	1.9%
Father's Occupation	Business	81	38.4%
	Private Job	82	38.9%
	Govt. Job	19	9.0%
	Professional	17	8.1%
	NA	12	5.7%
Family Type	Joint Family	47	22.3%
	Nuclear Family	164	77.7%
Economic Status	Below Middle Class	6	2.8%
	Middle Class	83	39.3%
	Upper Middle Class	78	37.0%
	Rich Class	7	3.3%
	Prefer Not to say	37	17.5%
State of Origin	Karnataka	45	21.3%
	Kerala	46	21.8%
	Tamil Nadu	57	27.0%
	North India	63	29.9%

Table 7.1 explains in detail the profile of the respondents. The data starts from the variable Gender, which clearly states that respondents are equally distributed between male and female categories. And it's evident with 1.9% of responses that fewer number were not preferred to reveal their gender. The highest percentage falls against the category of CBSE in terms of types of schools with 61%. Notably, there are 15% of respondents from state board schools are also part of this study. About 38% of students revealed that their father's occupation was Business and Private Jobs. 78% of the participants come from a nuclear family. When it asked about the economic status of the family, an equal percentage of respondents i.e. closer to 39% of respondents stated that they belonged to the middle class and upper middle class. A major pool of respondents is filled by students from three southern states of India and the remaining are from northern states of India.

ANOVA (Analysis of Variance) is a tool that helps in checking the effect of one or more factors by comparing means of various samples. It also guides to checking if experimented outcome/result is significant and to verify if the statement accepts or rejects the null statement.

Table 7.2 ANOVA comparing means of study constructs with demographic profile

Constructs	Fear Factor		Anxiety Factor		Shyness Factor		English Communication		Classroom Engagement	
Demographic Profile	Sig.	H_a Accepted	Sig.	H_a Accepted	Sig.	H_a Accepted	Sig.	H_a Accepted	Sig.	H_a Accepted
Gender	.872	No	.225	No	.365	No	.383	No	.585	No
Types of School	**.028**	**Yes**	**.007**	**Yes**	.762	No	.151	No	.186	No
Father's Occupation	.200	No	.447	No	.092	No	.204	No	.391	No
Family Type	.082	No	.082	No	.096	No	.345	No	.184	No
Economic Status	.110	No	**.002**	**Yes**	**.000**	**Yes**	**.001**	**Yes**	**.000**	**Yes**
State of Origin	.359	No	.858	No	.404	No	.606	No	**.023**	**Yes**

Table 7.2 depicts that, Types of School has a greater variance in Fear factor and Anxiety Factors with the significance of 0.028 and 0.007 respectively and accepts the alternate hypothesis. The economic status family showed significant variance among all the variables except the Fear factor with a significant value of less than 0.05. There is significant variance in the perception of respondents towards classroom engagement based on the Economic status and State of origin of the students.

Table 7.3 Hierarchical regression models on factors influencing classroom engagement

Models	1	2	3	4	5	6	7
(Constant)	2.32 ***	2.23 ***	2.24 ***	2.23 ***	1.84 ***	1.55 ***	1.43 ***
	−0.2	−0.21	−0.23	−0.19	−0.21	−0.23	−0.22
Fear	0.45 ***				0.29 ***	0.20 **	0.16 *
	−0.05				-0.06	−0.06	-0.06
Anxiety		0.46 ***			0.28 ***	0.24 ***	0.18 **
		−0.05			−0.06	−0.06	−0.06
Shyness			0.44 ***			0.20 **	0.09
			−0.06			−0.06	−0.06
English				0.47 ***			0.25 ***
				−0.05			−0.06
R Square	0.36	0.35	0.32	0.42	0.45	0.48	0.54
Adj. R Square	0.36	0.35	0.32	0.41	0.44	0.47	0.53
Num. obs.	140	140	140	140	140	140	140

*** $p < 0.001$; ** $p < 0.01$; * $p < 0.05$

Table 7.3 displays the hierarchical regression analysis consisting of seven different models. The first three models tested the impact of predictors on the dependent variable which is Classroom Engagement and the fourth model tested the influence on the dependent variable accounted by the mediating variable which is English Communication. All four variables compared with the dependent variable significantly influence the variations in the dependent variable which can be confirmed by the star symbol presented along with coefficient values. The standard error values are mentioned within the parenthesis below the coefficient values. Hence it could be interpreted that individually all these four factors are significantly influencing the classroom engagement of students in multicultural and multilingual classrooms. Each of the variable's coefficients is above 0.4 which represents the contribution value of predicting the variation in Classroom Engagement. Models from the fifth and

sixth describe the influence of independent variables on the dependent variable when they are combined. The sixth model denotes the changes in the impact of the predictor variables on Classroom Engagement when the inclusion of mediating variable. Here in each stage or model, one independent variable is added to the existing independent variable to test the combined influence on the dependent variable. In Model 7, the mediating variable i.e. English Communication added its combined effects are displayed; the mediating variable is significantly influencing the dependent variable with a coefficient of 0.25 which is said to be the highest in the overall model. When the mediating variable is introduced, it can notice that there is a change in the beta value and level of significance. Here it is revealed that the Fear factor is 0.16 and the Anxiety Factor is 0.18, but when it comes to the Shyness factor it is just 0.06 and it lost its significance in influencing the dependent variable. This means that the mediating variable of English Communication acts as an indirect mediator by reducing the negative impact of Fear and Anxiety factor towards Classroom Engagement. Moreover, English Communication act as a full mediator between the Shyness factor and Dependent Variable as the Shyness factor lost its significance after the inclusion of the mediating variable in model 7.

9. Discussion and Conclusion

The population of this research consists of students pursuing their graduation in selected universities in Bangalore which admit students from multicultural and multilingual backgrounds. The commonality among the above-mentioned population is that they belong to a different states of origin and have different languages as their mother tongue. Most of the respondents have English as the only common communicating language in the multilingual environment (Aldridge, Fraser, and Laugksch 2011). That means a larger amount of their communications are only in English. Hence, they have little space to communicate with others in their language. They have limited knowledge of the other languages or the local languages. Based on the current study it is found that three key factors influence classroom engagement in the multilingual classroom namely, Fear factor, Anxiety factor and Shyness factor (Deng, Kiramba, and Viesca 2021). Demographic factors such as type of school, family type, economic status and state of origin are the fundamental elements correlated with their level of engagement in the class. Nonetheless, when we draw attention to the factors impacting classroom engagement, it's visible that all three factors

are highly influencing. Among the three factors, anxiety is the factor found to be causing the highest impact on the dependent variable. Since most of the students belonging to tier II and III cities who were there in Bangalore for the first time were facing some cultural shock and which resulted in Fear, Anxiety and Shyness among the students. Fear factor includes the fear of making mistakes while communicating with others in a new environment or multicultural environment (Harvey, Baumann, and Fredericks 2019). Anxiety and shyness reflected comparing themselves with others and domineering characteristics of students from metropolitan and tier I cities (Fisher et al. 2022). Due to this phenomenon, they usually prefer subtle expressions.

Aspects like peer pressure and cultural bias are prevalent but if the communication skill of the student is good, they might not be affected by these factors but rather select their group and manage better ways in the situations (Sun et al. 2022).

A few suggestions according to these findings are, communication language is very crucial to aid students in a multicultural and multilingual environment, hence the students at a school level need to be trained in developing communication skills. Bridge courses and icebreaking sessions should be introduced to induce the habit of communicating and socializing with each other. An initial assessment system can be introduced to test the language and communication skills of the students and according to the insights gained, a mechanism of communication and cultural enrichment laboratory/environment can be established. In a multicultural and multilingual environment some students are so burdened by language anxieties that eventually they lose a sense of purpose, face identity crisis and do not achieve their potential. Here they may be identified and given counselling and mentoring which addresses their specific language barriers and help them to be on par with their peers and the demands of the multicultural environment. Educational institutions can benefit from this paper to build new tools and approach students from diverse backgrounds and small towns. Students with language anxieties have a greater chance of actualizing their potential in an inclusive and understanding environment. A common concern with such students is their lack of awareness about the ways of adopting and adapting to multicultural and multilingual spaces despite having a keen interest in the same. Education coupled with competitiveness is the major goal in today's world hence academic spaces should include systems and mechanisms to promote healthy competitiveness by enabling students to overcome fear, anxiety and shyness in their communication through English.

REFERENCES

1. Aldridge, Jill M., Barry J. Fraser, and Rüdiger C. Laugksch. 2011. "Relationships between the School-Level and Classroom-Level Environment in Secondary Schools in South Africa." *South African Journal of Education* 31(1):127–44. doi: 10.15700/saje.v31n1a407.

2. Deng, Qizhen, Lydiah Kananu Kiramba, and Kara Mitchell Viesca. 2021. "Factors Associated With Novice General Education Teachers' Preparedness to Work With Multilingual Learners: A Multilevel Study." *Journal of Teacher Education* 72(4):489–503. doi: 10.1177/0022487120971590.

3. Fisher, L., M. Evans, K. Forbes, A. Gayton, Y. Liu, and D. Rutgers. 2022. "Language Experiences, Evaluations and Emotions (3Es): Analysis of Structural Models of Multilingual Identity for Language Learners in Schools in England." *International Journal of Multilingualism* (May):1–21. doi: 10.1080/14790718.2022.2060235.

4. Harvey, Marina, Chris Baumann, and Vanessa Fredericks. 2019. "A Taxonomy of Emotion and Cognition for Student Reflection: Introducing Emo-Cog." *Higher Education Research and Development* 38(6):1138–53. doi: 10.1080/07294360.2019.1629879.

5. Mazer, Joseph P., Timothy P. McKenna-Buchanan, Margaret M. Quinlan, and Scott Titsworth. 2014. "The Dark Side of Emotion in the Classroom: Emotional Processes as Mediators of Teacher Communication Behaviors and Student Negative Emotions." *Communication Education* 63(3):149–68. doi: 10.1080/03634523.2014.904047.

6. Pérez-García, Elisa, and María Jesús Sánchez. 2020. "Emotions as a Linguistic Category: Perception and Expression of Emotions by Spanish EFL Students." *Language, Culture and Curriculum* 33(3):274–89. doi: 10.1080/07908318.2019.1630422.

7. Pickett, Linda, and Barry Fraser. 2010. "Creating and Assessing Positive Classroom Learning Environments." *Childhood Education* 86(5):321–26. doi: 10.1080/00094056.2010.10521418.

8. Qureshi, Habiba, Fareeha Javed, and Sana Baig. 2020. "The Effect of Psychological Factors on English Speaking Performance of Students Enrolled in Postgraduate English Language Teaching Programs in Pakistan." *Global Language Review* V(II):101–14. doi: 10.31703/glr.2020(v-ii).11.

9. Sun, Xiaojing, Marloes M. H. G. Hendrickx, Thomas Goetz, Theo Wubbels, and Tim Mainhard. 2022. "Classroom Social Environment as Student Emotions' Antecedent: Mediating Role of Achievement Goals." *Journal of Experimental Education* 90(1):146–57. doi: 10.1080/00220973.2020.1724851.

Note: All the figure and tables in this chapter were made by the authors.

Integrating Advancements in Education, and Society for Achieving Sustainability – Dimitrios A. Karras et al. (eds)
© 2024 Taylor & Francis Group, London, ISBN 978-1-032-70841-6

Testing the Nexus between Quality of Work Life and Continuance Commitment in Selected Companies of Western Maharashtra

Pravin Vitthal Yadav*, Umesh S. Kollimath, Shriram Shripad Badave

Anekant Institute of Management Studies (AIMS), Baramati, Dist-Pune (MAH) 413102 India

Dnyaneshwar Tukaram Pisal

SVPM's Institute of Management, Malegaon BK, Baramati, Dist-Pune (MAH) 413115 India

Sudarshan Arjun Giramkar

Parikrama College of Management, Shrigonda, Dist-Ahmednagar, 413701, India

Abstract: The current study's goal is to identify the nexus of "Quality of work life" and "Continuance commitment" among blue-collar workers of selected manufacturing enterprises in western Maharashtra. This aspect of blue-collar workers' "Quality of work life" and its association to "continuance commitment" were gauged by conducting survey. 390 samples were probed throughout the process. The "Pearson Correlation Coefficient (r)" statistical test was performed to understand the nexus. Pursuant to the findings, various aspects of work-life quality are found to be highly linked to employees' continuance commitment to the company. Employees that have a strong feeling of loyalty to the company do so because they believe they have few other options or that quitting would be expensive. They more likely to remain in the organization for longer periods of time.

Keywords: Continuance commitment, Quality of work life, Regression, Reliability

1. Introduction

The fast changing and unpredictable business environment have paved way to people with many job alternatives if one possesses specified skill set. On the other hand, people element is gaining paramount importance to every organization. Competitive advantage/s can be gained only through consistently reaching set goals, setting appropriate benchmarks for organizational performance, etc. Competent and skilled people are pivotal to this. This requires organizations to adopt strategic human resource management (Agarwala, 2010). Hence, a well-oiled organization implies necessity of top-tier talent, tethering the employees' hearts and minds to the organizational cause, and forging long-lasting teams (Garg, 2019). But it is not so simple to achieve this because; there is "war for talent". To win the war and gain competitive advantage "Quality of work

*Corresponding author: praveen26dec@gmail.com

DOI: 10.4324/9781032708461-8

life (QWL)" is key to success for business enterprises. Human resources are the "gears" of corporations, delivering the product or service that allows firms to fulfil their goals and survive in society through their talents, methods, knowledge, and labour. A healthy work life is the cornerstone for increased productivity. Organizational development is dependent on the people engaged in order to achieve their "task", "goals" and "outcomes".

But, the same is not universally true. People may stick to the organization for a prolonged period due to lack of other better options, personal problems/interests, anticipation of better compensation, etc. Therefore, 'continuance commitment' assumes importance in this context. Continuance commitment is associated with less desirable job conduct (Meyer et al., 2002). It's the employees willingly or deliberately staying with an organization as they don't want to risk the existing perks from the present occupation. An employee may believe that the personal consequences of leaving present organization would be adverse, compelling him/her to remain loyal to the present job. This form of loyalty is based on the benefits that an organization provides to its personnel. Employees manifest job-loyalty so that nothing of worth is lost. Employees who are committed to staying with an organization for a long time will be retained. Thus, it can be surmised that individuals with a long-term commitment stay with a company because of the money and other investments they've made during their present tenure and not because they believe in their organization's values.

The present article in essence, takes an effort to identify the connection of "Quality of work life" parameters and "Continuance commitment" with reference to blue-collar

workers of selected manufacturing industries in western Maharashtra.

2. Conceptualization of Variables

Quality of Work Life (QWL)

Usage viz., "quality of work life" originally emerged in the research publications and the print media in the United States during 1970s. This can be attributed to the early works of Louis Davis. The "International Council" for "Quality of Work Life" founded in nineteen seventy two & hosted the event in Toronto. (Bindu & Yashika, 2014). Fundamentally, Eight factors for employment, including aspects of a person's work history or work environment, were used to interpret QWL (Bhola, 2018). "Adequate and fair wages", "safe and healthy working conditions", "opportunity to use and develop human capacities", "opportunity for continued career growth and security", "social integration in the workplace", "constitutionalism in the work organization", "balanced role of work in total life space", and "social relevance of work" are the eight broad conditions of employment that Richard E. Walton defines (Fernandes et al., 2017).

Continuance Commitment

Allen and Meyer have identified the third level of organizational commitment as continuity commitment (1990) (Allen & Meyer, 1990) express 'perseverance commitment' as a mathematical function of the volume and/or quantity of individual investments (or side-bets), as well as a scarcity of alternatives. Individuals committing a considerable time and efforts to attain specialized skills in order to enhance their earning potential. Employees can only benefit from the greater pay by staying with the company in this situation. Continuance commitment is seen as a consequence of individual decisions to continue with a firm due to the pre-existing personal time and resources considerations, as much as the switchover costs to a different occupation. Thus, those who have considerably invested in their company are least interested to leave. Besides the fear of losing their investments in the present organization, (Allen & Meyer, 1990) claimed that individuals build continuation commitment due to a perceived lack of alternative. They imply that a person's loyalty to the organization is based on their personal appraisal of the opportunities available outside. An employee may presume that the abilities he/she has acquired cannot compete in the field. This leads to such employees getting stuck with their present organizations. Those who labour in environments where the training and skills acquired are organization-specific may fall in such traps. As a result of the monetary, psychological, social, and other expenses involved with quitting the business, the employee is compelled to remain loyal to the organization. Unlike affective commitment, which is founded on a connection to a feeling and emotions towards the organization, continuance commitment when an employee feels they have a lot to lose if they quit their company, they are more likely to remain committed to it.

3. Review of Literature

(Suliman & Iles, 2005) say that continuous commitment is an affirmative organizational behavior. It was also discovered that Gender and commitment to continue have a negative relationship. "Quality of work life" has a lesser but detrimental effect on

"Continuance commitment", according to (TADEMR AFAR, 2015). As "Quality of work life" improves, certainly this leads to a plunge in commitment to continue. At both state and foundation universities, the "Quality of work life" influences around 3% of the variation in academicians' "Continuance commitment". QWL and Continuance Commitment of Academic Staff were found moderate association (Farid et al., 2015). The findings contributed to efforts by (Ahmadi et al., 2012) to raise "Quality work life" and "Organizational commitment" among employees of public organizations in Kurdistan. Respondents identified growth and development as a crucial antecedent of all four forms of commitment studied: "affective", "normative", "continuance (alternatives)", and "continuance (maintenance) (cost)". Participation has a considerable positive link with pledges to continue (alternatives) and continue (cost). The chance for continuing progress, stability, and the social importance of work-life are all favorably associated to continuance commitment, according to (Yusoff et al., 2015). Employee loyalty, defined as their attachment to the company because they believe quitting is costly. The "Continuance commitment" was not found to be statistically considerable as a intermediary for the link in QWL factors and turnover intention in this study. Furthermore, the most important factor determining organisational commitment is the work environment (Adikoeswanto et al., 2020). As a result, high employee work life quality can lead to continued commitment, this is one of the factors that contribute to employees sticking with a company for a long time.

4. Research Hypothesis

H0: "There is no significant relationship between Quality of Work Life (QWL) and Continuance Commitment of Labours".

H1: "There is significant relationship between Quality of Work Life (QWL) and Continuance Commitment of Labours".

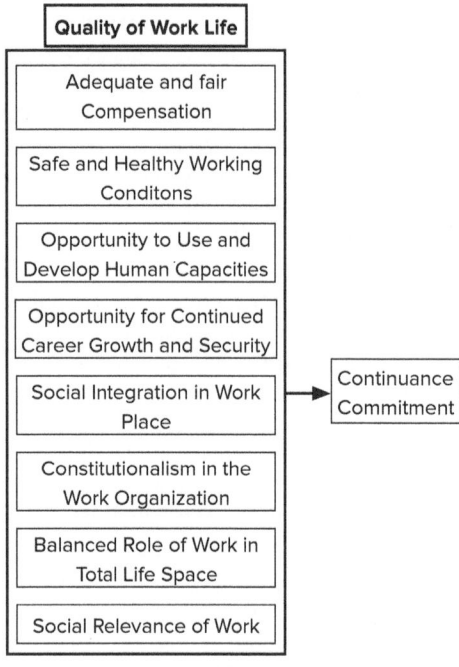

Fig. 8.1 Research model

Source: Authors

5. Research Methods

Design of the Study

Since the current study is a detailed inquiry that is driven by hypotheses, it is diagnostic in character. Because it focuses on the link and association between variables like QWL and Continuance Commitment, the diagnostic inferential research approach is employed. To evaluate the effect of QWL on Continuance Commitment, primary data in the form of feedback from a sample of blue-collar workers was gathered. The primary information from respondents was gathered using three different organized, closed-ended schedules.

Sampling Design

Workers, or blue-collar workers, from big industrial firms in western Maharashtra were taken into consideration and chosen as a sample. The sample was drawn using the simple random sampling approach. Due to the size of the sample universe, the researcher sought representative samples from each organization. We personally spoke with representatives from each firm to collect the replies. In western Maharashtra, there are 677 significant manufacturing enterprises, 114 of which were chosen because their combined workforce exceeds 500. 1,36,036 people made up the research's sample universe. The sample size was calculated using a common statistical procedure by the researcher. The exact sample size was determined by adding the samples from each firm. This study's real sample size is 390 respondents.

Measurement

Schedule I, II, and III's final data were subjected to a "Cronbach's Alpha" reliability study with SPSS. It was determined whether QWL has an impact on continuance commitment using the "Simple Linear Regression" and "Multiple Linear Regression" tests. To determine whether there was a relationship between QWL and continuance commitment, the hypothesis was put to the test using the "Pearson Correlation Coefficient (r)" test.

6. Data Analysis

To see if these parameters are present at work, quality of work life's and continuance commitment standard statements were tested on samples. A five-point rating scale is used to assess the status of "agreement" and "disagreement" specifically since I = "Strongly Disagree" to V = "Strongly Agree". These parameters are assigned ranks depend upon "mean score".

Table 8.1 "Quality of work life (QWL)" and "continuance commitment": A simple linear regression analysis

Model Summary				
Model	R	R-Square	Adjusted R-Square	Std. Error of the Estimate
1	0.944	0.892	0.892	0.09454

a. Predictors: (Constant), Quality of work life (QWL)
Source: Researcher's compilation

The given table illustrates "R, R-Square, Adjusted R Square and Std Error." R expresses the association between the dependent variable's observed and forecasted values. In the given table R = .944. The model summary and overall fit statistics are also shown in the table above. Researcher reveals the adjusted R^2 is .892 with the R^2 = .892 which state "linear regression" exhibits 89.2 % of the variance in the given data.

Table 8.2   Anova

"ANOVA"					
Model	Sum of Squares	df	Mean Square	F	Sig.
Regression	28.657	1	28.657	3205.953	0.00
Residual	3.468	388	0.09		0.00
Total	32.125	389			

a. Dependent Variable: Contiunance commitment
b. Predictors: (Constant), Quality of Work Life (QWL)

Source: Researcher's compilation

The F-test is shown in the given table. F-test in the "linear regression", the null hypothesis is that "there is no linear relationship between the two variables". This assessment is substantial in terms of F = 3205.953 and 389 degrees of freedom, therefore do accept the fact that the variables are in our model, "Quality of Work Life (QWL)" and "Continuance Commitment", have a linear relationship.

Table 8.3   Coefficients

Coefficients					
Model	B	Std. Error	Beta	t	Sig.
1 (Constant)	1.098	0.62	0.944	17.842	0.00
Q WL	0.779	0.014		56.621	0.00

a. Dependent variable: Continuance commitment
Source: Researcher's compilation

According to the tabular regression equation of Continuance Commitment is: Continuance Commitment = 1.098+.779 (Quality of Work Life (QWL)). The value of coefficient denotes for each additional mean of "Quality of Work Life (QWL)" the "Continuance Commitment" to rise by an average of .779.

Table 8.4   Effect of "quality of work life (QWL)" factors on "continuance commitment" of workers

Coefficients						
		B	Std Error	Beta	t	Sig.
1. Adequate fair compensation	X1	1.677	0.86	0.129	19.47	0.000
2. Safe & Healthy Environment	X2	0.50	0.011	0.208	4.41	0.000
3. Development of Human capacities	X3	0.066	0.09	0.264	7.35	0.000
4. Growth & Security	X4	0.112	0.010	0.24	10.757	0.000
5. Social Integration	X5	0.088	0.009	0.216	9.58	0.000
6. Constitutionalism	X6	0.075	0.010	0.222	8.62	0.000
7. Total Life Span	X7	0.08	0.011	0.191	8.158	0.000
8 Social Relevance	X8	0.123	0.014	0.223	6.497	0.000

Dependent variable: Continuance commitment
Source: Authors

A multiple regression was calculated in the above table to predict "Continuance Commitment" based on "Quality of Work Life (QWL)" characteristics. The effect of eight "Quality of Work Life" (independent variable) elements has been studied using a multiple regression equation devised by the researcher on "Continuance Commitment" (dependent variable).

$$CC = 1.677 + 0.050*X1 + 0.066*X2 + 0.112*X3 + 0.088*X4 + 0.075X5 + 0.080*X6 + 0.070*X7 + 0.123*X8$$

It is obvious from the table above that "Development of human capacities" has a greater influence on Continuance commitment which has a "Beta" value of 0.264 subsequently "Growth and security" & "Social relevance" which has a "Beta" values 0.240 & 0.223 correspondingly.

7. Hypothesis Testing

Table 8.5 The correlation between "quality of work life (QWL)" and "continuance commitment (CC)"

"Correlations"		"QWL"	"CC"
"QWL"	"Pearson Correlation"	1	.944**
	"Sig. (2-tailed)"		.000
	"N"	390	390
"CC"	"Pearson Correlation"	.944**	1
	"Sig. (2-tailed)"	.000	
	"N"	390	390

**. "Correlation is significant at the 0.01 level (2-tailed)."

Source: Researcher's compilation

Pearson's r indicates correlation between "Quality of Work Life" (QWL) and "Continuance Commitment (CC)" that is 0.944. this is extremely close to 1. Thus, it states that there is a significant and beneficial relationship between "Quality of Work Life (QWL)" and "Continuance Commitment (CC)".

"P" value is 0.000, and this number is <0.05, the null hypothesis is rejected, and the alternative hypothesis, which states significant link between the two variables, *i.e.* "Quality of Work Life (QWL)" and "Continuance Commitment" (CC) is accepted.

8. Discussion of the Findings

Following is the result of a consensus examination of numerous standard statements of continuance commitment: the highest agreement was found with "I continue with my present job as leaving asks for considerable personal sacrifice—another organization may not offer equivalent benefits", followed by "It wouldn't make much difference for me if I leave my present organization", and "One of the few serious consequences of leaving this organization would be that I would have to leave this organization", followed by "It wouldn't be too costly for me to leave my organization now", "My loyalty with present organization is a matter of necessity and not desire". "I believe, I have few options to consider leaving this organization", "I can't leave my present job right now, even if I wish to". The researcher observed that the F-test in linear regression contains the null hypothesis i.e two variables have no "linear relationship" after utilizing Simple Linear Regression to study the impact of two variables. The assessment is substantial through F = 3205.953 and 389 degrees of freedom, therefore one has to accept that variables in our model, *i.e.* "Quality of Work Life (QWL)" and "Continuance Commitment", have a linear relationship. Continuance Commitment has the following regression equation: Continuance

Commitment = 1.098+.779 ("Quality of Work Life" (QWL)). In this study coefficient suggests that the "Continuance Commitment" rises average of .779 for every incremental mean of "Quality of Work Life (QWL)". Effect of eight "Quality of Work Life" factors (independent variable) on "Continuance Commitment" has been studied using a multiple regression equation (dependent variable). The beta value of 0.264, "Development of human capacities" has the greatest effect on continuance commitment, followed by "Growth and security", and "Social relevance", through beta values of 0.240 and 0.223, respectively. The association between "Quality of Work Life (QWL)" and "Continuance Commitment (CC)" is 0.944, according to Pearson's r. The Pearson's r value is 0.944, which is quite near to 1. However, it was discovered that "Quality of Work Life (QWL)" and "Continuance Commitment" had a substantial and favorable link. Thus, null hypothesis is ruled out, whereas the alternative hypothesis, i.e there is a strong association amid "Quality of Work Life (QWL) and Continuance Commitment (CC)", is accepted because the 'P' value is 0.000, which is less than 0.05.

9. Conclusions and Future Implications

The findings of the study endorse the association between quality of work life and continuance commitment of the employees. The findings of the study indicate an employee has a feeling of devotion to the company because the employee believes that quitting would be too expensive or that there are few other options. many work-life aspects promoting long-term commitment as well as the development of human capacities that boost employees' goal-orientation and drives high levels of engagement among employees. It is found that the workers get loyal to an organisation since they believe in present work life quality as valuable and that it fulfils a meaningful organisational goal. High-continuance commitment employees are more likely to stay with the company for longer period of time. This research suggests that multiple elements of quality and meaningful work can significantly improve long-term commitment, which has practical consequences for organisations looking to re-strategize their commitment-boosting initiatives. Since then, there has been a pragmatic shift in organisational research, with a greater emphasis on positive elements of work and life; these concepts have begun to generate buzz among academic and organisational researchers, indicating the need for more research.

REFERENCES

1. Adikoeswanto, D., Eliyana, A., Sariwulan, T., Buchdadi, A. D., & Firda, F. (2020). Quality of Work Life's Factors and Their Impacts on Organizational Commitments. 11(7), 450–461.

2. Agarwala, T. (2010). Strategic Human Resource Management (9th ed.). Oxford University Press.

3. Ahmadi, F., Salavati, A., & Rajabzadeh, E. (2012). Survey Relationship Between Quality of Work Life and Organizational Commitment in Public Organization in Kurdistan Province. Interdisciplinary Journal of Contemporary Research in Business, 4(1), 235–247.

4. Allen, N. J., & Meyer, J. P. (1990). The measurement and antecedents of affective, continuance and normative commitment to the organization. 1–18.

5. Bhola, S. S. (2018). Assessment of Quality of Work Life with Reference to. November.

6. Bindu, J., & Yashika, S. (2014). Quality of Work Life with Special Reference to Academic Sector. 3(1), 14–17.

7. Farid, H., Izadi, Z., Ismail, I. A., & Alipour, F. (2015). Relationship between quality of work life and organizational commitment among lecturers in a Malaysian public research university. Social Science Journal, 52(1), 54–61. https://doi.org/10.1016/j.soscij.2014.09.003

8. Fernandes, R. B., Bruna, Martins, S., Pereira, R., Custódio, C., Da, G., Filho, C., Guilherme, Braga, A., Luiz, & Antonialli, M. (2017). Quality of Work Life: an evaluation of Walton model with analysis of structural equations Qualidade de vida no trabalho: uma avaliação do modelo de Walton, com análise de equações estruturais. Pág, 38.

9. Garg, P. (2019). Testing the reciprocal relationship between quality of work life and subjective well-being: a path analysis model Testing the reciprocal relationship between quality of work life and subjective well-being: a path analysis model Shivani Agarwal * Pooja G. July. https://doi.org/10.1504/IJPOM.2019.10022103

10. Klein, L. L., Pereira, B. A. D., & Lemos, R. B. (2019). Quality of working life: Parameters and evaluation in the public service. In Revista de Administracao Mackenzie (Vol. 20, Issue 3). https://doi.org/10.1590/1678-6971/eRAMG190134

11. Meyer, J. P., Stanley, D. J., Herscovitch, L., & Topolnytsky, L. (2002). Affective, Continuance, and Normative Commitment to the Organization: A Meta-analysis of Antecedents, Correlates, and Consequences. 52, 20–52. https://doi.org/10.1006/jvbe.2001.1842

12. Suliman, A., & Iles, P. (2005). Is continuance commitment beneficial to organizations? Commitment-performance relationship: a new look.

13. Taşdemir Afşar, S. (2015). Impact of the Quality of Work-Life on Organizational Commitment: A Comparative Study on Academicians Working for State and Foundation. ISGUC The Journal of Industrial Relations and Human Resources, 17(2), 45–75. https://doi.org/10.4026/1303-2860.2015.0278.x

Integrating Advancements in Education, and Society for Achieving Sustainability – Dimitrios A. Karras et al. (eds)

Analytics for Public Governance: A Case of Aspirational Districts Program of NITI Aayog

9

Pratham Parekh*

Assistant Professor, Nirma University, Ahmedabad, Gujarat, India

Abstract: In January 2018, NITI Aayog, initiated a Transformation of Aspirational Districts Program (TADP) for data-led development monitoring and transformation of 112 most backward districts. These districts covers 15% of India's total population. The uniqueness of the program is that this program is considered as world's largest ICT-enabled data-driven result-based governance program covering more than 250 million citizens.

The study attempts to make a qualitative assessment of the pros and cons of data-driven public governance in India. To substantiate arguments, the study uses descriptive analysis of data collected from 112 districts under themes like skill development, financial inclusion and basic infrastructure across 54 months. In anticipation, the injury highlights the status, progress and challenges for the data-driven public governance model by offering novel perspectives on TADP and opening up avenues for further detailed research in the field of data-driven public governance.

Keywords: Public governance, Public policy, Data driven governance

1. Introduction

In contemporary times, data has emerged as valuable resource after oil. In every sector of society, economy and polity data has become pivotal aspect for decision-making. Across the world, governments are have embraced big data for citizen behaviour informed decision-making. Data is considered as a backbone of governance models. With the striking rise in sophisticated information and communication technologies (ICTs), usage for data for public good has increased. Analytics derived from public data has encouraged governments to devise customized welfare and development solutions. Data collection, analysis and dissemitiation through complex algorithms has reduced operational costs for governments. On other side, governments also gained exposure for novel kind surveillance techniques that can reduce tax evasion, petty crimes etc.

*Corresponding author: pratham.parekh@nirmauni.ac.in

DOI: 10.4324/9781032708461-9

Governments across the world has both disadvantages and opportunities to optimize decision making through data driven models of governance. With data driven governance models, governments can simulate numerous scenarios for public welfare (Janssen & Hoven, 2015).

During 2017, a taskforce for Artificial Intelligence (AI) was get up by Ministry of Commerce & Industry to explore potential usages of AI for economic transformation of the country. In 2018, NITI Aayog published a draft national strategy for artificial intelligence followed by four major committees constituted by Electronics and Information Technology in 2019 to encourage usage of data driven practices and effective deployment of AI for identifying key policy interventions, mapping technological capabilities, cyber security, etc. across various sectors. These committees broadly recommended an AI and data resource platform at central level that be used for promotion of collaboration among academia, industry and governments. By 2020, India became founding member of Global Partnership for AI (GPAI). With this backdrop, NITI Aayog developed a National repository for AI based research, data analytics and knowledge assimilation called AIRAWAT (NITI Aayog, 2020). In India, state and central governments are now enabled to capture public behaviour trends and integrate it with political, economic and social decision-making models across all levels of governance and administration. In India, the efforts for harnessing potentials of data for governance is observed in recent efforts by the government.

Numerous efforts are being made for linking data driven governance with actual transformation of the conditions at grass root level. One of such effort is reflected in a flagship program of NITI Aayog popularly known as the Aspirational district programme. With demise of the Planning Commission, an effective programme called Backward Districts Grant collapsed but the need for addressing regional balanced development did not. This need is now being fulfilled by novel data driven Aspirational District Programme (ADP).

The Prime Minister and Chairman of NITI Aayog initiated aspirational District Programme (ADP) during January 2018. The object the program is to improve socio-economic outcomes of "backward" (termed as "aspirational") districts. This program re-imagined governance informed and driven by data. ADP vested ownership and accountability of development on district administration monitored and analysed through real time system at central level. The program envisage convergence and collaboration among various levels of administration, industry and civil society for development of 112 districts (initially there was 115 districts) across 28 states covering 8600 gram panchayats i.e. more than 20% country's population. The program synchronizes the efforts of central ministries, district governance and planning machinery of the Government of India through data collection, analysis and dissemination.

2. Brief history of District Level Public Governance in India

Post-independence, concept of empowering local governments through planning and monitoring emerges from the report of Grow More Food (GMF) Enquiry Committee in 1952. The committee recommended the Community Development Blocks as a solution to the challenges posed to variety of local governments that are observed to be working in silos or without any common shared objective (Mehta, 1957). Recommendations of this encouraged initiation of a Community Development Program (CDP). CDP primarily focused on the agriculture and allied economic activities that led to development of rural community. This enabled local governments to address issues of rural health, communication, rural education etc. The program was enthusiastically adopted and implemented in the newly independent country though there was not linkages with ICTs. During 1957, the Balwant Rai Mehta Committee recommended establishment of three layer Panchayati Raj Institutions that can be considered as an agencies of local governance organizations at the village, block(taluka) and district levels of administration. By 1964, coverage of Community Development Block was extended over the entire India, such vast coverage created a need for resource allocation at the local level of administration. This need was addressed by the Administrative Reforms Commission in 1967 by recommending that purposeful planning has to be undertaken for resource allocation for local-level administrations (Parekh, 2022). In the new century, the country's decentralized planning efforts started inclining towards allocation of financial resources to micro administrative departments and expenditure through local governments diluting the very objective decentralized planning. Brief list of major steps taken for strengthening district administration is as follow:

1964: Coverage of Community Development Block was extended over the entire country, such vast coverage created a need for resource allocation at the local level of administration

1978: Prof. M. L. Dantwala Committee Recommended that block level planning must serve as a link between village plans and district plans.

1984: The Reserve Bank of India (RBI) steps into development discourse by developing a credit plan for district-level planning.

1984: Hanumantha Rao Committee Recommended the decentralization of administrative functions, powers and finances through specialized district planning agencies and district planning cells.

1985: G. V. K. Rao Committee Made an effort for making district panchayats as managing and monitoring units for all development programs at a rural level.

1985 to 1990: 7th Five Year Plan Focused on decentralization before which planning processes, decision-making and strategy formulation remain concentrated at the state and union levels.

After recommendations of the Balwant Rai Committee though some states constituted panchayats, the panchayat remained a non-permanent feature government system. Development programs/schemes formulated by the central government (known as Centrally Sponsored Schemes or CSS) and were implemented through line departments, making the district-level planning process collapse or irrelevant.

1992 to 1993: 74th amendment Decentralized planning where districts became key units (known as Urban Local Bodies or ULBs) for regional as well as national planning processes. This constitutional amendment recognized districts as a planning unit for urban local demography. This 74th amendment also mandated the formation of District Planning Committees that would be responsible for drafting district plans based on plans received from panchayats and municipalities.

1992 to 1997: 8th Five Year Plan Stressed building and strengthening peoples' institutions and generating active participation within liberalization and privatization frameworks. During plan period, district plans focused on population control, environment preservation and development of infrastructure at the local level.

1997 to 2002: 9th Five Year Plan Focused on themes related to social justice and equity.

2002 to 2007: 10th Five Year Plan Advocated area approach and policy priorities for strengthening decentralized planning.

2004: 2nd Administrative Reform Commission Decentralized planning mandated the preparation of village plans based on the development needs of the village at block levels and was finally consolidated at the district level for preparing district plans. Such activity though mandated, remained merely paperwork in several states, resulting in regional imbalances in development. Regional developmental imbalances remained the major priority of planning in India (Mohan, 2005). Such mechanism created a confusing scenario that was studied

2007: Group of Ministers (GoM) Accepted recommendations for devolution of power to urban local governments under themes like school education, public health (administration of community health centres or area hospitals), traffic management, civil policing, urban environment, heritage maintenance, land management and registration etc.

2008: Expert Group of Ministry of Panchayati Raj Constituted an expert group that made recommendations on developing district and sub-district plans at all levels of Panchayats aimed to deliver basic needs to citizens at grass root levels. Based on this report by Shri V. Ramachandran-led expert group, the Planning Commission issued detailed guidelines for developing district plans in the 11th Five Year Plan (Ministry of Panchayati Raj, 2008). In same year, Integrated District Planning - Planning Commission for transforming vertical planning processes into the horizontal process. This initiation ensured the participation of district administration in the preparation of sectoral plans like a district health plan, district watershed plan, district education plan etc. this effort confirmed sectoral thrusts in development that fine-tuned the planning processes with increased specializations to address local problems (Planning Commission of India, 2014). Integrated District Planning since 2008 focused on specific roles of local bodies and line departments. This framework aimed to ensure that sectoral specialization is efficiently and responsibly utilized. In such a context, urban planning is left with scattered sub-optimal planning agencies and processes. The local governments then were empowered to be part of actual decision-making in district planning assuming them as a vital layer connecting regional planning and national planning.

3. The Study

The study uses data of 112 districts that is made public by the NITI Aayog on its dashboard of aspirational districts. Instead of ranking districts with delta rank or composite scores, the study divorce from NITI Aayog's methodology of monthly delta ranking and attempts to grasp progress made by 112 districts under Basic Infrastructure theme from March 2018 to March 2022. The study explores 8 indicators under the basic infrastructure theme and 16 indicators under the financial inclusion and skill development themes.

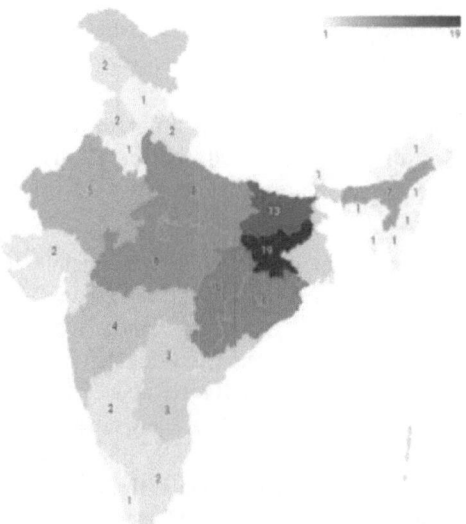

Fig. 9.1 Distribution of aspirational districts in among states of India

Source: Author's compilation

Out of 112 Aspirational Districts, majority districts are from Eastern zone (37.50%) followed by Central India districts (25%) and North Eastern districts (12.50%) and least districts are from Western zone of India (5.36%). Jharkhand has highest number of Aspirational Districts i.e. 19 districts (16.96% of total Aspirational Districts). Goa do not have any aspirational districts while districts of West Bengal has been "dropped off" by NITI Aayog.

The study is based on 23 indicators across three themes *i.e. Financial Inclusion (8 indicators), Skill Development (4 indicators)* and *Basic Infrastructure (11 indicators):*

4. Discussion

As a sub-discipline of sociology, public sociology emphasizes the expansion of disciplinary boundaries and allowing non-academic audiences (Burawoy, 2005). Public sociology advocates engaging sociology with issues of public policies, governance and political activism. This stem of sociology leverage over empirical methods (mostly statistics, mathematics or computational) linked with sociological theoretical frameworks to grasp not just "what is" or "what has been" but also "what might be". The majority of versions of public sociology have remained normative and political (Piven, 2007). The study grounds its rationale on such a theoretical perspective.

It is well known that the economic growth of India since its independence presents an applauding picture. This economic growth also comes with a shadowing cost of socio-economic inequalities and regional imbalances attaining development. Post-independence period Indian governments attempted to get rid of colonial exploitative effects on the economy and society that is still observed as an ongoing project. India's adoption of Five Year Plans always had economic growth, national development and regional development as dominating agendas. The role and priorities of the Planning Commission of India have remained fluctuated with changing political scenarios. By the 1990s, it became clear to governments that a narrow vision of linear economic planning and growth is not going to address the developmental issues of the country. Issues like social deprivation, regional disparities, evaluation of social inequalities etc. became priority issues during the 11th Five Year Plan that continued until the last five-year plan. Though the country is adopting and continuing with rigorous development planning at all levels of governance, India is unable to translate economic growth into human wellbeing.

The study thus focuses on development assuming national development as an accumulation of regional development and focuses on district-level governance (assuming districts as an aggregated unit of blocks and villages). Districts have always remained an important part of development planning in India. Districts as planning units has been debated a lot since independence. An important milestone in district-level development planning is observed in 2008 when the planning commission made an attempt to streamline previously recommended measures and initiated Integrated District Planning that focused on specific roles of local bodies and line departments to sectoral specialization. Due to this urban planning is left with scattered sub-optimal planning agencies and processes. The local governments then were empowered to be part of actual decision-making in district planning assuming them as a vital layer connecting regional planning and national planning.

By January 2018, newly formed national planning body of India launched Transformation of Aspirational Districts Programme. This program is meant to target rapid transformation of districts lagging in various development parameters. This programme axels over concept of 3Cs *i.e.* 'Convergence', 'Collaboration' and 'Competition'. It primarily makes an effort to strike convergence of centrally sponsored schemes and state government schemes through collaboration while the Prabhari Officer and District Collector/Magistrate are responsible authorities for improving socio-economic indicators through competitive federalism. Achievements of districts under this program is monitored through quantifiable outcomes across

sectors like health & nutrition, education, skill development, basic infrastructure, agriculture, water resources and financial inclusion considered as thematic areas of development. These thematic areas have been narrowed down into 49 key performance indicators. Initially the program included 117 districts that is reduced to 112 districts covering 28 states and more than 8000 local bodies/gram panchayats.

The program makes States responsible for empowering districts and building their capacities for recording and analysing strength, weaknesses and identifying low hanging fruits for swift progresses. The NITI Aayog at central level measures progress of each Aspirational District by implementing rigorous statistical models through ICTs. Based on output generated by such statistical models, NITI ranks all these districts that motivates further competition.

In terms of application of information technology, the NITI Aayog has developed a public Management Information System (MIS) that provides monthly progress and rank of each district under program.

5. Conclusion

After analysing public data of NITI Aayog on Aspirational Districts, the study attempts to grasp the distance of districts from the set benchmark/target and compounded annual growth rate of districts' eight indicators under the theme of basic infrastructure. Overall, it is observed that more than 45% are missing their annual targets for all eight indicators. Districts are observed to perform weakly in terms of indicators linked with financial inclusion and skill development. It can be assumed that aspirational districts are doing well in terms of road and water accessibility.

While household infrastructure like house and electricity needs reasonable focus at the implementation level. District-wise movement can be visualized on the dashboard developed by researcher at https://public.tableau.com/views/adr_16621297151650/Dashboard2?:language=en-US&:display_count=n&:origin=viz_share_link in terms of growth towards the attainment of the target, the compounded annual growth rate is calculated for 56 months and it is found that 45% of districts experienced negative growth rates i.e. their movement towards the target is stagnant. In addition to describing data and gauging the progress of districts, the study also observed qualitative insights.

The data-driven governance model demonstrated by the Aspirational Districts Program on first look appears to be a vital initiative for a country like India. But such systems tend to face several miss-management hurdles due to the cultural and political inclination of the scheme. The majority of indicators are receiving data directly from Districts. These districts focusing on competition may ascent on extravagating monthly data. The lack of granular data on selected monitoring indicators has to face impediments to administrative efficiency. Further, the indicators which are selected by NITI Aayog require more meticulous attention for instance from data captured for any given indicator is difficult to distinguish between output and outcome. The output may not necessary accumulate into the outcome, for example, an output indicator like "Percentage coverage of establishment of Common Service Centres at Gram Panchayat level" cannot represent the "quality of service" (outcome indicator) provided to citizens through such common service centres. This limitation can serve as an inception point for outcome-based monitoring and management systems.

Another important drawback of data-driven transformation reforms is that such reforms may miss the opportunity to pay detailed attention to the quality of implementation. Considering the Aspirational, districts program as a district database that enables and encourages district and state administration to diagnose needs and customize solutions for targeted interventions fails to align with an idea ranking districts. Competition rather creates a psyche of focusing on targets rather than focusing on solutions. Even while considering competition as a better mechanism for development, one needs to look at the incentives attached to competition. The induced competition of improving district ranks is forced on IAS officers serving as District Development Officers, District Collectors and District Magistrates. These officers serve their office generally for three years dealing with more than 100 ongoing schemes/programs. Such a scenario creates a sceptical viewpoint on the improvement of implementation quality as focusing on visible political priority rather than competing with fellow districts may be a rational choice for such officers. Additionally, a centralized monitoring system that aligns with politically inclined target settings is being monitored closely and realistically resisted.

The Aspirational District Program in such sense also fails to promote decentralization as rural local bodies as panchayats have not been given any role. It is clear that developing such systems takes the planning process to new levels but assuming data as a substitute for engrained administrative failures can rise an urgent need to focus on actual administrative reforms and developing implementation capacity. Many such intriguing questions raised during studying open an arena for undertaking sociologically driven research.

REFERENCES

1. Burawoy, M. (2005). For public sociology. *American sociological review, 70*(1), 4–28.
2. Janssen, M., & van den Hoven, J. (2015). Big and Open Linked Data (BOLD) in government: A challenge to transparency and privacy. *Government Information Quarterly, 32*(4), 363–368.
3. Mehta, A. C. (1957). *Report of the Foodgrains Enquiry Committee, 1957.* Ministry of Food and Agriculture, Government of India, New Delhi.
4. Mohan, K. (2005). *Addressing regional backwardness: an analysis of area development programmes in India.* Manak Publications.
5. NITI Aayog. (2020, January 9). AIRAWAT-Establishing an AI Specific Cloud Computing Infrastructure in India. India AI. Retrieved December 23, 2022, from https://indiaai.gov.in/research-reports/airawat-establishing-an-ai-specific-cloud-computing-infrastructure-in-india
6. Parekh, P. (2022). Sociological evaluation of basic civic infrastructure in 112 Aspirational Districts of India (in-press). *Gujarat Institute of Development Research, Ahmedabad.*
7. Piven, F. F. (2007). From public sociology to politicized sociologist. *Public sociology: Fifteen eminent sociologists debate politics and the profession in the twenty-first century*, 158–166.
8. Planning Commission. (2014). *Manual for Integrated District Planning: Report of the Task Force of the Planning Commission* (No. id: 5626).
9. Sen, D., Ozturk, M., & Vayvay, O. (2016). An overview of big data for growth in SMEs. *Procedia-Social and Behavioral Sciences, 235*, 159–167.

Integrating Advancements in Education, and Society for Achieving Sustainability – Dimitrios A. Karras et al. (eds)
© 2024 Taylor & Francis Group, London, ISBN 978-1-032-70841-6

10

An Improved Deep Learning Framework for Assessing the Proficiency Level of English Learners

R. Manikandan
Vel Tech Rangarajan Dr. Sagunthala R&D Institute of Science and Technology, India

Fredrick Ruban Alphonse*
Christ (Deemed to be University), Krupanidhi Degree College, India

B. Kiran Bala
K. Ramakrishnan College of Engineering, India

Gulnaz Fatma, Saba Tariq
Jazan University, Saudi Arabia

Abdulwahab Mohammed Saeed Mohammed
Jazan University Alardah University College, Saudi Arabia

Abstract: The deep learning technique for evaluating proficiency of verbal language is the subject of this research. Instead of read language, suggestions or inquiries that call for expressive language answers elicit more speech patterns, which more accurately represents a learner's competence level. Any automated process must manage disfluent, non-grammatical, verbal interactions with no understanding of the fundamental content, in as well as deal with extremely variable non-native, learner, and noisy real-world monitoring settings. A powerful deep learning-based speech recognition algorithm and a Gaussian Process (GP) grader are used in conjunction to handle these. The effectiveness of a variety of features extracted from the audio that use the recognition assumption is examined. On actual candidate inputs, it is demonstrated that the suggested approach can forecast grades at a level comparable to the original examination graders. Efficiency is further enhanced by interpolation with the exam ratings.

Keywords: English learners, Deep Learning, Recurrent Neural Network, Gaussian Process

1. Introduction

English is the primary language of communication used across borders. It makes it easier for individuals from many nations to connect with and comprehend one another. As a result of globalization, it is anticipated that employers would

*Corresponding author: fredrick.ruban@res.christuniversity.in

DOI: 10.4324/9781032708461-10

increasingly look for qualified individuals with strong interpersonal skills (Fenyvesi 2020). The number of English language learners enrolled in public schools across the country increased by more than 51% during the ten-year period (Peregoy & Boyle 2000). Numerous studies have brought up important questions concerning the fairness and equality of learning opportunities for learners. Helping English language learners develop appropriate English Language Proficiency so they can have better access to content learning and achieve academic success is essential to addressing these students' immediate needs as they deal with the simultaneous challenging task of learning educational content while working to develop their English language proficiency (Kunasaraphan 2015).

An assessment of proficiency would assess learners' broad English language skills based on everyday texts and language as well as their capacity to organize a cogent piece of writing. Given the requirement for students to participate in seminars, tutorials, and lectures, it would be perfect if it also included the productive skills of reading and listening. Higher education institutions often have English proficiency tests for admission since it is thought that having a particular degree of language ability is required for academic achievement. The evaluation of English language proficiency (ELP) is a crucial component of the academic careers of English language learners since the results of such assessments define and have an impact on their teaching, categorization, and advancement (Clark-Gareca, et al., 2020). Therefore, the most crucial factor in assessing their academic development is to provide accurate and genuine English Language Proficiency assessments. English

Language Proficiency evaluations based on dubious metrics might have serious academic repercussions. If English Language Learners aren't properly tested, their English skill levels can be misclassified, and they might get the wrong kind of education. They could even be mistakenly labeled as having learning difficulties, which would have a significant negative effect on their academic future. In order to assess English learning learners' readiness to take part in state content-based exams in subjects like reading/language arts, math, and science, it is crucial to include their level of English proficiency (Yang & Qian 2020).

2. Methodology

The research is based on the assessing the proficiency level of English learning, the methodology section is divided into two sections as Speech recognition and proficiency assessment. For the first part the data is collected form the AMI and DNN based Gaussian process is used for the speech recognition (Han 2022). At the second section RNN-LSTM is used for assessing the proficiency level of English learner. Finally, the finding was calculated based on the graders and the RNN algorithm (Malinin 2019).

3. Data Collection

The accessibility and dependability of corpora of non-native speech that have been transcribed by humans, especially those that include utterances, represent one of the main obstacles in the field of pronunciation evaluation. The majority of the nonnative language samples used in the non-native English pronunciation mistake detection articles examined in this chapter were evidently their own and intended for usage in the specific job. Low amounts (7–11) of non-native learners reciting pre-defined phrases or terms comprised the largest category of these. Only one native speaker from each of the articulation categorizations was used to collect the information. To use in the publications described in Table 10.1 beneath, data from more individuals were gathered.

Table 10.1 Pronunciation Error Detection

Dataset	Approach	Speakers	Natural
Custom	Phone Identification	100	N
LeaP	ERN	46	Y
ISLE	Phone Identification	46	N
iCALL	Native Contrast	305	N
C-AuDit	Native Contrast	94	N
CU-CHOLE	Phone Identification and ERN	210	N

4. Speech Recognition

A deep learning-based joint decoder method has been constructed in this research. The tandem GMM-HMM system and the stacking hybrid process of software up the joint technology. The hybrid and tandem algorithms are both trained for discrimination. Tandem and stacked hybrid systems both require transformed properties. On the AMI conference corpora, a bottleneck (BN) DNN is learned using context-dependent state targets (Han 2022). The AMI database was used for the DNN development since it contains a large - scale dataset with more (mainly) non-native English speakers and

is hand translated (Settles, et., 2020). The architecture of the BN DNN is $720 \times 10004 \times 39 \times 1000 \times 6000$. The 40-dimensional efficiency can be enhanced features from 9 image sequences with a delta added to each frame characteristic make up the DNN's input. As a result, the input vector size is 720. The Perceptual Linear Prediction (PLP) feature trained GMM-HMM system is used to pre-train the BN DNN using context-dependent goals produced by aligning the training data (Goldenberg 2020). After that, the pretrained model is adjusted using the CE (frame-level cross-entropy) standard. The data are then used to retrieve the 39 dimension limiting characteristics, which are then modified to use a global semi-tied linear combination. The 39-dimensional heteroskedasticity linear discriminant analysis (HLDA) projected PLP features with Δ, Δ^2, Δ^3 are supplemented with the modified BN features. Normalization of the cepstral mean at the speaker stage, (CMN) and cepstral variance normalizing (CVN) are used. As a result, a 78-dimensional per-frame extracted features is produced for use as input by the tandem and hybrid models.

Tandem GMM-HMM models are built in two pairs. One is a minimal phone error (MPE) criterion-trained speaker-independent (SI) classifier. The other uses Speaker Adaptation Training to create a speaker-dependent classifier (SAT). Constrained maximum likelihood regression analysis (CMLLR) on the input data is used to conduct SAT, which is then proceeded by discriminative training with the MPE criteria. A combination of 9 successively altered bottleneck and PLP texture features serves as the input to the layered hybrid DNN. This results in a 702-input size overall. The architecture of the DNN is $702 \times 10005 \times 6000$. The situationally states from the BULATS data are its throughput objectives. State linking uses the same number of stages and logistic regression as Parallel devices. Utilizing racially biased layerwise pre-training with context-dependent objectives produced by aligning the training examples with the Tandem SAT method, initialization is carried out. The frame-level CE requirement is then used to adjust the pretrained network.

5. Pronunciation Assessment

The goal is to evaluate the accent of an unidentified student's statement X_u. It is believed that a teacher-spoken referencing phrase X_y, already occurs. The method then generates a vector $U = \{u_i\}$ containing an evaluation of the pronunciation level for each of the A phonology in X_u. By purpose, the word quantity of student and educator statements is the same as $Y_u = Y_v$. For all investigations, the aforementioned phoneme sequence before remains the same (Mohammadi & Izadpanah 2019). Both the learner and educator feature sequencing could be divided into the sets corresponding to phones, presuming segmented knowledge for phoneme $Z = \{v_e\}$ as mentioned in the above equ (3) (Settles, et., 2020). R_u^e, R_v^e, the paired sets, typically have different lengths. The time is normalised to a given length L in order to assign equal priority to each phone occurrence. The most straightforward approach may be to set L = 1 and select, for instance, either the phone's central extracted features, $R_u^e = R^{e,central}$, or the mean of those feature vectors across each degree. Another option might be to reconstruct the original or linear interpolation the beliefs on each aspect to achieve features extracted with same unit Length, for example, $R_u^{e,T}$ and $R_v^{e,T}$ with L = 20, if the feature field is presumed to be constant along each dimension (such as in the backside features scope).

6. Result and Discussion

The previous sections of this work provided two methods for predicting a candidate's competency and native language by identifying rhythm cues from the matched phone sequence of the candidate's statements. TensorFlow is now being used to create the systems, which were trained and tested on actual data. The candidate replies to the spoken section of the Business Language Testing Service (BULATS) for foreign English language learners, offered by Cambridge English Language Testing, are used to generate the training and testing data. The five portions of the BULATS speaking test are all focused on business scenarios. Short answers to questions that were posed make up. Candidates read eight phrases aloud that were their own, often multi-sentence replies to a succession of oral and visual cues. The method is estimating the candidates' overall competence score, which is measured on a scale from 0 to 30. For the native language (L1) prediction objectives and human given proficiency level tasks (in the latter case, trained end-to-end), the shallow background features and the deep features presented are retrieved for each speaker in the data presented in Tables 10.2 and 10.3 show the outcome. For the native language (L1) prediction objectives and human given proficiency level tasks (in the latter case, trained end-to-end), the shallow background features and the deep features presented are retrieved for each speaker in the data presented in Tables 10.2 and 10.3 show the outcome.

Table 10.3 Grade prediction

	Correlation Factor	Mean Square Error
Baseline	0.779	17.7
RNN	0.785	15.9

Table 10.4 Proficiency Detection rate

	Accuracy (%)
Base Line	56
RNN	73

The division of L1 classification results by participants score is shown in Table 10.4 follows (grouped by CEFR level). There were two effects proposed to influence this association. Firstly, it ought to become difficult to detect the speaker›s L1 as their skill rises because the rhythm of more proficient presenters should resemble native speech in the L2. In contrast side, better speakers› statements are simpler for the ASR to grasp, therefore it is reasonable to assume that their length data will be more accurately aligned.

Table 10.4 Detection accuracy

	$<A_1$	A_2	A_3	A_4	A_5	C
Base	30	58	50	57	55	60
RNN	40	64	49	53	74	70

It is obvious that the longer an effective predominates for baseline features, and rhythmic characterization appears to be constrained by the precision of alignment data. This effect is less noticeable because the link between score and L1 detection accuracy is reduced when using deep features. This might be explained by the long short-term memory giving low quality length data more robustness, lessening the effect of

ASR faults on the efficiency of the end-to-end network.

7. Conclusion

A list of thirteen standard qualities was created, along with an analysis of the properties of rhythm that have been used in the literature. It was shown that using this strategy significantly improves score and L1 predictive accuracy. The phone length pronunciation features were determined to supplement both the baseline and deep characteristics, demonstrating that each set of features is capturing a distinct aspect of speaker proficiency. These characteristics were applied to a neural network structure to forecast the L1 based on expressive language by non-native applicants of a spoken English test and the user competence score. The replacement of the baseline characteristics with more extended, adjustable deep features that were extracted utilizing a learning algorithm across a recurrent network of neurons was then demonstrated. With end-to-end training, the score and L1 forecasting challenges are rerun with the additional characteristics. It was shown that using this strategy significantly improves score and L1 predictive accuracy.

REFERENCES

1. S. F. Peregoy and O. F. Boyle. (2000). "English Learners Reading English: What We Know, What We Need to Know," *Theory Pract.*, vol. 39, no. 4, pp. 237–247, Nov.

2. Fenyvesi. (2020). "English learning motivation of young learners in Danish primary schools," *Lang. Teach. Res.*, vol. 24, no. 5, pp. 690–713.

3. K. Kunasaraphan. (2015). "English Learning Strategy and Proficiency Level

of the First Year Students," *Procedia - Soc. Behav. Sci.*, vol. 197, pp. 1853–1858.

4. B. Clark-Gareca, D. Short, M. Lukes, and M. Sharp-Ross. (2020). "Long-term English learners: Current research, policy, and practice," *TESOL J.*, vol. 11.

5. B. Settles, G. T. LaFlair, and M. Hagiwara. (2020). "Machine Learning–Driven Language Assessment," *Trans. Assoc. Comput. Linguist.*, vol. 8, pp. 247–263.

6. Y. Yang and D. D. Qian. (2020). "Promoting L2 English learners' reading proficiency through computerized dynamic assessment," *Comput. Assist. Lang. Learn.*, vol. 33.

7. C. Goldenberg. (2020). "Reading Wars, Reading Science, and English Learners," *Read. Res. Q.*, vol. 55, no. S1.

8. H. Mohammadi and S. Izadpanah. (2019). "A Study of the Relationship between Iranian Learners' Sociocultural Identity and English as a Foreign Language (EFL) Learning Proficiency," *Int. J. Instr.*, vol. 12, no. 1.

9. A. Malinin. (2019). "Uncertainty Estimation in Deep Learning with application to Spoken Language Assessment," p. 234.

10. X. Han. (2022). "Investigation on Deep Learning Model of College English Based on Multimodal Learning Method," *Comput. Intell. Neurosci.*, vol. 2022, pp. 1–10.

Note: All the tables in this chapter were made by the Authors.

Integrating Advancements in Education, and Society for Achieving
Sustainability – Dimitrios A. Karras et al. (eds)
© 2024 Taylor & Francis Group, London, ISBN 978-1-032-70841-6

11

Exploring the Impact of E-Learning on the Morality of Undergraduate Students

S. Rakesh, Kethan Kumar K,
Fredrick Ruban Alphonse*, Mubashira V.
Krupanidhi Degree College, India

Banumathi P.
PKR Arts College for Women, India

Abstract: In the contemporary period, the education system has taken up a whole new approach and place. E-learning enables to enhance knowledge through various online content which can be delivered to the learner at any time and at anywhere. However, the idea of e-learning also brings an impact on students' moral behavior and ethical values. The present study elicits how such behaviors have had an impact and can bring an impact. The unethical behavior of individuals might affect their future goals both in educational and professional career and eventually their manners too. The present research paper examines the unethical behavior of the students in the virtual classrooms with regard to attendance, tests and assignments. It also investigates the factors that have led the students to take an immoral approach towards e-learning.

Keywords: Morality, E-learning, Technology, Psyche, Behaviour

1. Introduction

E-learning is a method of learning by utilizing technology for accessing the educational curriculum outside the traditional classroom. It involves formalized teaching with the help of e-resources, and this makes learning accessible anytime and anywhere. E-Learning has become a crucial tool for tutors around the world to provide consistent education. In 1999, the term 'E-Learning' was phrased by Elliott Masie in the Tech Learn conference at Disneyworld. However, the very first attempt to e-learning was taken in 1924 by Sidney Pressey, a professor at Ohio State University. He invented the very first electronic learning machine, which was referred to as the Automatic Teacher. E-learning has made its inerasable mark in the educational field. With the fast internet connections, the opportunities of Multimedia and other domains are at ease. Eventually social media has fixed it is roots in education, and it is evolving constantly.

*Corresponding author: fredrick.ruban@res.christuniversity.in

DOI: 10.4324/9781032708461-11

E-learning gives a baggage of required solutions for the problems such as poor infrastructure, pandemic, natural calamity and so on. Learning in an environment of physical presence has declined and now everything is compressed in the portable virtual world. However, researches have proved that e-learning has negative impacts on the moral values of university students. The objectives of the research is: to detect the interest of the students in E-Learning, to know how E-learning is causing a negative impact on the morality of the students, and to determine the psychology of the students behind preferring E-learning.

2. Theoretical Framework

Sigmund Freud conceptualizes that the moral development that occurs as a person's ability to set aside his/her selfish needs is replaced by the values of significant socializing agents (Cherry 2020). In the structural theory, Freud states that morality comes from the superego. This model explains three regions which are interdependent. They are the id, ego, and superego (Castelloe 2013). The 'id' is the fundamental and primal part of personality which is present ever since the birth of a person. It signifies the animal nature in human beings. The id comes from pleasure principle. Ego is based on the reality principle. It covers the defense mechanisms and reasoning capabilities. The function of reality principles is to satisfy the desires of id in a realistic manner. It weighs the total costs and merits of that action before deciding to act upon. In cases, the id's impulses would be satisfied by the process of delayed gratification wherein the ego would allow the behavior, but in a particular time and place.

Superego covers the feelings of guilt and conscience. It makes connection between the psychic health and moral action. It is a component of personality which encompasses the internalized moral standards that one acquires from his/her parents and environment. The Freudian concept of perfect self is encased in the superego. The theory states that the superego functions to suppress the urges of the 'id'. It also tries to tune the ego behave morally rather than realistically. Freud mentions that the superego is the main and last component of personality to develop. The morals values learnt from one's parents underpin to the formation of superego. In addition, the formation of superego is nurtured by the knowledge of right and wrong that one acquires from the society. The superego also helps in giving guidelines for making appropriate judgements. The superego is divided into two components, and they are namely conscience and ego ideal. It is recorded that the ego ideal is the main part of the superego. It carries the rules and standards for an ethical behavior. These factors add to the behaviour that has been approved by the parents and other authoritative personalities of the society. The theory states that following and obeying these rules render the feeling of accomplishment, pride and value. The person one who breaks these rules would be tormented by the feeling of guilt.

3. Method

The present study has used random sampling method. The researchers have selected Bengaluru college students ageing between 17–20. The experimental group consists of 86 learners and the group is homogeneous as the learners belong to the same peer group, place of origin and educational level. These 86 respondents were asked to respond to the questionnaire consisting of fourteen questions. The researchers implemented the survey technique using a mixed questionnaire and the participants responded explicitly to open-ended questions without any inhibitions.

4. Interpretation

The present research aimed at detecting the mode of learning preferred by the respondents. The first question placed in front of the respondents is a closed-ended question and the question is Do you prefer to attend the online class? The choices for the question are Yes and No. The preference of the respondents is apparent in the Table 11.1 pasted below.

Table 11.1 Frequency distribution

Variables	Items	Frequency	Percent	Mean	Std. Deviation
G.	Male	38	44.2	1.56	.500
	Female	48	55.8		
Q1	Yes	29	33.7	1.66	.476
	No	57	66.3		
Q2	Yes	35	40.7	1.59	.494
	No	51	59.3		
Q3	Yes	76	88.4	1.12	.322
	No	10	11.6		
Q4	Offline	44	51.2	1.49	.503
	Online	42	48.8		
Q5	Yes	50	58.1	1.42	.496
	No	36	41.9		
Q6	MS T.	77	89.5	1.19	.584
	G Meet	3	3.5		
	Zoom	5	5.8		
	C.Web	1	1.2		
Q7	Yes	34	39.5	1.60	.492
	No	52	60.5		
Q8	0	14	16.3	2.31	1.551
	1	16	18.6		
	2	13	15.1		
	3	22	25.6		
	4	14	16.3		
	5	7	8.1		
Q9	OP1	44	51.2	1.91	.966
	OP2	6	7.0		
	OP3	36	41.9		
Q10	OP1	50	58.1	1.66	.928
	OP2	21	24.4		
	OP3	9	10.5		
	OP4	6	7.0		

Source: Authors

From Table 11.1 it is evident that 33.70% of the students prefer to attend the online class and the remaining 66.30% of the students have registered that they do not prefer to attend the online class. There could be several reasons presumed for abstaining themselves from attending the online class and of all, the most substantiating reasons could be lack of interest in studies and network issue. Since the majority of the students have preferred to say that they do not prefer online class, the second question focused on determining their understanding about the online class. In order to find out the students' interest and relationship with the E–learning, the researchers have tried exploring the preference to attend the classes. The maximum percentage that is 59.30% of the students have chosen to say that the online class can be a substitute for the offline class, and on the other hand, 40.70% of the students have chosen to say that the online class cannot be a substitute for the offline class.

The third question placed in front of the respondents is 'Did the network glitch hamper your E-learning process?' and choices are Yes and No. This question has been framed primarily to detect the technical issues faced by the students. The Table 11.1 gives a transparent report on the technical glitch experienced by the students. It can be figured out from the chart that 69.80% of students have faced technical glitch and that has hampered their E-learning. Whereas 30.2% of students have recorded that technical glitch did not hamper their E-learning process. In the process of data collection, the researchers got access to listen to the difficulties faced by the students in the process of E-learning. They opened up their problems to the researchers and the most widely heard response was that the respondents had to trek for a few kilometers away from their home in certain areas to get proper signal and access to strong and stable internet connection. Due to the technical glitch, they could not attend the online class or submit their online assignments on time. On one hand the students and teachers living in remote areas have faced trouble due to the poor network connection, whereas on the other hand the city-dwellers too have found it challenging. The present research has detected that 69.80% of the students were facing network issues which has directly affected their E-learning and 30.20% of students have not experienced such difficulty. The research anticipated this result therefore it was curious to place a question pertaining to the mode of learning.

The fourth question in the questionnaire tried detecting the mode of learning preferred by the students. The given question is Which mode of learning do you prefer? and the choices are online and offline. As mentioned in Table 11.1, the study reveals that 76.70% of respondents have chosen offline mode of learning and 23.30% of respondents have selected online mode of learning. During the pandemic, universities and colleges had to use applications like MS Teams, Zoom, Google Meet, Cisco WebEx and so on to create virtual classrooms. These applications were perceived as lifesaving jackets by academicians. The Table 11.1 communicates that 89.5% of students have used MS Teams, 7.23% of students have used Zoom, 2.1% of students have used Google Meet and 1.17% of students have used cisco WebEx to attend the online classes. The data shows that majority of the students have used MS Teams to attend online classes. This result varies from college to college and university to university because colleges and universities had liberty to choose their online platform for e-learning. The Table 11.1 delineates that the majority group of students have preferred to attend offline class and the minority group of students have preferred to attend online class. From the interaction with the respondents, the present research could figure out the following points as the reasons for the students choosing offline mode of learning: a) The offline mode of learning helps the students to acquire knowledge thoroughly, b) It gives the students the space to get their doubts clarified, c) The offline mode of learning gives the students the space for face-to-face interaction, d) Immediate interruption to clarify doubts can be done in offline mode of learning. These were the points shared by the respondents as the reasons why they choose offline mode of learning. The research also found that the students who are less interested in studies preferred online mode of learning. They believe that online mode of learning helps them to attend the virtual classes without much effort and supports them to perform well in the exams without working hard. When the students registered that they are able to perform better in online exam, there is a serious need to test the integrity of the students. In order to examine the integrity of the respondents, the present research placed a direct question which will contribute to detect if the students have or have not involved in malpractice in the online exams.

The fifth question in the questionnaire is 'Have you indulged in malpractice in the exams/ tests?' and the choices are Yes and No. This question was framed basically to test the integrity and the ethical behaviour of the students. It also enables to detect how e-learning affects the morality of the students. The Table 11.1 gives a clear picture of the percentage of students who have involved in malpractice and who have been honest in taking up online exams. It is evident that the students accepted they have indulged in malpractice in exams. The Table 11.1 showcases that 58.1% of the students have indulged in malpractice and the remaining 41.9% did not involve in malpractice. During the pandemic, universities conducted exams online which created an opportunity for students to cheat. Therefore, universities look for techniques and methods that can be adopted to prevent the students from involving in malpractices. In this regard, the most searched question in google is different ways to cheat the invigilator in online exams. The biggest challenge in online class is attendance management. Therefore, the seventh question in the questionnaire aimed at exploring if the students have attempted giving proxy attendance. The framed question is 'Have you attempted giving proxy attendance in the online class?' and the choices are Yes and No. Students find out different methods to attempt proxy attendance and the above pasted Table 11.1 shows that 39.5% of students have been genuine in registering their attendance whereas 60.5% of students have indulged in registering proxy attendance. The majority group of students have involved in giving proxy attendance. This shows that e-learning has encouraged and given space to exhibit immoral behaviors of the college students. Since a major group of students have attempted giving proxy attendance, it is necessary to detect the effectiveness of e-learning mode. For this, the students were asked to rate the effectiveness of e-learning mode in the scale of 0-5. From the above given Table.1, it can be decoded that 8.1% of the students have selected 5 points on the rating scale and 16.3% of the students have selected 4 points on the rating scale. It is quite obvious that 25.6% of the students have chosen 3 points on the rating scale and 16.3% of the students have selected 1 point to point at the effectiveness of their e-learning. Likewise, 16.31% of students have chosen to say that e-learning mode is not at all effective therefore they have selected 0 point in the rating scale. It is very transparent from the Table 11.1 that the majority group of students has acknowledged the fact that e-learning has not been effective in their cases. Only the minority group of students have registered that e-learning has been effective in their cases.

Since e-learning has not been effective in the cases of the respondents of the present research, it is necessary to detect how did the students prepare their assignments. The question framed for this purpose is 'How do you prepare the online class assignments?' and the options are 'working sincerely' and 'copying'. The responses of this question can be read from the above table that 49.4% of students have prepared their online class assignments sincerely. They have not indulged in malpractice whereas 50.6% of the students have confessed that they have copied to prepare their assignements. A majority group of students have attempted copying, which shows that students lose integrity when they are beyond the monitoring of their teachers. In such case, students do not sharpen their skills and they put an end to their critical thinking skill and creativity. It is clear that a majority group of students have been lazy and inactive in the online class. To probe further into it, the present research has framed the question: 'How do you respond when you are asked to answer question(s) during the progress of the lectures?' Since it is a closed ended question, the students had to choose any one

from the four options. The options are 'I give my genuine answers', 'I pretend that I have network issue', 'I borrow answers from someone/google', and 'I do not reply when the teacher asks question'. The Table 11.1 mentioned above showcases the responses of the respondents. It is clear that 57.70% of the students have chosen to say that they give genuine answers for the questions asked by their faculties in the online classes. There are 22.30% of students who have confessed that they do not reply when the faculties ask questions and 12.90% of students have agreed that they borrow answers from someone or google. The above pasted table depicts that 7.10% of students pretend to say that they face network issue. A minority group (42.30%) of students who are found to be insincere in answering the questions are seen as morally corrupted.

5. Conclusion

The present research has found that majority of the respondents have indulged in malpractice, and they do not prefer online class. The students agreed that e-learning is indeed making a path to cultivate immoral behavior. It also revealed that the respondents have untuned psyche. In the light of Freud's structural theory of psyche, the present research has discovered that the majority group of students who have indulged in malpractice and involved in exhibiting unethical behaviour have failed to cultivate superego in them. As a result, their conscience was not active to feel guilty for the unethical behaviour. Hence, the present research suggests that the superego must be fine-tuned to be active and matured amidst the students.

REFERENCES

1. Manjusha, M.P. (2021). Examining the interest and adaptability of undergraduate rural. *Ilkogretim Online - Elementary Education Online*, 20 (5): pp. 2078–2084.
2. Gudmundur Heidar Frímannsson. (2017). Moral and Historical consciousness. *Historical Encounters: A Journal of Historical Consciousness, Historical Cultures, and History Education*, 4(1), pp. 14–22.
3. Molly, S. Castelloe. (2013). On the Origins of Morality-Rethinking Freud's vision of good and evil".
4. Sibi, K. J. (2020). Sigmund Freud and Psychoanalytic Theory. *Langlit: An International Peer-Reviewed Open Access Journal*, Special Issue, pp. 75–79.
5. James Hansell, Joshua Ehrlich, Wendy Katz, Howard Lerner & Katherine Minter. (2008). *Psychoanalysis & Psychodynamic Psychology*. American Psychological Association.
6. Zhang, S. (2020). "Psychoanalysis: The Influence of Freud's Theory in Personality Psychology." *Advances in Social Science, Education and Humanities Research*, 433, 229–232.
7. Freud, Sigmund. (1920). *A General Introduction to Psychoanalysis*. New York: Boni and Liveright.
8. Odokuma, Ese. (2010). Explanations of Freud's Psychoanalysis Theories on the Lives and Works of Some Western Artist: An African Perspective. *An International Multi-Disciplinary Journal, Ethiopia*, 4 (3a), 227–233.

Integrating Advancements in Education, and Society for Achieving Sustainability – Dimitrios A. Karras et al. (eds)
© 2024 Taylor & Francis Group, London, ISBN 978-1-032-70841-6

12

Impact of Personality on Entrepreneurial Intention: The Mediating Role of Self-efficacy

Md Rezaul Haque and Manjit Kour*

Chandigarh University, Mohali, Punjab, India

Abstract: This study was conducted on university students in India to examine the role of personality traits on Entrepreneurial intention (EI) and the mediating part of self-efficacy. The relationship between the variables was measured using Smart pls 4.0 software. Extraversion, openness, and conscientiousness were discovered to influence entrepreneur intention. There is a negative relationship between neuroticism and agreeableness. Entrepreneur intention is strongly related to self-efficacy. It serves as a bridge between extraversion, openness, conscientiousness, and entrepreneurial intent. This study will add to the existing literature, and future researchers can use it as a foundation for their research.

Keywords: Extraversion, Openness, Neuroticism, Conscientiousness, Agreeableness

1. Introduction

Many experts believe that an entrepreneurial mindset, or entrepreneurial intention (EI), is a necessary first step toward reducing reliance on traditional employment. Successful entrepreneurship takes a significant investment of time. Thus, the motivation to start a business can be the driving force behind the first crucial action in establishing a pattern of sustained commercial activity. Because of this, encouraging and enabling entrepreneurial endeavors is no longer a luxury but a necessity. There needs to be more literature concerning the significance of macro-level factors in the entrepreneurial process. The degree to which a nation has developed economically influences the extent to which its residents are inspired to take the entrepreneurial plunge. Most research on entrepreneurial aspirations has been done in advanced economies; comparative evidence from developing nations needs to be more extensive (Ahmed et al., 2020). This research aims to determine if there is a correlation between a person's character and interest in starting their own business. Moreover, analyze how one's sense of self-efficacy influences the connection between entrepreneurial aspirations and personality.

*Corresponding Author: Mdrezaulhaque18@gmail.com

DOI: 10.4324/9781032708461-12

Career is determined by the interaction of environment and individual personality (Salameh et al., 2022). Personality refers to an individual's relatively stable thinking, feeling, and acting patterns. When an entrepreneur has the right personality traits, they are more likely to take the initiative, tolerate risk, and persevere through failure, all of which are essential for a startup to sustain a culture of innovation, renewal, and forward momentum long after the initial success has faded (Xie et al., 2018). Individual uniqueness is revealed in the degree to which a collection of personality traits shape their thought, emotion, and behavior. Entrepreneurs need to secure necessary resources like money, supplies, and networks to see their businesses grow and prosper. Given these parameters, it stands to reason that prosperous business owners share certain defining traits (Xie et al., 2018). Developing entrepreneurial skills is one of the primary objectives for advancing society and increasing its members' employability (López-Núñez et al., 2019).

2. Literature Review and Hypothesis Development

Personality characteristics conducive to a business's launch and success include those who are optimistic and risk-taking. In contrast, those that are pessimistic and pessimistic make it harder (Xie et al., 2018). Researchers have revealed that high EI students have a similar personality type to business owners. They have high levels of extraversion, conscientiousness, and openness, with low agreeableness and neuroticism, characterizing this persona. (López-Núñez et al., 2019). Entrepreneurial self-efficacy is an excellent way to predict whether or not someone will act or want to act like an entrepreneur. Both directly and indirectly, self-efficacy affects EI (Doanh, 2021).

One's agreeability can be measured by observing how one acts and reacts around others. Those who score high on the agreeableness scale tend to be loyal, helpful, cooperative, and humble. They care deeply about the feelings of others and will often give in when others demand it. A person who is difficult to get along with is likely manipulative, self-absorbed, suspicious, and cruel (Zhao et al., 2010).

H_1a: *Agreeableness has a positive effect on EI.*

Conscientious people are highly organized, dedicated, hard-working, and motivated in their pursuit of success (Zhao & Seibert, 2006). Entrepreneurial Intentions are significantly influenced by "conscientiousness" (Ahmed et al., 2020). Entrepreneurs are more likely to be hard workers, think things through, make plans, establish goals, and maintain high standards (Salameh et al., 2022).

H_1b: *Consciousness has a positive effect on EI.*

People who went into business for themselves were likelier to be extroverted and vocal than the average citizen. Extraverted people have a lot of talking, hanging around, and making friends. Their outer appearance is vivacity, activity, assertiveness, and domination, while their inner temperament is optimism, good spirits, and a desire for novelty and challenge (Zhao et al., 2010). Interest in entrepreneurial pursuits is correlated with extraversion. (Zhao & Seibert, 2006).

H_1c: *Extraversion has a positive effect on EI.*

Anxiety, anger, depression, and shame are all examples of neuroticism, a broad category of negative affect. High-neurotic people tend to avoid challenges they perceive as having a low chance of success (Salameh et al., 2022). There is a strong correlation between neuroticism and risk aversion (Ahmed et al., 2020). They take criticism and failure very personally and are easily disheartened. They may worry, feel hopeless, or even panic when faced with adversity (Zhao et al., 2010).

H_1d: Neuroticism has a positive relationship with entrepreneurial intention.

Entrepreneurs performed better in openness (Zhao & Seibert, 2006). Entrepreneurship is a mindset that places opportunities ahead of threats. The ability to be open to new experiences revealed a critical function in opportunity recognition. Entrepreneurship entails taking risks and dealing with adversity. Individuals with a high openness dimension are not afraid of new challenges, are versatile and imaginative, and frequently demonstrate a high level of creativity (Zhao & Seibert, 2006).

H_1e-Openness has a positive effect on EI.

Self-efficacy is a crucial trait for entrepreneurs because it boosts goal-directed behavior and the thirst for knowledge that fuels innovation (Dorcas et al., 2021). The confidence in one's abilities to carry out the many responsibilities inherent in starting and running one's own business is known as entrepreneur self-efficacy. There was a positive correlation between entrepreneur self-efficacy and the intent to launch a business. Students who rated themselves as more capable business owners were more likely to want to do so (Chen & Greene, 1998).

H_2a: Agreeableness has positive relationship with SE.

H_2b: Consciousness has positive relationship with SE.

H_2c: Extraversion has a positive relationship with SE.

H_2d: Neuroticism has a positive relationship with SE.

H_2e: Openness has a positive relationship with SE.

H_3: SE has a positive relationship with EI.

H_4a: SE mediates the relationship between Consciousness and EI.

H_4b: SE mediates the relationship between Agreeableness and EI.

H_4c: SE mediates the relationship between Extraversion and EI.

H_4d: SE mediates the relationship between Neuroticism and EI.

H_4e: SE mediates the relationship between Openness and EI.

3. Research Methodology

The research was carried out on university students. Students have the potential to become entrepreneurs. They could pursue entrepreneurship as a career. The survey questionnaire was given to 190 university students. The students' personalities were evaluated using the Five-Factor Model (FFM). With this model's help, researchers can categorize a wide range of personality characteristics into a manageable framework, wherein causal relationships can be more easily identified (Zhao & Seibert, 2006). agreeableness, conscientiousness, *Extraversion*, openness, and neuroticism are the five components of the FFM structure. The data was gathered using the questionnaire developed by (Salameh et al., 2022), which has 21 items on a seven-point scale. Both EI and self-efficacy were assessed by five times on a five-Likert scale. The entrepreneurial intention was measured by (Linan & Chen, 2009) questionnaire. Smart PLS 4.0 structural equation modeling software was used to test the model's empirical validity and investigate support for the proposed hypotheses.

4. Results

The survey questionnaire was distributed to 190 students, but only 175 responded. The response rate is 92%. 46.3% of the students were female, with the remainder being male. Most students are between the ages of 21 and 24, unmarried (90.3%), and have no prior work experience (68.6%). We received responses from students with postgraduate (40%), graduate (26.3%), and undergraduate (31.4%) qualifications and whose major was science (38.6%), commerce (46.3%), and humanities (14.9%).

In (Table 12.1) All factor loadings are statistically significant. Goodness-of-fit statistics point to a well-fitting model. The AVE values were above the recommended threshold of 0.50, ranging from 0.53 to 0.75. Composite Reliability (CR) output was also considered, obtaining values over 0.70, therefore adequate. The Cronbach's alpha (α) coefficients test the validity of each variable. It comes from .62-0.92, which shows that the variables are reliable.

Table 12.1   Measurement and operationalization

Construct	Items	Loading	CR	AVE	A
Agr	A1	1.483	0.773	0.673	0.619
	A2	1.218			
	A3	1.071			
	A5	1.336			
Con	C1	1.844	0.799	0.527	0.687
	C2	1.993			
	C3	1.127			
	C4	1.429			
Ext	E1	2.065	0.907	0.663	0.872
	E2	1.648			
	E3	2.850			
	E4	2.661			
	E5	1.938			
EI	EI1	2.623	0.921	0.745	0.886
	EI2	1.806			
	EI3	2.772			
	EI4	2.480			
SE	SE1	2.180	0.938	0.753	0.918
	SE2	2.835			
	SE3	2.982			
	SE4	2.228			
	SE5	2.077			
Neu	N1	1.740	0.753	0.523	0.918
	N2	1.153			
	N3	1.923			
	N4	1.462			
Ope	O1	2.293	0.920	0.741	0.884
	O2	2.695			
	O3	2.251			
	O4	2.119			

Note: Agreeableness: Agr; Conscientiousness: Con; Extraversion: Ext; Openness: Ope; Neuroticism: Neu; Entrepreneurial Intention: EI; Self-efficacy; SE

Heterotrait-Monotrait (HTMT) ratios were used to evaluate the discriminant validity. The HTMT ratios are shown in (Table 12.2). The bias-corrected and accelerated bootstrap confidence intervals technique was used with 5,000 resembling and two-tailed tests at a 95% significance level. The results showed that all the HTMT scores were less than the benchmark scores of 0.85 and 0.90, indicating that the study variables had discriminative validity.

Table 12.2 HTMT ratios

	AGR	CON	EI	EXT	NEU	OPE	SE
AGR							
CON	0.820						
EI	0.432	0.317					
EXT	0.810	0.755	0.168				
NEU	0.833	0.842	0.168	0.349			
OPE	0.842	0.656	0.373	0.748	0.399		
SE	0.829	0.703	0.505	0.719	0.350	0.760	

Table 12.3 describes the direct and specific effect between the variables. We discovered a correlation between entrepreneurial intention and conscientiousness, extraversion, and openness after examining the beta, standard deviation, T statistics, and p-value. However, there is little correlation between agreeableness, neuroticism, and the intention to start a business. Entrepreneurial intent and contentiousness, as well as openness and extraversion, are all mediated by self-efficacy. But fall short of mediating the contrast between agreeableness and neuroticism.

Table 12.3 Direct and specific effect measure

Hypothesis	Path	beta	Standard deviation	T statistics	P values	Remark
H_1a	Agr -> EI	0.214	0.159	1.350	0.177	Rejected
H_2a	Agr-> SE	0.111	0.098	1.136	0.256	Rejected
H_1b	Con-> EI	0.171	0.160	1.072	0.003	Accepted
H_2b	Con-> SE	0.197	0.087	2.256	0.024	Accepted
H_1c	Ext -> EI	−0.280	0.111	2.529	0.011	Accepted
H_2c	Ext -> SE	0.194	0.073	2.651	0.008	Accepted
H_1d	Neu-> EI	0.102	0.128	0.796	0.426	Rejected
H_2d	Neu -> SE	−0.006	0.070	0.081	0.935	Rejected
H_1e	Ope -> EI	0.314	0.120	2.608	0.009	Accepted
H_2e	Ope -> SE	0.372	0.090	4.117	0.000	Accepted
H_3	SE -> EI	0.514	0.095	5.428	0.000	Accepted
H_4a	Con -> SE -> EI	0.101	0.049	2.067	0.039	Accepted
H_4b	Agr -> SE -> EI	0.057	0.054	1.061	0.289	Rejected
H_4c	Ext -> SE -> EI	0.100	0.043	2.333	0.020	Accepted
H_4d	Neu -> SE -> EI	−0.003	0.036	0.080	0.936	Rejected
H_4e	Ope -> SE -> EI	0.191	0.062	3.086	0.002	Accepted

5. Discussion

Entrepreneurship refers to a person's behaviors that include identifying economic opportunities, taking the initiative, being creative and inventive, organizing social-economic systems to put resources and situations into action, and accepting the risk of failure. To succeed as an entrepreneur, one needs to have certain traits. Studies across all fields and professions have found that conscientiousness is the best predictor of professional success. To put it simply, extroverts are the most fun to be around because they are the most outgoing and social among us. An individual with the trait of openness to experience is naturally inquisitive and eager to learn about and try out novel situations and ways of thinking. According to the current study, conscientious, extraverted, and open people are more likely to become entrepreneurs in the future.

Personal disparities in coping and emotional stability are reflected in people's levels of neuroticism. A high Neuroticism score indicates an individual is more likely to experience anxiety, irritability, hostility, sadness, introspection, impulsivity, and vulnerability (Zhao & Seibert, 2006). A high Agreeableness score indicates that a person is highly dependent on pleasant connections and committed to cooperative principles. At the bottom of the spectrum, you'll find people who are manipulative, egocentric, distrustful, and downright cruel (Zhao & Seibert, 2006).

One's belief in their abilities can play a role in whether or not one launches their own business. Self-efficacy influences behavior through self-regulation processes, such as boosting motivation, goal-setting prowess, and the ability to exert behavioral control in pursuit of desired outcomes. According to the results, self-efficacy cannot mediate the connection between entrepreneurial intent and the personality traits of neuroticism and agreeableness. For agreeableness, sociability, and openness, however, it does not.

6. Conclusion

Entrepreneurship is essential to create jobs, alleviate poverty, foster innovation, advance society, and boost economic competitiveness. In this context, gaining a better understanding of the main determinants of entrepreneurship is unquestionably beneficial to economies and society as a whole. We investigated the role of personality traits (agreeableness, conscientiousness, extraversion, openness, and neuroticism) in influencing entrepreneur intention and the role of self-efficacy as a mediator. Extraversion, openness, and conscientiousness all impact an entrepreneur's intention. Where there is a minor relationship between neuroticism and agreeableness, entrepreneur intention is strongly related to self-efficacy. It serves as a bridge between extraversion, openness, conscientiousness, and entrepreneurial intent.

REFERENCE

1. Ahmed, M. A., Khattak, M. S., & Anwar, M. (2020). Personality traits and entrepreneurial intention: The mediating role of risk aversion. *J Public Affairs, February*, 1–15. https://doi.org/10.1002/pa.2275
2. Chen, C. C., & Greene, P. G. (1998). Does Entrepreneurial Self-efficacy Distinguish Entrepreneurs from Managers? *Journal of Business Venturing, 9026*(97), 295–316.

3. Doanh, D. C. (2021). *The moderating role of self-efficacy on the cognitive process of entrepreneurship: An empirical study in Vietnam. 17*(1), 147–174.

4. Dorcas, K. D., Celestin, B. N., & Yunfei, S. (2021). Entrepreneurs Traits/Characteristics and Innovation Performance of Waste Recycling Start-Ups in Ghana: An Application of the Upper Echelons Theory among SEED Award Winners. *Sustainability 2021, 13*(5794). https://doi.org//10.3390/su13115794

5. Linan, F., & Chen, Y.-W. (2009). Development and Cross-Cultural application of a specific instrument to measure entrepreneurial intention. *ET&P.*

6. López-Núñez, M. I., Rubio-Valdehita, S., Aparicio-García, M. E., & Díaz-Ramiro, E. M. (2019). Are entrepreneurs born or made? The influence of personality. *Personality and Individual Difference, 154*(May). https://doi.org/10.1016/j.paid.2019.109699

7. Salameh, A. A., Akhtar, H., Gul, R., Omar, A. Bin, & Hanif, S. (2022). Personality Traits and Entrepreneurial Intentions: Financial Risk-Taking as Mediator. *Frontiers in Psychology, 13*(July), 1–11. https://doi.org/10.3389/fpsyg.2022.927718

8. Xie, X., Lv, J., & Xu, Y. (2018). The Role of the Entrepreneurial Personality in New Ventures. In *Inside the Mind of the Entrepreneur, Contributions* (pp. 91–108). https://doi.org/10.1007/978-3-319-62455-6

9. Zhao, H., & Seibert, S. E. (2006). The Big Five Personality Dimensions and Entrepreneurial Status: A Meta-Analytical Review. *Journal of Applied Psychology, 91*(2), 259–271. https://doi.org/10.1037/0021-9010.91.2.259

10. Zhao, H., Seibert, S. E., & Lumpkin, G. T. (2010). The relationship of personality to entrepreneurial intentions and performance: A meta-analytic review. *Journal of Management, 36*(2), 381–404. https://doi.org/10.1177/0149206309335187

Note: All the tables in this chapter were created by the authors based on their analysis of the data.

*Integrating Advancements in Education, and Society for Achieving
Sustainability – Dimitrios A. Karras et al. (eds)*

13

Multi-component Intervention for Problematic Video Gaming Usage Among Adolescents: Scoping Study

Sterin Joseph* and Tony Sam George

Christ (Deemed to be) University, Bengaluru, Karnataka, India

Abstract: Video game is a prominent choice for entertainment, socialisation and relieving stress. Video games can be negative coping for escaping reality and unpleasant emotions. Due to their stage of development, adolescents are more susceptible to problematic use. Identifying the dysfunctional elements that contribute to problematic use is essential, and teaching adolescents various inventive ways of coping with their vulnerabilities are necessary. Recovery has no currently accepted definition, and there may be several different approaches to achieving it. A multi-component system enables us to see beyond the limitations of a single strategy. The component model of addiction (CMAT) model enumerates that there can be multiple pathways to overcome problematic usages. The following databases: JSTOR, PROQUEST, APA Psynet, and EBSCO will be searched to find publications for the scoping review. Through this study, we study whether the multi-component interventions help to manage the maladaptive strategies of the problematic video game.

Keywords: Problematic video game, Intervention, Adolescents

1. Introduction

Video games' high levels of involvement may result from vulnerabilities. Video game maltreatment may be a maladaptive coping mechanism for uncomfortable emotions. Problematic players conceal the problems of real life by retreating into a safe virtual space. Individuals who have specific vulnerabilities or impairments are more vulnerable than others. Problematic thinking patterns can result in illogical, often harmful behaviours, including problematic usage. Exposures might not exist.

According to research, problematic gaming is linked to various emotional issues. There are several traits that all harmful usages share. It is essential to address each of these elements to deal with the problematic use of video games. We must be familiar with several theories and open to the idea that we can unify them in meaningful ways.

*Corresponding author: sterin.joseph@res.christuniversity.in

DOI: 10.4324/9781032708461-13

No one psychotherapy has all the answers for all the clients. The multi-component intervention technique might have more excellent treatment utility than a solitary behaviour intervention. Through the scoping study, we can comprehensively view various components of the intervention and provide a valuable contribution to the field of psychotherapy.

2. Background of the Study

The imminent shift to maturity marks the end of a crucial stage in the life cycle during which growth is experienced on the biological, cognitive, and social-economic levels. Children who endure stress from abusive parenting are more likely to utilise tactics to manage or express their uncomfortable feelings that are problem-focused, emotion-focused, or both. More effective coping techniques are required to address psychological issues adequately. Video games are a fantastic kind of entertainment. Still, they also have the potential to promote harmful behaviour or a lack of self-control that may harm people's daily life. This behaviour has begun to draw notice. Social, economic, and behavioural aspects influence people's gaming habits and cognitive processes. Problematic gaming usage needs to be addressed on numerous levels for an intervention to be successful.

3. Method

The objectives include the scoping assessment and outcomes summary of a multi-component intervention for problematic video game use.

Source of information and search strategy: Searching the following database yielded articles for the scoping review. Ebsco, ProQuest, APA Psynet, and JSTOR. It was modified to make the search mechanism compatible with the various databases. Every database was searched for peer-reviewed full-text publications from 2013 to 2022. The articles were added after the 2013 publication of the updated Diagnostic and Statistical Manual of Mental Disorders (DSM-5). To do this, the concept of a multi-component intervention for problematic video game use was defined. Those articles were included if problematic video gaming was the topic under investigation rather than addiction. The PRISMA-ScR search method found 1080 papers with unique titles. Ten full-text articles were accessed through the titles and abstracts screening.

Search string "multi-component" OR "multi-stage", OR "multi-element" AND "interventions" OR "treatment" OR "management" AND "problematic video games" OR "problematic computer games" OR "problematic electronic games" AND "adolescents" OR "teenagers" OR "school children".

4. Inclusion Criteria

Articles on multi-component intervention for problematic video gaming. Articles on intervention for non-clinical populations. Articles on problematic video games of adolescents of non-clinical populations

5. Exclusion Criteria

Articles on intervention for video gaming addiction. Articles on interventions for psychiatric comorbidities due to gaming. Articles on adult video game usage

6. Findings

The study shows that psychological treatments work better than passive control to curtail unhealthy gaming behaviours. For persons with online gaming disorder who have maladaptive firm views about the social benefit of gaming in social anxiety, activating behaviour through real-world socialising may be beneficial. According to Gonzalez-Bueso et al. (2018), cognitive behaviour therapy (CBT), motivational interviewing (MI), or a combination of psychological therapies have been employed as an intervention component of a more comprehensive therapeutic approach in the majority (80%) of studies on treating Internet gaming problems. Individual and group strategies have been used in studies; the most effective techniques include CBT and mindfulness treatment (González-Bueso et al., 2018). CBT is helpful, but studies have shown that it works far better with other therapies.

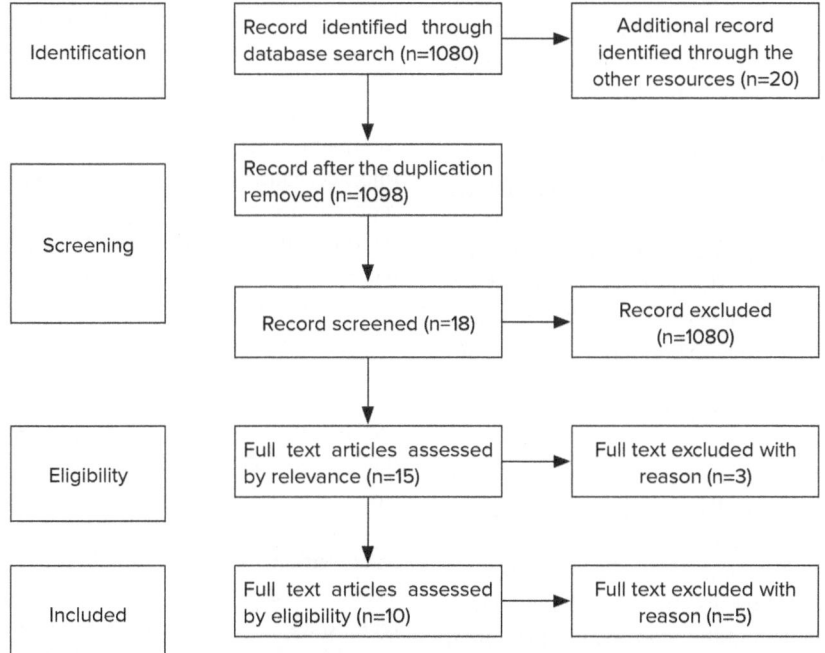

Fig. 13.1 Prisma flow diagram

Source: PRISMA Flow chart of scoping review (based on framework by Arksey & O' Malley, 2005; Moher et al., 2009)

In 2015, Pallenson et al. looked at the efficacy of a manualised approach that included CBT, short family therapy, solution-focused therapy, and motivational interviewing techniques. The results showed that mothers' symptoms associated with online gaming had significantly decreased.

Young (1999) said that family groups emphasising family communication and teaching parents about addiction would be a vital part of such therapy. Familial solid ties among the clients may assist them in getting over their problematic video game use. According to studies, alternative therapies are now employed in treatment to assist clients with gaming concerns. These therapies include psycho-educational training, sleep hygiene, and virtual reality therapy (Costa & Kuss, 2019). It could be beneficial to look at the often-employed therapeutic practice to develop and validate dependable recommendations.

After Parent Management Training for Game Addiction (PMT-G) intervention, it was predicted that parents would have more excellent knowledge of the gaming course, better handle issues, and experience less family conflict. The intervention allowed for more empathic communication among family members (Chanvit Pornnoppadol, 2018). Due to the calming benefits of the new connection, the home environment was more conducive to changing treatment components (individual counselling, family therapy, group mentoring and social support, and alternative activities were generally helpful). The counsellors successfully engaged the teenager because they acted as positive role models. Practitioners may want to rethink how they engage with adolescents suffering from problematic gaming in a therapeutic setting by using technology, particularly video games. The Programa Individualizado Psicoterapéutico para la Adicción a las Tecnologías de la información y la comunicación (PIPATIC) programme comprises numerous intervention areas and is divided into six parts. Family, intrapersonal, interpersonal, and lifestyle changes are all included in psycho-educational therapy. The programme attempts to lessen the symptoms of online gaming addiction and enhance the well-being of teenagers (Torres-Rodríguez et al., 2018). Motivational interviewing techniques, stimulus control, learning adaptive coping skills, self-monitoring techniques, cognitive restructuring, and addiction-related problem-solving are among the techniques discussed by Hung et al. (2010) and King et al. (2011).

According to a meta-analysis of randomised control trials (RCTs), experimental subjects who got counselling during the post-test substantially impacted the level of addiction in the control group (Kim & Noh, 2019). With a decrease in two or three measures of game time, Pallison and the other 13 sessions were investigated. CBT, Family therapy, MI, and solution-focused therapy for adolescents were among the therapeutic modalities often employed (Zajac et al., 2020). Another study discovered a connection between 13 sessions of eclectic psychotherapy that included CBT, Family therapy, MI, and solution-focused treatment and a substantial decrease in internet gaming disorder (IGD) symptoms reported by parents (Zajac et al., 2017). At three-month follow-ups, teenagers who participated in a nine-day self-discovery camp showed a much-reduced rate of time gain. 14 CBT sessions, eight sessions of individual counselling, three lectures on medicine, and a workshop on gaming, engagement, and healthy non-gaming activities make up the camp experience (Zajac et al., 2017). The cognitive restructuring emphasis of the Acceptance and Cognitive Restructuring Intervention Program (ACRIP) module, which incorporates CBT and mindfulness techniques, was deemed pertinent and valuable (GeorgekuttyKuriala,2020). Due to the ideas of self-awareness, self-acceptance, detecting cognitive distortions and their course combined with CBT techniques that are most suited to prevent and transform this problematic thinking pattern, adolescents

7. Limitations of the Study

The study was conducted after 2013, when the DSM-5 was updated, and video games were included as a "condition needing further study." It is challenging to decide when and how much therapy is required to increase the natural recovery rate since more studies must be conducted on the condition's ordinary course.

8. Conclusion

The long-term impact of assuming significance to bring about positive change is the essential component of the multi-component method. The multi-component intervention will assist the teenager in embracing new strategies for addressing numerous vulnerabilities. The pressing need for psychological treatment for excessive gaming requires we address each exposure to validate the intervention. This research will back helpfully therapies for treating problematic video gaming and is recommended for more extensive research

REFERENCES

1. Chanvit Pornnoppadol, W. R.-a. (2018). A Comparative Study of Psychosocial Interventions for Internet Gaming Disorder Among Adolescents Aged 13–17 Years. International Journal of Mental Health and Addiction.
2. Costa, S., & Kuss, D. J. (2019). Current diagnostic procedures and interventions for Gaming Disorders: A Systematic Review. Frontiers in Psychology, 10, 578.
3. González-Bueso, V., Santamaría, J. J., Fernández, D., Merino, L., Montero, E., Jiménez-Murcia, S., Del Pino-Gutiérrez, A., & Ribas, J. (2018). Internet Gaming Disorder in Adolescents: Personality, Psychopathology and Evaluation of a Psychological Intervention Combined With Parent Psychoeducation. Frontiers in Psychology, 9, 787.
4. Georgekutty Kochuchakkalackal Kuriala, M. E. (2020, March). Efficacy of the Acceptance and Cognitive Restructuring Intervention Program (ACRIP) on the Internet Gaming Disorder Symptoms of Selected Asian Adolescents. Journal of Technology in Behavioral Science.
5. Joseph, S. (2022). Problematic Gaming Among Adolescents Within a Non-Clinical Population: A Scoping Review. In ECS Transactions (Vol. 107, Issue 1, pp. 14665–14673). https://doi.org/10.1149/10701.14665ecst
6. Kim, J., Lee, S., Lee, D., Shim, S., Balva, D., Choi, K.-H., Chey, J., Shin, S.-H., & Ahn, W.-Y. (2022). Psychological treatments for excessive gaming: a systematic review and meta-analysis. Scientific Reports, 12(1), 20485.
7. Kim, S., & Noh, D. (2019). The Current Status of Psychological Intervention Research for Internet Addiction and Internet Gaming Disorder. Issues in Mental Health Nursing, 40(4), 335–341.
8. Torres-Rodríguez, A., Griffiths, M. D., & Carbonell, X. (2018). The Treatment of Internet Gaming Disorder: a Brief Overview of the PIPATIC Program. International Journal of Mental Health and Addiction, 16(4), 1000–1015.
9. Zajac, K., Ginley, M. K., & Chang, R. (2020). Treatments of internet gaming disorder: a systematic review of the evidence. Expert Review of Neurotherapeutics, 20(1), 85–93.
10. Zajac, K., Ginley, M. K., Chang, R., & Petry, N. M. (2017). Treatments for Internet gaming disorder and Internet addiction: A systematic review. In Psychology of Addictive Behaviors (Vol. 31, Issue 8, pp. 979–994). https://doi.org/10.1037/adb0000315

*Integrating Advancements in Education, and Society for Achieving
Sustainability – Dimitrios A. Karras et al. (eds)*
© 2024 Taylor & Francis Group, London, ISBN 978-1-032-70841-6

The Narratology of the Lesbian Protagonist in Jeanette Winterson's *Oranges are Not the Only Fruit*

14

G. Ranjitha[1]
Assistant Professor, Kalasalingam Academy of Research and Education, India
C. Jothi[2]
Assistant Professor II, Kalasalingam Academy of Research and Education, India

Abstract: Narratology is a literary technique that helps to identify and state the meaning of the narratives in fiction. Through narratology and its interpretation, the nature of the story is identified and clearly analysed. Oranges Are Not the Only Fruit is a novel written by Jeanette Winterson. The novel exemplifies and portrays the life of a female protagonist who identifies herself to be a lesbian. The lesbian identity of the character is identified through the storytelling narrative. The characters' portrayal and voice also play a major role in segmenting the novel to be a queer one. The novel is written from the Protagonist's point of view which helps to identify the innocence and issues regarding budding sexuality. The paper attempts to explore the various narrative techniques in the novel and thus aims to categorise the novel under LGBT literature.

Keywords: Narratology, Lesbian identity, Point of view, Character, Voice

1. Introduction

Oranges Are Not the Only Fruit is Jeanette Winterson's semi-autobiographical novel. In order to communicate and convey the emotional experience of her childhood the author wants to be an escapist so that she can escape herself from the crude realities that happen around her. In order to do the same, the author uses storytelling techniques where she includes Fairy tales, parables, and stories of King Arthur Knights. Like the protagonist Jeanette, the author Jeanette Winterson is also grown under an adoptive mother in an evangelical household.

Identification of the various narrative structures in the novel helps to find the insight of the author. The relevant finding of the novel as an LGBT literature is also analysed by dissecting and exploring the narratology. The narrative characteristics which the study attempts to analyse are the characteristics of portrayal and the voice. The author's use of a young girl's point of

[1]g.ranjitha@klu.ac.in, [2]c.jothi@klu.ac.in

DOI: 10.4324/9781032708461-14

view is also examined in order to find the leading of sympathy of the character when she explores her sexuality.

2. Methodology

Gerard Gennett's book *Narrative Discourse* (1980) is considered to be an important work regarding narratology. In this book, he claims that narrative discourse and analysis have the capacity of transforming the story itself. The book has given a special focus on Fabula Time which indicates the progress of narratives and happenings throughout the story. He implies that the existence of Fabula Time is in two forms namely objective and subjective. Objective time is the real-time and subjective time is the depiction of time in the minds of the characters. The characters' anticipations, lives, and memories can be got by analysing subjective time (Narrative Discourse, 1980).

Patrick O' Neill in his book *Fictions of Discourse: Reading Narrative Theory* (1996) has clearly figured out that narratology itself is a form of narrative (3). The primary focus of his narrative theory is literary narrative. His conceptual idea is based on narratology which he considers to be a branch of narrative theory. Through narratology, a literary text is analysed in order to bring a contextual and intertextual relationship.

O Neill's next ideology is based on "Game Theory" in narratology. He says that "all narratives are a form of semiotic game, presenting particular and particularly effective arrangements and interrelationships of real or invented events for reception and interpretation by known and/or unknown audiences" (Fictions of Discourse: Reading Narrative Theory, 26). He states that the serious context of an earthquake can be given

a non-serious and lucid one by the author. It is the responsibility of the reader to analyse the context by witnessing the narratology. The readers can take it either way.

The concept of "Textuality" is also contributed by Patrick O'Neill. Through the connection between Intertextual and extratextual agents in order to understand the narrative, Textuality identifies the real author and the real reader (Fictions of Discourse: Reading Narrative Theory, 38).

Lucy Irigaray in her book *The Sex Which is Not One* idealized female discourse. She exemplifies that the female discourse is characterized by simultaneity, fluidity, and multiplicity. She wants to break down the conventional monolithic masculine perceptions of linearity, unity, and oneness through this conceptual idea.

The paper attempts to apply the methodology of interpreting narrative techniques with the mentioned theorists. In specific, it tries to analyse the progress of time by applying Gerard Gennette's perspective on Fabula time. Patrick o' Neill's ideology of analysing narratives is also interpreted in order to bring contextual and intertextual relationships. Lucy Irigaray's conceptual ideology framework of investigation of the women's discourse as a non-conventional and non-linear work which is applied in the select work *Oranges are not the only fruit*.

3. Discussion

The novel Oranges are not the only fruit is told by the female protagonist Jeanette. The story's narration clearly states that Jeanette is grown under a domineering, paranoid, arguably unbalanced, evangelical Christian adoptive mother. The mother and daughter relationship is exemplified by the gap which is between them. Both mother and daughter are committed to God and the daughter Jeanette is even surrendered to become a missionary as her mother prophesized. Her mother's over-dominant attitude and reluctance to befriend anybody, Jeanette ultimately realizes that she has a sexual attraction toward girls. The sympathetic narration of the protagonist Jeanette helps to realize the suffering of the young girl and pity falls on her.

Character identification plays a very important role in the narration. The narration of the protagonist clearly helps to identify the protagonist and the antagonist. Because of the intensive purposes, Jeanette is the hero in the story and through her narration about her dominant mother portrays her mother to be the villain. Jeanette's mother is very stern in what she believes and the narrator also attributes that her mother has no mixed feelings. She categorizes the people who are with her to be either friends or enemies. The description of her mother's domineering attitude is portrayed while narrating the details about making Jeanette prepare tea every morning while her mother is in prayer, "As soon as 'Vengeance is mine saith the Lord' boomed through the wall into the kitchen, I put the kettle on. The time it took to boil the water and brew the tea was just about the length of her final item, the sick list. She was very regular" (Oranges 4).

Point of view in a narrative is a very prominent and fundamental element. According to Janet Burroway, point of view is "a complex and specific concept, dealing with the vantage point and addressing the question: Who is standing Where to watch the scene?" (Imaginative Writing: The Elements of Craft, 44). She also points out that the point of view clearly depicts

the voice of the narrator and thereby helps to understand the diction. In the novel, *Oranges Are Not the Only Fruit* Jeanette Winterson Chooses to tell the story from Jeanette's point of view. The point of view supports the identification of Jeanette's suffering under a dominant mother. It helps to recognize actions and dialogues and also the thoughts and feelings of the young girl protagonist. Through Jeanette's point of view, the sexuality exploration is also justified and provides supporting elements because of the suffering element that she has been undergoing. Thus, it clearly helps to find out the struggle and her journey to identify herself.

The character's voice is also a vital element in a narrative. Burroway states that voice "is a chosen mimicry and is one of the most rewarding devices of imaginative writing, a skill to pursue in order to develop rich characters both in their narratives and in their dialogue" (Imaginative Writing: The Elements of Craft, 42). Voice helps to detect the characterization of the narrative. The author has given a voice of an immature child. Through that innocence is exemplified. The autobiographical tone of the author is narrated through Jeanette's voice as a child. In order to represent the innocence of the protagonist being herself the author has used a child's persona.

The innocence of Jeanette's voice is attributed to her complete surrendering and belief in what her mother says. She is a young and innocent girl like whoever belongs to her age. Jeanette completely believes every prophecy and story by her mother. Her mother has claimed and made her believe that she has been born and brought up only to save and help the world with God. "My mother, out walking at night, dreamed a dream and sustained it in daylight. She would get a child, train it, build it, dedicate it to the Lord…We stood on the hill and my mother said, 'You can change the world'" (Oranges 10). Jeanette completely trusts her mother in doing that so she wholeheartedly accepts to become a missionary as her mother's wish.

The voice also helps to witness how indifferently and strangely Jeanette's mother has handled impeding sexual knowledge to her daughter. Mrs. White is a friend of her mother who frequently visits her mother at their home. On a Sunday, When Jeanette's mother and Mrs. White are there at home the people at home clearly can hear the neighbours having sex. Jeanette's mother has completely become agitated. Jeanette describes the situation in a much of a tense situation. She innocently narrates that at that time her mother immediately starts to play piano and asks all to sing hymns louder in order to drown out the noise which comes because of the sex. The voice clearly reveals that this act of her mother and agitated situation makes sex between a male and a female a kind of crude and unacceptable reality. The situation also makes her lose and lack knowledge regarding sex which can be observed through her voice. "I didn't know quite what fornicating was, but I had read about it in Deuteronomy, and I knew it was a sin. But why was it so noisy? Most sins you did quietly so as not to get caught" (Oranges 54). The voice of the child questions that any act of sin is committed in silence and made to be hidden, But this sinful act of sexual activity is done with a huge noise and also without any repentance. It clearly shows the confusion seen in the child's voice.

Jeanette when found out about her homosexuality it is treated as an innocent act through the child's voice in the narration. She exhibits her very first activity of sexual relationship with her friend Melanie. The voice of Jeanette's innocence is revealed when she expresses her love towards her is equalized with love for God which consists of more profound meaning. "We bustled through the kitchen and I stood on the stairs to kiss her. 'I love you almost as much as I love the Lord,'" (Oranges 103-4).

The other important aspect which is seen in the narration is the usage of fables by the author. It is a semi-autobiographical fiction but consists of several fairy tales, Biblical myths, and a few Arthurian tales too told by the narrator Jeanette. The fables are told as an account of escapism by the protagonist. The protagonist could not tolerate the crude and dominant realities that happen around her. So, she wants to escape by narrating fables in order to draw a line between reality and imagination. The stories are told by the narrator as a form of interlude.

Jeanette as an innocent child who was under a dominant abusive mother is proscribed to take decisions on her own and do what she likes. She always has to do and involve herself in satisfying society's needs and especially her mother's. In relief of the same, she uses fairy tales in her narratives to illustrate her inner thoughts and desires. Jeanette's mother has instructed her regarding the distinctions between the "inner world and outer world" (Oranges 32). The inner world is Jeanette's thoughts and expectations and the outer world is her behaviours which society witnesses. The stories help her to bridge the gap between the inner and the outer world.

In the opening narratives when she is so innocent, she brings a story of a hunch bag man who adopts a perfect girl and provides her duties to milk the goats, to educate the people, and sing songs for the festival. This narrative is correlated and compared with her childhood of being under her mother who adopts her and gives her duties to be a missionary and servant to God.

The church pastor while giving sermons regarding leading a flawless life the narrator tells the story of a king searching for a flawless wife. In order to hide the guilt that she bears because of the homosexual activity she correlates the fable with her life through her storytelling narration. The Arthurian king beheads every woman who does not satisfy his expectations which she associates with her being a fault-done homosexual. The story makes her feel pressurized in order to be a perfect girl.

As the title indicates "Oranges" are provided to be the only fruit for Jeanette. The novel's narrative framework helps to identify the facts of compulsory Heteronormativity. It acts as an extended metaphor that exemplifies that no options are available other than heterosexuality. "Elsie Norris and me ate an orange every day" (Oranges 39). Jeanette is really aware of these facts of being in a choiceless situation. Once when Jeanette's sexuality is being explored and was made humiliated by her mother in front of all the church members, she finally decides to move out of her house. As soon as she moved out redemption is achieved by her. In her narrative, she delivers the freedom she got after she moved out of the house. This time she uses the extended metaphor of telling that "oranges are not the only fruit" (Oranges 219). She also has the freedom to eat pineapples. The narration

helps to identify not only the sexuality being explored but also Jeanette's identification of individuality and her achievement of freedom.

4. Conclusion

The paper attempts and achieves the findings of various narrative structures in the novel in order to understand the author's clear perspective. The novel through characterization, voice, and the point of view of the protagonist Jeanette reflects the author's inner self since it is a semi-autobiographical novel. The storytelling technique enables to identify the protagonist's inner emotions and their connections to the outer world which is society and her mother. The transition of Jeanette from an innocent child to a homosexual explored girl is also revealed through her evolutions in the narrative. Ultimately the author's aim of achieving individuality and freedom is also clearly exemplified by the meticulous narrative perspective that Jeanette Winterson intends. The finding the protagonist Jeanette is a lesbian, is achieved through the narratology of the novel *Oranges Are Not the Only Fruit*.

REFERENCE

1. Irigaray, Luce. (1985). *The Sex Which Is Not One*. Trans. Catherine Porter and Carolyn Burke. Ithaca: Cornell UP.
2. Winterson, Jeanette. (1987) *Oranges Are Not the Only Fruit*. New York: Atlantic Monthly P.
3. Genette, Gérard (1988). Narrative Discourse Revisited. Ithaca, N.Y. Cornell University Press P.
4. Patrick. O. Neill (1996). Fictions of Discourse: Reading Narrative Theory (Theory/Culture). University of Toronto Press P.

5. Xiowei, Chen. Peeling the Orange (2014). An Intertextual Reading of Oranges are Not the Only Fruit. Comparative Literature and Literary Theory.

6. Amy Benson Brown PhD. (2008) Inverted Conversions, Journal of Homosexuality, 33: 3-4, 233–252, DOI: 10.1300/J082v33n03_11

7. Emma Hutchison (2010). Unsettling Stories: Jeanette Winterson and the Cultivation of Political Contingency, Global Society, 24:3, 351–368, DOI:10.1080/13600826.2010.485561

8. Delaney, Daniella (2022). Discuss Jeanette Winterson's exploration of sexuality and identity in Oranges Are Not The Only Fruit. LBU Review, 1(1).

9. Richard M. Wafula& Chris L. Wanjala (2017). Narrative Techniques in Chinua Achebe's Things Fall Apart, Journal of Social Sciences, 50:1-3, 62–69, DOI: 10.1080/09718923.2017.1311740

Integrating Advancements in Education, and Society for Achieving Sustainability – Dimitrios A. Karras et al. (eds)
© 2024 Taylor & Francis Group, London, ISBN 978-1-032-70841-6

15

'The Computer has Become an Object to Think with': Advancement in Speech Therapy for Children

Rohini R., Manali Karmakar*

Vellore Institute of Technology, India

Abstract: Language is one of the pivotal tools through which human experiences are formed. However, children diagnosed with neurodevelopmental disorders face challenges in developing normative communication skills, leading to discrimination and marginalization in a neurotypical society. This paper offers a review of a few current research work on Human-machine interaction (HMI) training programs that could be used for developing communication and pragmatic skill in children. Through the lenses of a technology-driven society, the paper discusses the status of research on speech development for children with communication disorders. With the advancement in HMI, training a child can improve language production. The paper's novelty lies in its ability to show how individual focus given to a child through HMI can accelerate early diagnosis and establish personalized treatment. Therefore, this paper attempts to justify the requirement of a mobile application that will lay individual focus on every child and diagnose the part of language production that the child struggles with.

Keywords: Human-machine interaction (HMI), Speech disorder, Mobile application, Language production

1. Introduction

According to the National Institute on Deafness and Other Communication Disorders (NIDCD), approximately 18.5 million individuals have a speech, voice, or language disorder. Among the general population mentioned above, children (the age between 3 and 17) with communication disorders are identified to suffer from a range of difficulties when it comes to language production. The inability to produce language spontaneously or at the right age creates significant problems, such as the feeling of exclusion and demotivation. The communication disorder also discourages the child from joining family gatherings and school activities.

*Corresponding author: manali.karmakar@vit.ac.in

DOI: 10.4324/9781032708461-15

Over the decades, speech pathologists have been playing the role of moderators to address this issue of difficulty in language production in children. However, the first half of the 20th century strongly marked the steady replacement of human beings with machines, robots, and humanoids. The human mind began to think through a computer. With the advancement in Human-machine interaction (HMI), training a child becomes possible to improve language production via mobile applications.

This paper culminates valuable inputs from recent research based on the inclusion of Artificial Intelligence (AI) in the lives of children who struggles with specific language production disorders. This paper would also like to review the development that has happened so far in diagnosing and treating through HMI to children with speech disorders.

AI and Language Development in Children: Artificial Intelligence (AI) is rapidly altering the present and the future world, where speech pathologists are effortlessly being replaced by speech therapy technological mobile applications. According to Microsoft, current researchers are gaining positive results and are paving the way to create innovative tools to approach speech difficulty problems in children. The initial groundbreaking achievement of creating tools to improve a child's speech disorder through pre-equipped tasks and activities gained an impactful response from the parents and the children. In order to emulate the understanding of a speech pathologist and a child with the disorder, the AI tool must be able to comprehend the speech styles and impairments (Bull, 1997).

It can be argued that before and during the catastrophe of COVID-19, children continued to be the victims and were yet to receive medical attention from speech pathologists due to long waiting lists (McLeod et al., 2020). In addition, the unavoidable work-from-home (WFH) condition transformed parents to adopt HMI as a reality that naturally limited human-human interactions (Sahu & Karmakar, 2022). Research proves that without required services, children with speech disabilities might become vulnerable to lifelong educational and social limitations (McCormack et al., 2009).

As the Massachusetts Institute of Technology (MIT) reported, early intervention in children with speech and language disorders can yield excellent results in their academic and social skills later in life (Hardesty, 2016). To erase the situation of late treatment or not giving required individual attention to the child, it is crucial to produce a mobile application that will grant individual, one-on-one attention to the child, thereby diagnosing the specific pattern of the child's speech and language production. McLeod et al. (2020)'s study proved that with the help of technological solutions, speech improvement in children was noticeable and plausible compared to face-to-face speech therapy. Therefore, his paper asserts the importance of a personalized mobile application that must provide individual attention to the child in the place of speech pathologists.

An AI Tool as a Speech Pathologist: An outright comment has been given by the University of Texas at Austin stating, "We think that AI can evaluate a child's language skills "better" than their parents or a speech pathologist might." The usage of such a tool will give customised suggestions at any time of the day, ultimately giving attention to the child throughout the day, which will yield rapid results. This assertion can be justified by the statement released by a researcher named Jordan Green in the year 2016, who himself is a speech pathologist at Massachusetts General Hospital's Institute of Health Professions, "Better diagnostic tools are needed to help clinicians with their assessments." With technology in hand, clinicians depend on tools to keep track of the child's development as well as to provide an advanced level of assessments.

The question of whether tailor-made suggestions and assessments can be put into practice has been asked by researchers such as Jordan Green and Tiffany Hogan from MIT, Sharynne McLeod and others from Charles Sturt University, and Aubrey O'Neal from the University of Texas. Since the area of speech and language disorders are in heavy research, the availability of age-specific speech development in children makes it possible to understand the different sounds children make during different ages. If the child does not produce the right sound or word at a specific month/age (See Fig. 15.1), it can be understood that the child is finding it challenging to produce language. When it comes to prototyping information into a tool, research proposed by O'Neal (2018) suggests that creating an AI tool to address speech difficulty in children will not be possible without data corpus, voice recognition, and automatic transcription. Thus, for a personlised tool to be created, it is vital to lay individual focus on each child and document the regular speech conversations between the child and parents.

Signs of language problems include:

Birth–3 months	Not smiling or playing with others
4–7 months	Not babbling
7–12 months	Making only a few sounds. Not using gestures, like waving or pointing.
7 months–2 years	Not understanding what others say
12–18 months	Saying only a few words
1½–2 years	Not putting two words together
2 years	Saying fewer than 50 words
2–3 years	Having trouble playing and talking with other children
2½–3 years	Having problems with early reading and writing. For example, your child may not like to draw or look at books.

Fig. 15.1 Sign language problems

Source: Early Identification of Speech, Language, and Hearing Disorders. American Speech-Language-Hearing Association.

Evidence-based diagnostic solutions can improve accessibility which implies low-cost screening for n-number of children by decreasing the number of children with speech disorders (Source: Researchers at the Computer Science and Artificial Intelligence

Laboratory at MIT and Massachusetts General Hospital's Institute of Health Professions). Using machine learning to attend particular classifications with the impartation of training sets will eliminate false results, a statement confirmed by MIT researchers, John Guttag and Jen Gong.

Researchers at The University of Texas at Austin and MIT are in line to run a technological test to create a tool that will help parents to undertake the suggestions provided by the tool to be used on children. Although there might be varying signs of neurodevelopmental disorders in children, the availability of case studies and the determined neurodevelopmental disorders can be primarily used to create the tool. One of the advantages of using AI in the medical health sector is its ability to adapt to the repetitive causes and responses from various subjects (Kay & Hartmans, 2021). Based on the cumulative data, the tool will be able to present a suitable assessment of the child's disorder (which the tool would have encountered previously in another child).

The Relevance of an AI Tool in Language Production: Any tool that will emulate or replace human efforts must be of any advantage, collide with the pre-existing system, yield positive results, and be easily used (Rogers, 1962, p.1). In this case, numerous pre-existing mobile applications are used to improve the child's speech disorder generically by giving basic activities. Instead, the need of the hour is a tool that will diagnose the child's disorder in particular and extensively train the child in language production. Therefore, the juxtaposition of bringing a new tool to life is relatively high.

The most important task of creating a tool is the pre-input that must be given to the tool before producing it to a larger audience. With an initial data corpus, the tool can function independently. As mentioned, the data from multiple case studies can be given as input to the tool (O'Neal, 2018), but this will not be the end of the experiment. An innovative, most viable product (MVP) can be worn by the volunteer (child/parent) through which the software tool can analyze the conversations of the child.

By listening to the child's conversations, the tool can locate the specific issue in the cognitive, linguistic, and neural parts of speech production. It is based on the hypothesis that a learned machine has a rich set of knowledge to determine the correct training type to be given to the child (Mishra, 2021). Results gained by the tool will provide a real-time diagnosis of what the child is struggling with and offers tailor-made sessions and activities instead of providing basic training to the child when the impairment is identified. It is also interesting to note that the same tool can record the development of the child's language production. Due to the usage of technology in eliminating the impairment, it is believed that this process can comparatively speed up the child's speech development. From the literature available, this paper reveals the unspoken truth that AI can potentially diagnose a child's speech disorder and provide personalised treatment in the near future.

2. Conclusion

In conclusion, what is essential is that speech training for children must be individual-centric rather than providing general training. A tool operating like a speech pathologist (from diagnosing to training) is still a dream for many tech wizards as well as parents. The tool's features must focus on the child's language production, and gradually, while in use, the tool will learn the pattern of development required in the child's communication skills.

Since machine learning is rapidly developing, the data stored will only multiply in the sense of machine knowledge and enhance the learning experience of the upcoming generation of children with disorders. Overall development of such a tool will be time-consuming. For example, Amazon's Alexa (a voice-controlled virtual assistant) was created and tested for a period of six days a week for six months, during which Alexa collected all the questions and sent them to the team (Kay & Hartmans, 2021). After a successful data collection and input process, Alexa leaped into the market with its extraordinary ability to talk back and respond to the query of the person on the other side. Notably, creating such a tool exploits time, but the output will transform the generation into a tech-supported human world. This paper would like to conclude its argument by stating that a tool that comprehends the needs of a child in the context of language production and speech development by diagnosing the specific difficulty will remain a groundbreaking invention of all times.

REFERENCES

1. Bull*, S. (1997). "Promoting effective learning strategy use in CALL." Computer Assisted Language Learning 10(1):3 39.
2. Hardesty, L. (2016). | MIT News Office. "Automated Screening for Childhood Communication Disorders." MIT News | Massachusetts Institute of Technology, https://news.mit.edu/2016/automated-screening-childhood-communication-disorders-0922.
3. Kay, G. and Avery Hartmans. (2021). "Amazon Trained Alexa in Secret by Hiring Unsuspecting People to Ask...," https://www.businessinsider.in/tech/news/amazon-trained-alexa-in-secret-by-hiring-unsuspecting-people-to-ask-questions-in-a-room-filled-with-hidden-prototypes/articleshow/82557489.cms
4. McCormack, J., McLeod, S., McAllister, L., & Harrison, L. J. (2009). "A systematic review of the association between childhood speech impairment and participation across the lifespan." Int. J. of Speech-Language Pathology. 11(2):155 170.
5. McLeod, S., Ballard, K. J., Ahmed, B., McGill, N., and Brown, M. I. (2020). "Supporting Children With Speech Sound Disorders During COVID-19 Restrictions: Technological Solutions." Perspectives of the ASHA Spl. Interest Grps. 5(6): 1805 1808.
6. Mishra, A. (2021). "Self-Learning System for Child Development Using Conversational AI and Natural Language Processing (NLP)." Impact of AI Technologies on Teaching, Learning, and Research in Higher Education. IGI Global. 124 133.
7. O'Neal, A. (2018). "AI for Child Language Acquisition." Advanced Design for Artificial Intelligence. https://medium.com/advanced-design-for-ai/ai-for-child-language-acquisition-45582f07236
7. Rogers, E. M. (1962). Diffusion of Innovations. Free Press.
8. Sahu, O. P. and Manali Karmakar. (2022) "Disposable culture, posthuman affect, and artificial human in Kazuo Ishiguro's Klara and the Sun (2021)." AI & society. 1 9.

*Integrating Advancements in Education, and Society for Achieving
Sustainability – Dimitrios A. Karras et al. (eds)*
© 2024 Taylor & Francis Group, London, ISBN 978-1-032-70841-6

16

Analyzing the Personal Barriers of the Teacher Trainees Towards Grammar Learning: An Exploratory Study

P. Jayakumar*

Assistant Professor of English,
St. Joseph's College of Engineering, Tamilnadu, India

Aravind B. R., Harishraj A. N., N. Devadhas Prabu

Assistant Professor of English,
Kalasalingam Academy of Research and Education, Tamil Nadu, India

Abstract: Before instructing the learner on anything, it is crucial to comprehend their attitude. The students' feedback is useful for identifying learning opportunities and ways to advance their education. The most difficult aspects of learning a language for beginners are its grammar and spelling rules. They have a huge variety of unpredictable and unexpected, erratic regulations. This study makes an effort to identify the individual hurdles that learners face when learning grammar. A questionnaire was given to 60 B.Ed. students in the Chennai district as part of the study's data collection. After their pre-test, the data was gathered to create an appropriate syllabus and improve the teaching methods. According to the comments, the issues were quickly discovered and resolved in the post-test.

Keywords: Feedback, Grammar, Syllabus, Questionnaire and teaching methods

1. Introduction

Mishra, P. (2010) stated, "Teaching of Grammar has always remained a controversial subject as the method and material adopted in teaching it." The foundation of verbal and written communication is grammar. If the feedback is gathered depending on the subject taught, it will be advantageous to the students. Moses, R. N., & Mohamad, M. (2019) explored at the challenges both teachers and students encountered when acquiring and imparting writing skills in elementary schools. The initial step in the learning process is to analyse and identify the learners' difficulties. This study aims to investigate the difficulties B.Ed. students have in studying grammar.

The study of errors is one of the important platforms in teaching-learning process. In the error analysis study, it is required to record the kind of problems faced by the learners. In second language learning, it is a major concept for the learners and

*Corresponding author: jaikmabed@gmail.com

DOI: 10.4324/9781032708461-16

researchers. According to Khansir, A. A. (2012), "The basic task of error analysis is to describe how learning occurs by examining the learner's output and this includes his/her correct and incorrect utterances." Lennon, P. (2008) highlighted the view of Pit Coder and stressed the value of error analysis. Pit Coder in his paper (1967) "The significance of learners' errors" viewed "The learner is engaged in a process of discovering the language. In this view errors are not only an inevitable but also, very importantly, a necessary feature of learner language, without which improvement cannot occur."

2. Brief Review of Literature

Akhtar, R., Hassan, H., Saidalvi, A., & Hussain, S. (2019) discussed the difficulties and solutions faced by ESL learners when writing academically. They discovered through the study that there are difficulties with language proficiency, teaching strategies, and students' attitudes toward English. To resolve the problems, they advised using a few strategies. They suggested to employ some methods and approaches to overcome the issues.

Bhuvaneswari, G., Swami, M., & Jayakumar, P. (2020) studied the perspective of the UG students towards digital learning. The paper also analyzed the challenges and opportunities of online classrooms.

Jayakumar, P., & Ajit, I. (2016) & (2017) conducted the studies on teaching sentence structures and pedagogical implications in grammar learning.

Scott, C. M., & Balthazar, C. H. (2010) Examined applications for language problems in older school-age children and adolescents, as well as for screening and intervention.

Myhill, D., & Watson, A. (2014) A fully-theorized understanding of grammar in the curriculum was the subject of a survey of the literature on the teaching of grammar and its place in the curriculum.

Erdoğan, V. (2005) explored the role that error analysis plays in the linguistic and methodological aspects of teaching English as a second language. The study stressed the contribution that Teachers can identify the causes of errors and adopt pedagogical measures against them by using error analysis. As a result, it has become crucial to analyse learner language in order to address certain issues and offer solutions.

Jobeen, A., Kazemian, B., & Shahbaz, M. (2015) made a study on EA in teaching and learning. In order to comprehend the tactics and procedures employed in the process of second and foreign language acquisition, the purpose of this research is to investigate mistakes made by second and foreign language (L2) learners. The study draws the conclusion that their first language's rules have a significant influence on them (L1).

3. Aims and Objectives

The aim of the study is to diagnose problem faced by the learners to overcome their errors in grammar learning on the proposed area.

The study covers the following objectives:

- To identify the exact problem of the learners
- To know the learning style and preferences
- To change the practice in their learning

4. Methodology

The current study adopted the exploratory approach to identify the learning barriers of the students in the field of grammar. This study gathered data from 60 B.Ed. students who concentrated in English during their undergraduate years in order to get information from the learners. As this was an experimental study, the researcher gave the intended audience members pre-intervention and post-tests. The target audience's data was gathered using a questionnaire approach.

The study includes a rank correlation relating to the learners' personal barriers and challenges. In order to frame a syllabus and an effective teaching approach, the study's goal was to investigate their difficulties. The study then completed a mixed teaching strategy for grammar using a cognitive code approach and an Android application during the intervention period.

5. Rank Correlation

Table 16.1 Problems from the personal parameters in learning grammar

S. No.	Statements	Scores										Total
	Numerals	1	2	3	4	5	6	7	8	9	10	
1	U	4	3	7	3	4	2	7	3	7	20	60
2	D	4	6	6	8	6	5	8	4	9	4	60
3	L	12	6	6	8	6	2	3	5	8	4	60
4	M	5	10	7	9	3	8	10	4	1	3	60
5	C	6	6	10	9	6	9	1	5	7	1	60
6	P	9	7	9	1	6	9	3	7	3	6	60
7	LOT	3	7	4	7	4	5	9	9	5	7	60
8	UE	4	2	9	7	12	5	9	4	6	2	60
9	DD	2	6	1	2	10	9	7	12	6	5	60
10	PT	11	7	1	6	3	6	3	7	8	8	60
	Total	60	60	60	60	60	60	60	60	60	60	60

It is an attempt to understand the target learners' hierarchy of mental states. The following variables are identified in order to rank the learning opinion of grammar learning.

The students were asked to rank the various opinions by themselves in order to determine the issues arising from their particular parameters in learning grammar. The students' rankings of various opinions on studying grammar are shown in Table 16.1. The students ranked ten distinct types of opinions regarding learning grammar in accordance with their preferences. The students' priorities regarding grammar instruction were presented in order to rank their opinions. The scores listed below were used to calculate the weighed arithmetic mean.

I Rank   = 10 Points
II Rank   = 9 Points
III Rank = 8 Points
IV Rank   = 7 Points
V Rank   = 6 Points
VI Rank   = 5 Points
VII Rank = 4 Points
VIII Rank = 3 Point
IX Rank   = 2 Point
X Rank   = 1 Point

The total score of each stage was determined based on the points awarded and ranks attained. To determine the mean score, the total score thusly calculated was divided by the total number of students for each phase. The statement with the highest mean score out of the eight on the schedule was chosen by the majority of students as the action they took to deal with challenges. These steps are used to calculate weighted arithmetic means.

Having collected the data in the questionnaire, the researcher analyzed it both qualitatively and quantitatively. The main goal of the data analysis on personal barriers to grammar learning is to identify the factors that prevent people from successfully learning grammar in a second language. The following table shows the mean score of the data collected from B.Ed. learners. 10 problems were selected and arranged in order. Further it has been ranked and recorded as per the preferences of all 60 learners who majored in English. The provided data will help the other learners to overcome the learning difficulties in the field of grammar and to set a suitable pedagogy for the same.

Table 16.2   Weighted arithmetic mean score

S. No.	Contents	W.A. Mean	Rank
1	Uninterested	4.15	10
2	Disuse the grammar rules and patterns	5.25	7
3	Lack of practice	6.18	3
4	Memorize the rules without understanding	6.23	1
5	Confusions due to the similarity of grammar rules	6.2	2
6	Problems in exceptional rules in grammar	5.86	4
7	Lack of thorough knowledge in basic grammar	5.05	8
8	Understanding error	5.63	5
9	Distractions due to other disciplines	4.8	9
10	Postponing the studies	5.46	6

Based on the mean score shown in Table 16.2, the students' personal learning difficulties with grammar were ranked.

Table 16.2 shows the degrees of steps in terms of arithmetic mean, illuminating the issues arising from individual grammar learning settings. The first rank is given to Memorize the rules without understanding followed by Confusions due to the similarity of grammar rules, Lack of practice, Problems in exceptional rules in grammar, Understanding error, Postponing the studies,

Disuse the grammar rules and patterns, Lack of thorough knowledge in basic grammar, Distractions due to other disciplines and Uninterested.

6. Results and Discussion

Following the data collection, the study found that the learners have various problems understanding grammar. Based on the feedback provided by the students in the pre-test, the study created 10 different issues to determine the precise issue.

The major findings are as follows:

- The students found that memorize the rules without understanding is their first problem in their teaching-learning process. They felt in schools that grammar teaching was like memorizing the rules. They were taught rule-centered learning.

- Secondly, it is understood from the data, Confusions due to the similarity of grammar rules placed the second problem in their mind. It is evident from their opinion that they experienced grammar learning without differences.

- Lack of practice was addressed as their next issue. Grammar learning is more practice oriented. The students should develop their application in their learning. They have to apply the learned the rules in their spoken and written communication.

- Further, Problems in exceptional rules in grammar ranked as forth problem. The teaching grammar is a complete one when the teaching-learning process ended with teaching exceptional rules. It is meaningless if it is not entertained in the classroom teaching.

- Understanding error was ranked as fifth one. If the learner misunderstands the concept, he/she makes error for sure. Followed by Disuse the grammar rules and patterns can be one the strongest reason. One can forget the learned rules unless it is in practice.

- Naturally, postponing the studies is a problematic one for the learners not only in the field of grammar learning but also in the other fields.

- Grammar rules should be used after being learned. The students are required to use the rules they have learnt in the workbook or take a quiz on the material they have learned. The grammar rules would vanish from his or her active memory if they were not applied.

- Grammar lessons connect fundamental to advanced concepts. One must learn and grasp the fundamentals before moving on to the advanced level.

- Distractions from learning are a major issue for digital learners. Distractions occur in psychology if the subject matter is uninteresting.

- Basic grammar, distractions and uninterested are the minor problems felt by the students in their response.

7. Recommendations

Based on the findings the study recommends the following for the benefits of the learners who interested in learning grammar.

- As per Bloom's Taxonomy, the learner should keep it in mind that one should start from remembering, understanding, applying, analyzing, evaluating and creating.

- In order to avoid confusions, the teacher must facilitate the learners with multiple examples and practices.
- The teacher should study the attitude of the learners and employ a suitable method and approach for teaching the grammar subjects.
- For confusions, the teacher should teach the rules with differences. For example, the letter s used as plural when we add it to noun (Books are there), the letter s used as singular when we add it to verb (He plays cricket), the letter used in apostrophe (This is Ramu's pen).
- To complete the grammatical area, it is necessary to teach its exceptional rules to the learners since they are ESL learners. In voices, the learners are exposed only in active and passive. The teacher should also teach the kinds of passive and middle voices to make it a complete one.

8. Conclusion

The acquisition of additional number sentence patterns in daily usage will be made possible by the learners' grasp of grammar when studying a second language. For all English language second-language learners, the art of writing is essential. The requirement is necessary since it is mirrored in the nation's internal, end-of-semester, competitive, and other exam formats. Writing assignments from students will be used to assess their expertise. Knowing the language's grammar is necessary to develop strong writing abilities. The learners can create a broader variety of sentence forms with various patterns with the aid of their grammatical skills. Grammar and typography are the fundamental elements of successful writing in every language. The initial phase in the teaching-learning process is error analysis. The study emphasized the value of documenting students' difficulties with grammar instruction. The study also offered a few recommendations to overcome such problems while teaching grammar to ESL learners.

REFERENCES

1. Akhtar, R., Hassan, H., Saidalvi, A., & Hussain, S. (2019). A systematic review of the challenges and solutions of ESL students' academic writing. *International Journal of Engineering and Advanced Technology*, 8(5), 1169–1171.

2. Bhuvaneswari, G., Swami, M., & Jayakumar, P. (2020). Online classroom pedagogy: Perspectives of undergraduate students towards digital learning. *International Journal of Advanced Science and Technology*, 29(04), 6680–6687.

3. Binu Sahayam D., Bhuvaneswari G., Bhuvaneswari S., Thirumagal Rajam A. (2022) Stress-Coping Strategy in Handling Online Classes by Educators During COVID-19 Lockdown. In: Hamdan A., Hassanien A.E., Mescon T., Alareeni B. (eds) Technologies, Artificial Intelligence and the Future of Learning Post-COVID-19. Studies in Computational Intelligence, vol. 1019. Springer, Cham. https://doi.org/10.1007/978-3-030-93921-2_4

4. Jayakumar, P., & Ajit, I. (2016). Android app: An instrument in clearing Lacuna of English grammar through teaching 500 sentence structures with reference to the verb eat. *Man in India*, 96(4), 1187–1195.

5. Jayakumar, P., & Ajit, I. (2017). The pedagogical implications on the root and route of English basic verbs: and extensive study through android application. *The Social Sciences*, 12(12), 2244–2248.

6. Mishra, P. (2010). Challenges and Problems in the Teaching of Grammar. *Language in India*, 10(2).

7. Moses, R. N., & Mohamad, M. (2019). Challenges faced by students and teachers on writing skills in ESL Contexts: A literature review. *Creative Education*, 10(13), 3385–3391.

8. Myhill, D., & Watson, A. (2014). The role of grammar in the writing curriculum: A review of the literature. *Child Language Teaching and Therapy*, 30(1), 41–62.

9. Scott, C. M., & Balthazar, C. H. (2010). The grammar of information: Challenges for older students with language impairments. *Topics in Language Disorders*, 30(4), 288.

10. Khansir, A. A. (2012). Error analysis and second language acquisition. *Theory and practice in language studies*, 2(5), 1027–1032.

11. Lennon, P. (2008). Contrastive analysis, error analysis, interlanguage. Bielefeld Introduction to Applied Linguistics. A Course Book. Bielefeld: Aisthesis Verlag, 51–60.

12. Erdoğan, V. (2005). Contribution of error analysis to foreign language teaching. Mersin Üniversitesi Eğitim Fakültesi Dergisi, 1(2).

13. Jobeen, A., Kazemian, B., & Shahbaz, M. (2015). The role of error analysis in teaching and learning of second and foreign language. *Education and Linguistics Research*, 1(2), 52–62.

Note: All the tables in this chapter were made by the author.

Integrating Advancements in Education, and Society for Achieving Sustainability – Dimitrios A. Karras et al. (eds)
© 2024 Taylor & Francis Group, London, ISBN 978-1-032-70841-6

17

Reflection of Sustainable Development Goals in the Selected Political Discourse

Fredrick Ruban Alphonse and Arya Parakkate Vijayaraghavan
Christ (Deemed to University), India

R. Manikandan
Vel Tech Rangarajan Dr. Sagunthala R&D Institute of Science and Technology, India

Anuradha S.
Sri Sai Ram Engineering College

Mohd Aarif
Global Research Network, India

Samrat Ray
Peter The Great Saint Petersburg Polytechnic University, Russia

Abstract: Sustainable development goals are the blueprint to achieve a sustainable globe for living beings. There are 17 SDGs set up by United Nations General Assembly in 2015. The significance of sustainable development goals has reached the pinnacle in recent times. They are the global goals and a much-stressed factor in academia and political setup. Therefore, it is necessary to study how academic and political discourses work in disseminating the knowledge of SDGs. The present research paper aims to trace the latent sustainable development goals from the selected political discourse. The political discourse produced within the context of Tamil Nadu has been considered for the present study. It applies Teun A. van Dijk's political discourse analysis model to examine the selected political discourse. The study mints to look at the idea of sustainability through nature and social justice.

Keywords: Sustainable development goals, Natural resources, Equality and social justice

1. Introduction

The present research intends to explore the sustainable development goals latent in the political discourse of Seeman and Dr. Thol. Thirumavalavan. They are eminent political actors in terms of Dijk's political discourse analysis theory. Thol. Thirumavalavan is a scholar and activist from the southern Indian state of Tamil Nadu. He is the president of

*Corresponding author: fredrick.ruban@res.christuniversity.in

DOI: 10.4324/9781032708461-17

Viduthalai Chiruthaigal Katchi (VCK), who emerged as a politician in the late 20th century. He emphasizes unity among the Tamils and calls the Tamils to give away the disharmony that prevails in the name of caste. Seeman, the Chief-coordinator of *Naam Tamilar Katchi*, is another social activist who emerged in the 21st century. Seeman was enthralled by the Dravidian ideology at an early age. He dealt with Periyar's ideology and caste abolition in films. Presently, he voices for the voiceless Tamils around the globe. His politics touches upon a wide range of topics and issues, especially foregrounding the welfare of Tamils.

2. Method and Methodology

The political discourse of Seeman and Thol. Thirumavalavan has been selected as the primary source. The primary text is examined using Teun A. van Dijk's political discourse analysis model. This model functions to analyse and critique the political discourse in terms of theme and rheme. The word discourse indicates "the spoken word, or to all utterances written and verbal, or to a particular way of talking vocabularies" (Griffin, Gabriele 2013). Political discourse is one among its type that "is fundamentally argumentative and primarily involves practical argumentations" (Fairclough, Isabela, et al., 2017). The discourse that is produced within the political context is called political discourse. Dijk considers political discourse as a critical enterprise thus mints to understand it from a critical perspective. For Dijk, analysis of discourse can be classified as language use, discourse and verbal interaction, in which communication belongs to "the microlevel of the social order" and "power, dominance, and inequality between social groups are typical terms that belong to

a macrolevel of analysis" (Dijk, Teun A. van 1998). At microlevel analysis of discourse, the syntax, semantics, stylistics and rhetoric of a text are examined. The global meaning of the text, theme and rheme is analyzed at the macrolevel analysis of discourse. The present study analyses the selected discourse at macrolevel to find the underlying similarities between the selected political discourse and the seventeen substantial development goals.

3. Nature for Sustenance

In this part of discussion, the research deals with the discourse of Seeman with reference to "Seeman's Speech about Mother Nature". It is a 03:22 minutes video accessed from *Seeman for Universe* channel. The entire video is divided into two parts: engaging with nature in the past and engaging with nature in the present. The first part of the discourse focuses on 'engaging with nature in past'. It answers the questions: what has been preserved by the ancestors and what has been left as a legacy to future generations. An extract from the text is mentioned below.

> Didn't our ancestors build houses? Did they empty the river sand or break the mountains? Did they save millions of money in the banks for you? No! They preserved the lakes, the rivers, the trees and the mountains also for you. Why? The world is not just yours, it belongs to the future generations, your children and their children. You're just a tenant who lives here for rent. "Seeman's Speech about Mother Nature" (2:50-3:02)

In the above excerpt, Seeman sheds light on the natural life led by the ancestors of Tamils. He throws certain contemplative questions amidst the audience so that it can influence them. Seeman tries to teach them that the earth is not one's individual property, it is a space granted to all the living things, but it has been mutilated by human beings in the name of development. Generally, people believe that constructing huge concrete houses for their living is development. For this reason, they extract soil from the river and break mountains. They are also carried away by the fancy thought that owning a car is a mark of rich status in society, but they forget to realise how many liters of water are being utilized to prepare a car. Hence, he throws the question did our ancestors preserve the natural resources for us or constructed houses for us. Seeman attempts to convey that the ancestors of Tamils have protected and preserved nature and its resources for future generation. They have preserved mountains, rivers, lakes and trees for the use of future generations. This idea very well connects with the substantial development goals. Juxtaposing the substantial development goals and Seeman's idea makes it is evident that his notion correlates with three substantial development goals. The idea of 'water management' reflects in goal six, "sustainable consumption and production patterns" (United Nations) has been covered in goal twelve and "halt and reverse land degradation and halt biodiversity loss" (United Nations) has been dealt in goal fifteen. The crux of these three goals corresponds with the didactic text of Seeman. Water must be preserved for the need of the living beings. It has to be used for sustainable use and wastage of water must be controlled. Water can be preserved in many ways and saving rivers and mountains are pivotal act in this regard. Mountains contribute to cloud formation thereby bringing rain. Likewise, preserving river soil helps in retaining rivers and they contribute

immensely to the sustenance of land and agriculture. Agriculture is impossible without water. Lack of agricultural activities leads to starvation and poverty. This fact is explicit in the text of Seeman. He has understood the chain reaction that happens in the land and therefore educates his audience to discipline their behaviour towards nature like the previous generation. The previous generation was mindful of the upcoming generation; hence, they could preserve the natural resources prudently and hand over it to the present generation. But unfortunately, the present generation mishandles it, which is the chief charge of Seeman against the Tamils.

In connection to this, Seeman adds the behaviour of the people of contemporary society towards mother nature. 'Engaging with nature in the present' is the core idea of the second part of the discourse. He adds that "We cut down the trees that are hands of mother earth, we sucked the rivers that are blood vessels of earth. Now we break the head of the earth that are mountains. We emptied and sold the river sand now the mountains have been targeted. We kill the mountains to get mountains sand which is shortly called 'M Sand!'. (Seeman's Speech about Mother 1:22-1:42). These lines explicate the cruelty caused by the people of contemporary society to nature. People cut down trees and break mountains to fulfil their selfish needs. Seeman mentions that cutting trees is similar to cutting the hands of mother earth. His discourse attributes human qualities to the earth thereby personifying the earth with femininity. The word 'mother' generally reflects femininity, and in the present context, it is perceived as a figure engaging in reproduction and nurturing sustenance. Similarly, the earth is seen as a figure that nurtures the sustainability of human lives. The earth incessantly functions in reproducing natural resources and keeping them available for people's needs. There are certain natural resources like mountains which the earth does not reproduce. Hence, Seeman teaches his audience to preserve natural resources for the sustainability of all living things. This idea of Seeman connects with goal fifteen of SDG. It communicates the idea of protection and restoration of the terrestrial ecosystem.

4. Social Justice for Sustenance

In this part of the discussion, the research paper engages with the political discourse of Thol. Thirumavalavan. An excerpt from "Thol. Thirumavalavan Speech on Social Justice" is given here: "Majority of the people lagging too far to reach even the most basic needs that cannot be met. Where does social justice begin? Social justice begins where inequality is being rectified." (04:02-04:46). Thirumavalavan mentions that the majority of people are unable to meet their basic needs. It projects them as less privileged in society thus keeping them in a marginalized state. He tries to figure out the cause for the existence of inequality and attempts to suggest a solution to the problem. The excerpt mentioned below would give a better understanding of his notion of social justice:

> Social justice should be perceived ideologically. When it is not perceived ideologically, people tend to see social justice as merely offering financial assistance or a seat under reservation. They tend to see it as a privilege offered out of empathy. Thol. Thirumavalavan Speech on Social Justice (04:46-05:44)

Thirumavalavan explicates the notion of social justice within the context of contemporary Tamil society and attempts to redefine and bring a paradigm shift in perceiving it. He states that social justice has to be looked at ideologically and not from the point of view of reservation. The idea of social justice is always seen through the word reservation. An attempt to look at social justice through reservation does not give a complete understanding of it. Such a perception of social justice will only give a narrowed understanding of it. Hence, Thirumavalavan tries to redefine the notion of social justice. When social justice is merely seen through reservation, it is seen as an action connected to empathy and emotions, which is a narrowed approach to social justice.

To explicate social justice, Thirumavalavan places the marginalized communities at the forefront because they need to be recognized and encouraged to move forward. Only when this marginalized community is recognized and aligned with the mainstream society, the country will see an inclusive growth. For Thirumavalavan, unless an inclusive approach is employed, it is unfeasible for the country to notice harmony, equality, and development. His discourse on social justice connects with three sustainable development goals. Goal eight says that "promote sustained, inclusive and sustainable economic growth, full and productive employment and decent work for all" (United Nations). Thirmavalvan's social justice encompasses the economic growth of society. Providing a job for an individual cannot be treated as social justice because it does not benefit the entire society but rather develops only the individual's family. An approach that develops only an individual cannot be treated as a fair approach. Hence, an inclusive approach needs to be taken to focus on the development of the entire community.

Goal ten of SDGs deals with reducing inequality, which aligns with the idea of Thirumavalavan. It emphasizes reducing inequality within a country and among the countries. For the present discussion, it is sufficient to look at the first part of the statement. Reducing inequality within a country is important because a country has millions of divisions based on caste identity, religious identity, linguistic identity and so on. Such instances give birth to the concept of victimization. As a result, the powerful people dominate the weaker section through their economic power and political power. This creates exclusive communities wherein each community would insist on its seminal identity, and this would give rise to conflict among the communities. Thirumavalavan adds that today communities are disintegrated and fragmented based on their economic status. To align a marginalized community with the mainstream society, it is necessary to economically strengthen the community to reduce poverty and inequality. Goal sixteen focuses on "peace, justice, and strong institutions: promote peaceful and inclusive societies for sustainable development, provide access to justice for all and build effective, accountable, and inclusive institutions at all levels" (United Nations). This goal aims at promoting peace, justice and inclusive societies for sustainable development. The same ideas have been stressed by Thirumavalavan as well. He verbalizes that "decentralization of power is the fundamental argument of social justice. Empowerment of marginalized sections is the soul of social justice." (Thol. Thirumavalavan Speech on Social Justice

10:10-10:51). Thirumavalavan argues for social justice in the aforementioned lines. He insists on decentralization of power to witness social justice. Power and social justice are interconnected as power can pave a way to social justice. Power can be approached both positively and negatively. A positive approach to power brings social justice to a land and a negative approach to power can result in polarizing people. Therefore, Thirumavalavan believes that to bring peace among the nations and within a nation, it is necessary to decentralize the political power so that everyone in the committee gets benefited. Decentralization of power refers to sharing of power with people working under different capacities. It underpins to construct a strong inclusive society. It does not permit an individual to come up with the final decision instead it allows everyone in the managerial committee to contribute, which is an advantage of decentralization of power. When power is decentralized the degree of domination gets reduced. It also erodes prejudice and disparity by emphasizing the importance of each community that is being represented. Decentralization of power gives space and opportunity to the people of privileged and unprivileged communities. Such an approach would underpin to promote the social standard of the marginalized community. Hence, Thirumavalavan insists on the idea of decentralization of power and social justice for sustainable development.

5. Conclusion

The present research has explored that the political discourse of Seeman and Thirumavalavan reflect the sustainable development goals. The selected political discourse of Seeman engrains goal six, twelve and fifteen, and the selected political discourse of Thirumavalavan reflects goal eight, ten and sixteen. From the above discussion, it is lucid that the discourse of Seeman and Thirumavalavan emphasizes distinctive ideas. The former insists on protection of nature and the latter foregrounds social justice. However, the political discourse of Seeman and Thirumavalavan converges on the notion of sustainability.

REFERENCES

1. Amaglobeli, G. (2017). Types of Political Discourses and Their Classifications. Journal of Education in Black Sea Region. 3(1):18 24.

2. Dijk, Teun A. van. (1998). What is Political Discourse Analysis? Belgian Journal of Linguistics, 11(1):11 52.

3. Fairclough, Isabela. and Norman Fairclough. (2017). Political Discourse Analysis: A Method for Advanced Students. Routledge.

4. Griffin, Gabriele. (2013). Research Methods for English Studies. Edinburgh UP.

5. United Nations. *Transforming Our World: The 2030 Agenda for Sustainable Development.* Sustainabledevelopment. un.org, no publication date.

6. Seeman's Speech about Mother Nature (English Version)/Naam Thamizhar Katchi. (2019). YouTube, uploaded by Seemanism for Universe, https://www.youtube.com/watch?v=dTyt8_LkGyc.

7. Thol. Thirumavalavan Speech on Social Justice. (2021). YouTube, uploaded by Red Pix 24x7, https://www.youtube.com/watch?v=A8xS8K1PPiY.

8. Thol. Thirumavalavan Speech on Seminar on Social Justice/NEET/nba24x7. (2018). YouTube, uploaded by nba24x7, https://www.youtube.com/watch?v=cgx5ZmH_sIk.

*Integrating Advancements in Education, and Society for Achieving
Sustainability – Dimitrios A. Karras et al. (eds)*
© 2024 Taylor & Francis Group, London, ISBN 978-1-032-70841-6

An Analytical Study on Enhancing ESL Learners' Classroom Interaction for Second Language Acquisition

Aravind B. R.*

Assistant Professor in English,
Kalasalingam Academy of Research & Education, Tamilnadu, India

Jayakumar P.

Assistant Professor in English,
St. Joseph's College of Engineering, Tamilnadu, India

N. Devadhas Prabu and Harishraj A. N.

Assistant Professor in English,
Kalasalingam Academy of Research & Education, Tamilnadu, India

Abstract: The present study aimed to enhance students' talk time in classroom interaction for second language acquisition. Also, to identify the change in classroom interaction participation. Interaction analysis was used in the study. Additionally, analytics for classroom interaction was framed to observe the talk time of students and teachers. Purposive sampling was adopted for the research. A total of 70 students majoring in Engineering were used as the sample of the study. The research findings revealed that the increase in students' talk time was beneficial in effective classroom interaction. The result also confirmed that students' participation has changed over a period of the course. All the findings were analysed using SPSS 26 software package. Particularly, multiple regression was used to study the progress of participants and the differences among the participants.

Keywords: Student Talk Time, Teacher Talk Time, Interaction Analysis and Second Language Acquisition

1. Introduction

Our classroom atmosphere is entirely built on rote memory, and the classroom learning at the university level is too inadequate. Learners who are offered very little opportunity for active engagement and interaction are not provided any opportunities to enhance their intellectual and cognitive abilities. In the class, the instructor appears to have a highly powerful position. Sadly, poorly organised classes rapidly degenerate into pointless wasters of time.

*Corresponding author: aravind.abur@gmail.com

DOI: 10.4324/9781032708461-18

Students are expected to actively utilise the target language as part of the language learning process (Nunan, 1999). Therefore, a successful learner-cantered L2 classroom should offer a setting where learners may participate in instructional experiences and enhance the utilization of the language (van Lier, 2001).

Except for a few students who have experience speaking English at home or in other settings, most students in an English as a Second Language (ESL) programme are still unfamiliar with utilising the language in everyday conversation. As a result, the instructor has certain guidelines to help them learn new languages, particularly as role models. The instructor must foster an interactive learning environment in the language classroom in order to carry out those responsibilities (Aravind & Bhuvaneswari, 2023). The important term for language teachers is interaction. "Interaction is a collaborative exchange or ideas between a teacher and learners or a learner and other learners resulting in reciprocal effect on each other" (Brown, 2001). Language class interaction is distinct from other types of interaction. Interaction is the process of language acquisition in a language classroom.

According to Brown (2001), students can improve their language skills through interaction by participating in group projects, problem-solving activities, discussions, reading real language content, or even contributing to dialogue notebooks. Students can communicate in real-life situations using all of the language they know and have acquired casually. They pick up how to take advantage of language's flexibility even at the most basic levels in this way. Interaction is one of the crucial elements that must be present for the teaching and learning process to be fruitful.

Interaction is the cooperative interchange of thoughts, feelings, or information involving two or more individuals (Yanfen and Yuqin, 2010). It plays a significant part in the instruction of English. Students can broaden their language repertoire and apply the languages they already know through interactions with academics. As a result, grouping students can assist foster interaction in the classroom and increase student speaking time. Students gain teamwork skills, the ability to listen to others, and the ability to support their arguments.

The effective management of the process of interaction enhances the learning experience for students and motivates bored, disengaged pupils to perform and generate ideas. For efficient second language learning, a range of interactive strategies must be employed. The study of how students engage in a classroom is essentially the study of how communication works. In the context of classroom research, interaction analysis often refers to the process of analysing spoken language in the context of how it is utilised in a classroom by the instructor and the students.

2. Research Objectives

- To enhance students' talk-time in the classroom interaction for second language acquisition.
- To identify the change in the classroom interaction participation.

3. Research Questions

- RQ1: How to enhance ESL learners' classroom interaction for effective second language acquisition?

- RQ2: Was there any change in the participation of ESL learners' classroom interaction?

4. Significance of the Study

Due to time constraints, teachers try many different ways to get their pupils to communicate in class, but they typically fail. Student speaking time is increased, which is a remarkable benefit of classroom engagement. Students benefit from this interactive exercise by improving their vocabulary, understanding grammatical concepts, and overcoming shyness. It is seen as a healthy learning technique when students begin to extend their learning beyond the classroom while maintaining the same level of vibrancy and interacting with a similar set of individuals but in a new environment. Students benefit from peer feedback in the classroom because they may listen and make necessary adjustments to their knowledge level while they are in front of the teacher. A second attempt can be simply done if the instructor believes that a student's presentation might be improved.

Giving pupils performance evaluation is another advantage of employing classroom interaction. The instructor might devote extra attention to the class time to the subject that may be particularly difficult for the learners. When students wish to listen to or make the required adjustments in the course of practise, the teacher can provide them with advance material that can be accessed anytime they choose. It is definitely feasible for students to get along with one another and have a collaborative learning environment. The mind of a youngster undergoes considerable development as a result of the interactive

methods that mould his or her personality in every facet.

5. Research Methodology

The method of study that was used in this investigation was called interaction analysis. One of the methods for assessing classroom interaction is called interaction analysis. This approach entails doing a discursive analysis of instructional interaction. Research on the teaching and acquisition of second languages, the data for which are collected in whole or in part through careful observation or assessment of the teacher's and students' performance in the classroom. This study aimed to enhance and analyse teacher talk and learner talk in the context of classroom interaction, using a sample size of 70 students majoring in Engineering as the subject. In order to determine who would take part in the study, we used a method called purposive sampling. The research involved the use of classroom observation as a method for collecting the data. According to Cohen, Manion, and Morrison (2007), one of the primary benefits of observation as a method of research is that it provides the opportunity for an investigator to collect data from naturally produced interactions. Quantitative data analysis was performed on the information gathered for the study.

6. Research Design

The research plan of the study is interaction analysis where the participants' classroom interaction is analysed through a set of analytics for observation as shown in Table 18.1. All the 70 participants were grouped randomly and instructed to present a group presentation of own choice. Researcher gave a detailed instruction and steps to be followed during and after each presentation. For a period of three weeks, 18 classes were recorded. The classroom interaction was calculated and divided as student talk time and teacher talk time. Then the classroom interaction data was calculated through multiple regression using SPSS 26 version tool to find the difference value in students' interaction. The interaction analysis is implemented in order to investigate the research questions that are statistically proven using SPSS software.

Table 18.1 Analytics for Classroom Interaction

S. No.	Analytics for Classroom Interaction
	Student Talk Time
1	Group Presentation
2	Questioning
3	Requesting clarification
4	Offering opinion
5	Collaborative interaction
6	Providing feedback
	Teacher Talk Time
7	Directional questioning
8	Verifying
9	Correcting
10	Offering comments & opinion

7. Results and Discussion

Based on the observation findings of the study, it is found that analytics for observation items in classroom interaction have been applied and significant result was evident by the researcher. In total, the average value of student talk time is 11.33 in the 1st week, 17.16 in the 2nd week and 22.83 in the 3rd week. Similarly, the average value of teacher talk time is 24 in the 1st week,

17.7 in the 2nd week and 12 in the 3rd week respectively as shown in Table 18.2.

Table 18.2 Findings of the talk time analytics

Analytics for observation	Classroom Interaction		
	1st Week	2nd Week	3rd Week
Student Talk Time	11.33	17.16	22.83
Teacher Talk Time	24	17.7	12

The researcher knows how to make students involve in classroom interaction through natural activities. The analytics for observation representing positive attitudes towards student talk time had higher mean score those comparing the previous week mean scores. This shows that analytics for student talk time had prominent transformation that occurred after successive weeks with same procedure for classroom interaction. The findings belonging to teacher talk time mean score stated that respondents developed better attitude towards classroom interaction. Hence the mean score decreases from 1st week to 3rd week. In a way the researcher succeeds in engaging the students in effective classroom interaction for second language acquisition.

The analytics for observation of classroom interaction played a crucial part to second language acquisition and they interact more with students. Group presentation is advantageous in the classroom interaction. Because participants in the group has different experiences, views, ideas that they can bring to share for a successful presentation. The group participants bounce ideas off of each other which paved way for classroom interaction. Using questioning in classroom interaction to open conversations, enhance deeper thoughts and most importantly it promotes student-to-student interaction.

Because it facilitates language acquisition through active discussion, speaking and listening. Also, requesting clarification helps students to build critical thinking skills and respect other people's opinion and explains the concept clearly.

Likely, offering opinion is equally important. But students should be taught the way they share the opinion for meaning classroom interaction. In this study, participants engage in giving opinion not to prove others wrong. Rather participants share the opinion to understand other's opinion and able to see the different perspective to have an understanding which will give way for classroom interaction.

Collaborative interaction makes students gain their self-confidence and interact well in front of the class. The researcher created an atmosphere with the analytical interaction pattern for students' communication. Finally, student talk time analytics for observation ends with peer feedback. It helped the participants to gain an understanding and figure out areas of development through classroom interaction.

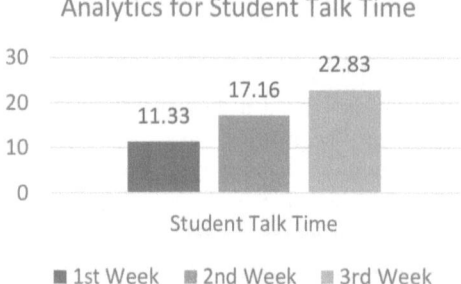

Analytics for Student Talk Time

Fig. 18.1 Analytics for student talk time

On the other hand, teacher talk time was intentionally reduced during the course of time. All the teacher talk time analytics were designed just to lead the classroom interaction lively. Here the role of the instructor is sustained to make classroom atmosphere as friendly as possible. Directional questioning tends to generate further interaction in the class. The findings also confirmed the exposure to discussion in the class. Verifying, correcting and teachers' comments made many initiations during classroom interaction. Participants' responded teachers' analytics which contributed in building the effective classroom interaction for students especially. Because the talk time of interaction is in proration of teachers' responses and feedbacks. If the teacher asks questions, students will response more as a result student talk time increases for classroom interaction.

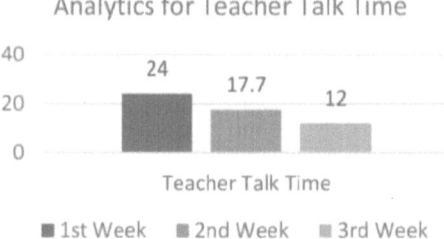

Analytics for Teacher Talk Time

Fig. 18.2 Analytics for teacher talk time

The above findings of student talk time and teacher talk time show that classroom interaction recognized as constructive in ESL classrooms. Teachers need to have more knowledge about the use of classroom interaction in more adequate way. It is reflected through students talk time related to teachers' analytics to pattern about how to interact effectively in classrooms. So, teachers should pay much attention to the appropriate use of talk time in the class. The present study suggested from the result that teachers should minimize teacher talk time and provide more student talk time to speak or interact in the classroom. Because from the result of the interaction analysis, patterns of classroom interaction in ESL class have shown great differences in performance and participation. Therefore, a good teacher maximizes student talk time and minimizes teacher talk time.

The result showed majority of students has developed the desire for talking and participating in the classroom interaction. They wish to participate in discussion and chance to interact in the classroom. It was also noticed that students wish to answer questions. The classroom atmosphere and the teacher should be friendly and not authoritative. There should be no distinction of superiority among the learners especially ESL learners for second language acquisition.

Participants in the study showed great interest and enthusiasm while performing classroom interaction during 3rd week. The researcher felt that PowerPoint presentation would encourage and motivate to initiate interaction in the class. Even participants felt that group presentation along with some steps instructed by teacher for interaction created an enjoyable learning experience. Since, peer level learning participants were comfortable and relaxed. Also, the level of anxiety of learning is reasonably reduced as evident in the result. Focusing on the volume and substance of student talk time may increase students' L2 acquisition.

8. Conclusion

The present study analysed the talk time of classroom interaction. The research aimed to enhance the students talk time to optimize the classroom interaction more dynamic. The results revealed that the participants talk time was largely increased over the course duration. Ultimately, teacher talk time deliberately reduced for responding the instructors' questions and comments. Referring to the findings of this study, the teachers were recommended to create interactive atmosphere in the class. The teacher must be a good motivator to students in initiating the interaction in the class.

REFERENCES

1. Aravind, B. R., & Bhuvaneswari, G. (2023). Utilizing Blogs on ESL learners' vocabulary learning through social constructivist theory: A descriptive study. MethodsX, 10, 101970.
2. Brown, H. D. (2001). Teaching by principles. An interactive approach to language pedagogy. New York: AW Longman.
3. Cohen, L., Manion, L., & Morrison, K. (2005). Research methods in education (5th ed.). New York, USA: Routledge Falmer.
4. Nunan, D. (1999) Second Language Teaching and Learning. Heinle and Heinle.
5. van Lier, L. (2001) 'Constraints and resources in classroom talk: Issues of equality and symmetry'. In Candlin and Mercer (2001) 90–107.
6. Yanfen, L., & Yuqin, Z. (2010). A Study of Teacher Talk in Interactions in English Classes. Chinese Journal of Applied Linguistics (Foreign Language Teaching & Research Press), 33(2).

Note: All the figures and tables in this chapter were made by the author.

Integrating Advancements in Education, and Society for Achieving Sustainability – Dimitrios A. Karras et al. (eds)
© 2024 Taylor & Francis Group, London, ISBN 978-1-032-70841-6

Integrated Learning Tools that Aide Education System to Face Challenges in ELL During COVID-19

Shruthi S. and Jothi C.*

Department of English Kalasalingam Academy of Research and Education, Krishnankoil, India

Abstract: Throughout the Covid-19 pandemic, all the institutions remained closed, and that caused to slow down English Language Learning (ELL). English is considered one of the most important languages as it plays a primary role in communication on a global scale. The widespread Covid-19 pandemic created new social norms in our life with education being one of the major sectors that was affected. To transcend knowledge, e-learning became the order of the day. A new approach had been implemented in the 21st century to improve students' learning abilities. As a result, this imparts a review of the article on the difficulties encountered by ELL. The learning tools like Google Classroom, Zoom, Teams, etc., were used to face the challenges. It was found critical to develop the English language through the intervention of technology. Thus e-learning managed to help to learn the English language rather than imparting effective knowledge through these tools.

Keywords: Covid-19 pandemic, English language learning, E-learning, Technical tools

1. Introduction

The COVID-19 (Coronavirus Disease-19) a pandemic, which is derived from the Greek words pan which means "all," and demos which means "local people". The Pandemic is an outbreak of an infectious illness that has spread over a considerable area such as several continents or the whole world and has affected a significant number of people. Pandemic is an unparallel diseases that has had an impact on a large population out of which a considerable number of people surmounted it. Seasonal influenza recurrences are one of the common pandemic diseases that has an impact on the consistent number of diseases that has affected many distinct places of the world at once rather than expanding internationally.

There have been several pandemics of illnesses for far like smallpox, flue, chickenpox, etc., throughout in the human

---

*Corresponding Author: c.jothi@klu.ac.in

DOI: 10.4324/9781032708461-19

history. The Black Death, also known as The Plague, claimed the lives of an estimated 75–200 million people in the 14th century and was the worst pandemic in recorded history. Numerous pandemics have recently occurred, including TB (Tuberculosis), HIV (Human Immunodeficiency)/AIDS (Acquired Immune Deficiency Syndrome), Asians, Russian, Spanish, flu (influenza), and COVID-19 are few instances.

In 2019, An urgent crisis brought about several changes in a variety of spheres of life, including commerce, trade, transportation, and education. To combat the Corona Virus's spread, collaborators were required to be responsive and make quick judgments. Infections of the upper respiratory tract are attacked by the COVID illness. Corona Virus was first discovered in Euhum, China, in December 2019. The World Health Organization declared a public health emergency of international concern on January 30, 2020 (WHO). It was believed that social isolation may end the Covid-19 transmission cycle. On March 17, 2020, the Pandemic emerged. Online learning gradually replaced face-to-face instruction. In the middle of March 2020, the educational policymakers made the decision to put into effect a regulation that would switch from offline to online learning days Diana et.al. (2021), For the protection of school and college students from the risk of COVID-19, the policy of remote distant learning was indicated.

1.6 billion pupils' education was impacted and disrupted because of COVID-19. The government decided to close schools and universities, so students started learning from home with the assistance of teachers using online techniques says Rosalina et al., (2021). In several nations during COVID-19, remote learning was used as an emergency reaction. Due to the conflict that had broken out among all of the educational institutions, they switched from in-person instruction to online instruction. In comparison to face-to-face learning, online learning has a different pedagogical impact kholili, A. (2021). For online classrooms, it was common practice to utilize online tools like Google Classroom (GC), Google Forms, Google Meet, Webex, and MS Teams, among others. Both benefits and drawbacks of online English language instruction exist. Although it attempted to reach every kid, this learning was not always successful. The stress that students experienced during the COVID-19 Pandemic, also had a detrimental effect on their ability to study. Given their lack of facilities and access to online education, it was difficult for students and instructors to convert quickly. Students had the ability to study at any time and from any location. With the aid of effective technology, the students were able to speak with their instructors while they were learning online.

Since many students in remote locations lacked a reliable internet connection, they were unable to participate in online classes. The Covid-19 pandemic presented several difficulties for both pupils and instructors. Remote instruction had emerged as the go-to method for continuing education during the pandemic period says Saudi et al., (2021), Virtual classrooms were introduced in all educational institutions and that was a success. According to Blanka Klimova et al., (2021) the relationship between students and instructors is seen as a benefit during the Covid-19 pandemic and the success of learning through online was attributed to the fact that the quality of education had increased over time. As a result, they gave amusement some thought. The effectiveness of the course might have increased during the Covid-19 pandemic. Statement on understanding the challenges and advantages of English language education at a distance. For students to be prepared for their job, where English is an essential course.

Diana et al., (2021) considers that GC is one of the best application to enhance learning. As it has set of advanced features that make it the ideal tool for use with students. Taufik et al., (2021) views the advantages of GC application which has effectiveness and adaptability in learning. GC is connected with Gmail, drive, hangout, YouTube, and calendar events.

2. Usage of Tools to Support Pandemic Learning

Walidaini's report in the year 2021 reveals that pupils were able to access the WhatsApp application during the Covid-19 pandemic in order to learn English online. WhatsApp was utilized for writing and reading abilities during online English learning which was advantageous during the pandemic in year 2019. In the study of Mu'awanah, (2021) online courses offered by Zoom are popular for supporting distance English learning and investigating the benefits and drawbacks of Zoom usage. The author demonstrates that Zoom was more efficient for practicing and studying English online since Zoom provides a chat application, video conference sessions, recording possibilities, and many other beneficial features for learning through online during the pandemic at 2019.

The article by Diana et al., (2021), shows that the Google Classroom application was used for online English language learning, and the majority of students emerged from the Covid-19 pandemic with a positive perception. Only a small percentage of students reported having trouble finding words to use in English language learning during online classes at Google Classroom. In Suadi's (2021), study on, Zoom and WhatsApp two internet tools, which were used to facilitate online learning during remote learning. This was confirmed favourably, and the students stated that using the two online tools had improved their language skills and reduced their shyness while interacting with other users who were in online classes.

According to Nehe's (2021) study, Google Classroom had positive relationships between lecturers, as well as amongst students themselves. The entire class participated in a speaking activity. During pandemic learning through online, students were also content, fearless, secure, and confident. There was no difficulty in using GC in English for speaking skills through online. The GC improved English language learning during the Covid-19 pandemic. The Students of English literature responded favourably to the learning process in GC.

The research by Abiky (2021), which looked at the influence of WhatsApp during the pandemic time on English learning resources, and Students found that WhatsApp was beneficial for their online study during the Pandemic. According to the research by Rosalina et al., (2021), there were several challenges with the network problem and students' lack of comprehension, when using Google Classroom to learn English online during the pandemic time. The Google Meet application was utilized for English language learning classes. The study of Kholili (2021), states that students, before and during the Covid-19 Pandemic, were exposed to English language learning through the hybrid method.

Figure 19.1 illustrates that most of the student's preferred WhatsApp and Google Meet as their most favourite tool to aide them during pandemic to enhance their English language learning. The above graph shows 1,2,3,4,5 as 1-Strongly disagree, 2-disagree, 3-neutral, 4-agree, 5-strongly agree.

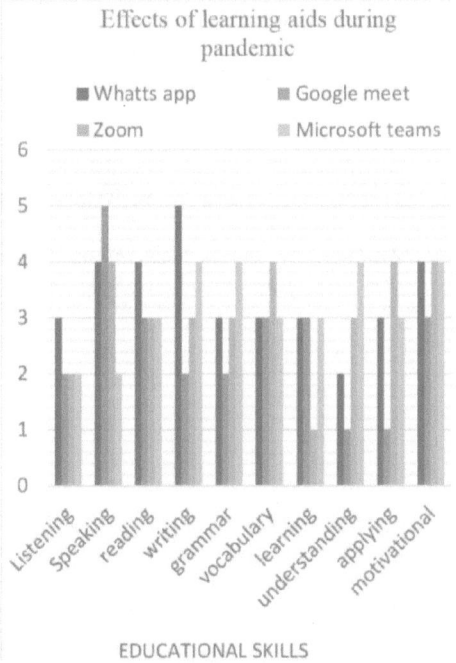

Fig. 19.1 Effects of learning aids during pandemic

Source: Made by the Author

3. Conclusion

The current study focuses on the overviewing of the pandemic English Language Learning concept. The review of literature acquired from several papers published in Google scholar and scientific journals indexed in the Scopus, provides various dimensions of English Language Learning during Pandemic. The study illustrates that students had enough training on English Language Learning skills with the help of learning tools line Whats App, Google Classroom, Zoom and other online tools. English Language Learning was effective, most of the students felt neutral in English Language Learning and few students were not satisfied with online English Language Learning. Students were not satisfied because they were from rural areas and were financially poor to equip themselves with the necessary gadgets to attend online classes.

The study thoroughly explains that the Pandemic online English Language Learning has certain limitations, as it only looks at articles in the ScienceDirect (SCI) journal and Google Scholars, excluding journal articles from other databases such as Academic search, Web of Science, and so on. Furthermore, only online English Language Learning during COVID-19 the Pandemic is included, which may allow researchers to investigate diverse subjects linked to English online Learning during the Pandemic in the future.

REFERENCES

1. Nehe, B. M. (2021). Students' Perception on Google Meet Video Conferencing Platform during English Speaking Class in Pandemic Era. *Journal of English Education, 10*(1), 93–104. https://doi.org/10.25134/erjee.v10i1.5359
2. Suadi, S. (2021). Students' Perceptions of the use of Zoom and Whatsapp in ELT Amidst Covid19 Pandemic. *SALEE: Study of Applied Linguistics and English Education, 2*(1), 51–64. https://doi.org/10.35961/salee.v2i01.212

3. Diana, N., Yunita, W., & Harahap, A. (2021). Student' Perception and Problems in Learning English Using Google Classroom During the Covid-19 Pandemic. *Linguists: Journal Of Linguistics and Language Teaching, 7*(1), 10. https://doi.org/10.29300/ling.v7i1.4274

4. Anisahril Walidaini, F. (n.d.). *The Use of WhatsApp Application for Learning English online the use of Whatsapp Application for Learning English online During Corona Virus Disease-19 Pandemic.*

5. Mu'awanah, N. (2021). Using Zoom to Support English Learning during Covid-19 Pandemic: Strengths and Challenges. *Jurnal Ilmiah Sekolah Dasar, 5*(2), 222–230.

6. Al Abiky, W. B. (n.d.). Days without schools: The effectiveness of WhatsApp, as an English learning tool, during COVID-19 pandemic. *Revista Argentina 2021, XXX*, 774–781. https://doi.org/10.24205/03276716.2020.2074

7. Rosalina, D., Ita Purnama, Y., & Ratih Tirtanawati, M. (2021). Analysis of English Online Learning During Covid-19 Pandemic Through Google Meet. *Jurnal Pendidikan Edutama.*

8. Kholili, A. (2021). *Prior to and in the Course of Covid-19 Pandemic: Exploring Learners' Experiences of Learning English through Narrative Lens. 3*(3), 195–204.

9. Klimova, B. (2021). ScienceDirect An Insight into Online Foreign Language Learning and Teaching in the Era of COVID-19 Pandemic. *Procedia Computer Science, 192*, 1787–1794. https://doi.org/10.1016/j.procs.2021.08.183

10. Taufik, M., Samsu Rijal, A., Dahniar, D., & Apriani, E. (2021). The Effectiveness of Learning English Using LMS Google Classroom during Covid-19 Pandemic. *AL-ISHLAH: Jurnal Pendidikan, 13*(2), 960–970. https://doi.org/10.35445/alishlah.v13i2.706

Integrating Advancements in Education, and Society for Achieving Sustainability – Dimitrios A. Karras et al. (eds)
© 2024 Taylor & Francis Group, London, ISBN 978-1-032-70841-6

20

Conceptualisation of Beauty: Theoretical Perspectives of Aesthetics and Black Aesthetic

Fredrick Ruban Alphonse*
Christ (Deemed to University), India

R. Manikandan
Vel Tech Rangarajan Dr. Sagunthala R&D Institute of Science and Technology, India

S. Barathi
SASTRA Deemed to be University, India

Vimochana M.
Vel Tech Rangarajan Dr. Sagunthala R&D Institute of Science and Technology, India

Gulnaz Fatma
Jazan University, Jazan Saudi Arabia

Yaisna Rajkumari
National Institute of Technology, Hamirpur, India

Abstract: Aesthetics is widely studied in philosophical and scientific perspectives. In a philosophical stance, aesthetics emphasizes the creation and appreciation of beauty, and deals with taste, beauty, and the nature of art. It is scientifically claimed as the study of sensory-emotional values and judgments of sentiment and taste. In addition, aesthetics is the basis of all human experiences which distinguishes humankind from the lower order creations. It works with the inexpressible realm of imagination and feeling. This article intends to delve into the idea of beauty that is dealt under the label Aesthetics and Black Aesthetic. It adopts a theoretical approach to find out the congruent and incongruent notions that are underlying in Aesthetics and Black Aesthetic. The perception of beauty is looked at from multiple dimensions to determine how Aesthetics perceives beauty and how Black Aesthetic understands beauty.

Keywords: Beauty, Black Aesthetic, Aesthetics, Taste, Judgments

1. Introduction

The present article aims to explore the notion of beauty that is dealt under the domain Aesthetics and Black Aesthetic. It employs a theoretical approach to investigate and trace the congruent and incongruent notions that are underlying in Aesthetics and Black

*Corresponding author: fredrick.ruban@res.christuniversity.in

DOI: 10.4324/9781032708461-20

Aesthetic. Further, the article examines the different notions of beauty that are perceived from multiple dimensions. Specifically, it attempts to determine how Aesthetics perceives beauty and how Black Aesthetic understands beauty.

The philosophical term 'aesthetic' originated from the Greek word *aisthanomai*, meaning "to apprehend through the senses" (Brown 5). It was derived and introduced by the German philosopher Alexander Baumgarten in 1735. Aesthetics is a branch of philosophy which studies the principles of beauty and centers "around understanding beauty in art and nature" (Bearleant 2). The term aesthetics is extensively applied to judgments, experiences, attitudes, objects, qualities and values. Aesthetics is the philosophical notion of beauty and is neither epistemological nor ethics. Therefore, Baumgarten defined aesthetics as "the science of sensory knowledge directed toward beauty and regarded art as the perfection of sensory awareness" (qtd. in Berleant 2). Until the emergence of Aesthetic movement, an English literary and artistic movement of 1890s, the term aesthetics was solely associated with Philosophy. Aesthetic movement or Aestheticism is a European phenomenon that has its headquarters in France. It was culminated in 1890s with Oscar Wilde as its exponent and Walter Pater as its acknowledged philosopher. Aestheticism developed out of the art and philosophy of Romantic movement which strived to liberate art from the rules of Classicism. "Art for art's sake" (Cuddon 12) is the chief slogan and tenet of aestheticism, which is an English version of French "ʃ'art pour ʃ'art" (Child & Fowler 2).

Evolution of Aestheticism created an explicit relation between aesthetics and literature.

It broke the Classical and Victorian views of art: art should teach moral values and the dominance of scientific thinking. Aestheticism is "a genuine search for beauty and a realization that the beautiful has an independent value" (Cuddon 13). It "attaches a high value to 'form' in art, the value of a work of art being dependent on form rather than on subject-matter" (Johnson 14). Writers of Aesthetic movement adopted an aesthetic style that emphasized art must carry refined sensuous pleasure rather than instilling sentimental messages. They professed that art must not take didactic approach instead it has to be beautiful. Moreover, they treated nature as crude when compared with art. In this connection, Pater states that "for art comes to you proposing frankly to give nothing but the highest quality to your moments as they pass, and simply for those moments' sake" (239).

2. Concept of Beauty

The primary characteristics of aesthetic style are using symbols, sensuality, synaesthetic and suggestion rather than statement. Furthermore, "dominance of form over content" (Lapierre 35), "celebration of beauty" (Lapierre 21) and emphasis on "power of imagination" (Lapierre 21) are the seminal values of aestheticism, which have been borrowed from Romanticism. For Kant, imagination is the "ability to represent in thought the features experienced in the sense perception of the external world" (Pluhar 505). In general, the values of aestheticism are commonly treated as the tenets of aesthetics.

In Literature, aesthetics is instilled through its genres with the writers' choice of technique. Usually, an author use rhythm, structure, humor, fantasy and imagery to carry out his aesthetic taste to the readers. Primary concern of aesthetics is in the "manner in which values, whether colors or tones or even words for the poet, are formally arranged in space" (Fernandez 357). Aesthetic values are emphatically personal and they "must be felt as one's own." (Parker 6). Denis Dutton has discovered the seven universal signatures of aesthetics:

1. Expertise or virtuosity. Technical artistic skills are cultivated, recognized, and admired.
2. Nonutilitarian pleasure. People enjoy art for art's sake, and don't demand that it keep them warm or put food on the table.
3. Style. Artistic objects and performances satisfy rules of composition that place them in a recognizable style.
4. Criticism. People make a point of judging, appreciating, and interpreting works of art.
5. Imitation. With a few important exceptions like music and abstract painting, works of art simulate experiences of the world.
6. Special focus. Art is set aside from ordinary life and made a dramatic focus of experience.
7. Imagination. Artists and their audiences entertain hypothetical worlds in the theater of the imagination. (Pinker 405)

An "aesthetic idea is a presentation of the imagination which is conjoined with a given concept and is connected, when we use imagination in its freedom" (Pluhar 533). Concept "is a device for taking the measure of the content of the Imagination" (qtd. in Cohen 2). When objects are judged in terms of concepts, the presentation of beauty is lost. According to Kant, judgment "is the ability

to think the particular as contained under the universal" (Pluhar 504). Universal refers to rule, principle and law. He proclaimed that no aesthetic judgment can "rightfully lay claim to everyone's assent" (Pluhar 510). Aesthetic judgment is not arbitrary action because some elements of nature can induce pleasure, and some cannot. Aesthetic pleasure is not based on the principles of one's will. Thus, the necessity of an aesthetic judgment is neither a theoretical necessity nor a practical necessity. Theoretical necessity means the judgment of knowledge and practical necessity refers to the normative judgment. Aesthetic judgments can be divided into empirical and pure. The former asserts the way of presenting an object is agreeable or disagreeable. An object is presented as agreeable in relation to sense. When something appears pleasing, then it is called as good. Both agreeable and good point out the power of desire. The latter articulates the object is beautiful, and it is the proper judgment of taste.

Taste is a natural ability to judge an object and is based on senses. Taste refers to one of the five senses which "provides gustatory discrimination and enjoyment" (Korsmeyer 195). It is unavoidably connected with "pleasure or displeasure" (Korsmeyer 195). Taste can be learned because it has its roots in education and elite cultural values. It differs according to education, class and cultural background. When taste gives a negative reaction to an object, the reason for it cannot be explained. Taste is the principle of itself, and it requires absolute autonomy. Hutcheson perceives taste as a sense whereas Hume observes it as a unique discrimination, but Voltaire defines taste as a part of redefined good manners. Taste aesthetically refers to the feeling of pleasure. Taste fosters to transit from sensible charm to the habitual moral interest. One who has limited palate may not bother about the food that he or she eats is a well-prepared meal or not. But the one with a fine sense of taste will be able to discern even mere ingredients that went into it while preparing. Thus, sense of taste is a metaphor used to appreciate objects that belong to art and nature. Properties of nature and art are judged only when one has first-hand experience with the properties. They arouse appreciative pleasure as a signal of the discernment of aesthetic quality. Therefore, "one needs taste to discern the aesthetic quality and cannot infer it from the presence of the non-aesthetic qualities" (Korsmeyer 199).

For Kant, taste is the ability to judge an object or a manner of presenting it by means of liking or disliking it. Thus, liking an object is called beautiful. Judgment of taste requires everyone to like the same object without liking being based on a concept. If an object is judged in terms of concept, then the presentation of beauty is lost. Kant disputes that "a pure judgment of taste cannot be based on pleasures of charm or emotion, nor simply on empirical sensations such as charming colors or pleasing tones, nor on a definite concept, but only on formal properties" (Crawford 55). The two cognitive faculties which work in the judgment of taste are "the Understanding and the Imagination" (Cohen 2). The task of imagination is "to 'mould' the object, to arrange and align its constituents" (Cohen 1-2). To decide whether an object is beautiful or not, rather than understanding, imagination is used because understanding is "the power of concepts" (Pluhar 505) and imagination is "the power of a priori intuitions" (Pluhar 505). According to Kant, the judgment of taste discerns that "something is or is not beautiful" (Crawford 52) and it "cannot be

proved since they do not rest on concepts or rules" (Crawford 55). Judgment of taste is neither cognitive nor logical, rather based on the feeling of pleasure therefore, it is aesthetic and subjective.

Beautiful is that which pleases without a concept and universality. Kant affirms that "beautiful prepares us to love disinterestedly something" (qtd. in Crawford 62). It is "cognized without a concept as the object of a necessary satisfaction" (qtd. in Atalay 44). Human beings call the beautiful objects of nature or art by their names, which presuppose that they are judge morally. Buildings and trees are called majestic and magnificent, trees as cheerful and gay, colors as humble, tender and innocent because they induce sensations that are analogous to the consciousness. Kant has structured four views on the beautiful: (i) the beautiful is liked directly, (ii) the beautiful is liked having no interest, (iii) the freedom of imagination is used while judging the beautiful, and (iv) the subjective principle is used to judge the beautiful as universal, which means valid for everyone. He proclaimed that "there is no beautiful thing; rather, beautiful is what we call something that gives us a certain feeling of attunement between the subject and the object-world" (Brown 6).

The German and British philosophers have represented beauty as the key element of art and aesthetic experience because it "is something valuable" (Graham 12). There are two types of beauty: free beauty and accessory beauty. Free beauty does not presuppose the concept of what the object is to be, whereas the accessory beauty presupposes the concept of what the object is to be. If a person does not experience pleasure while perceiving an object, then the person does not perceive the beauty of the object. Thus, judgment of beauty is sensory, intellectual and emotional.

Aesthetic appreciation of beauty is the perfect reconciliation of the rational and sensual parts of human nature. An aesthetic experience can be explicated using two terms: beauty and sublime. Sublime is diametrically high and found in the formless object. It is related to nature and is concerned with what is unlimited or even formless. Sublime deals "with things which are so big that they initially make us feel small" (Shelley 43–44). "Something is sublime if it pleases immediately by its resistance to the interest of the senses" (qtd. in Shelley 44). Sublime "prepares us to esteem something highly even in opposition to our own (sensible) interest" (qtd. in Crawford 62). For Kant, sublime is an object of liking, means feeling of pleasure. However, the object is not called as sublime but as attunement, refers to the mental sensation that attends perceiving an object.

For Hutcheson, the term beauty purports the idea that "certain qualities of things evoke in the mind" (Pluhar lii). Beauty stimulates "a pleasurable response" (McMahon 232) and "is a symbol of morality" (Pluhar 36). It is objective and universal because certain things appear beautiful to everyone, therefore no rule can enforce one's recognition of beauty. Beauty in nature is much concerned about the form of an object and "gives no individual result" (Archambeau 35). To perceive beauty therefore one must perceive the nature or structure from which beauty results. It is unfeasible "to perceive the beauty of an object, without perceiving the object, or at least conceiving it" (Reid 761).

3. Judgment of Beauty

The concept of beauty is assimilated and judged on different criteria. In America, the concept of beauty is evaluated according to the standard of Whites. The difference "between whiteness as beautiful (good) and blackness as ugly (evil) appears early in the literature of the Middle Ages-in the Morality plays of England. Heavily influenced by both Platonism and Christianity" (Gayle 40). Gerald, Gayle and Fanon have proclaimed an identical view in perceiving black and white colors. Carolyn Fowler Gerald enumerates the common standard adopted by Whites to evaluate beauty: "white as the symbol of goodness and purity; black as the symbol of evil and imparity" (Gayle 85). For Gayle, "To be white was to be pure, good, universal, and beautiful; to be black was to be impure, evil, parochial, and ugly" (40). In this connection, Fanon states that in "Europe, whether concretely or symbolically, the black man stands for the bad side of the character" (Markmann 146). He adds to the point that whiteness "has become a symbol of purity, of Justice, Truth, Virginity" (Sardar xiii) and blackness implies the diametrically opposite: "ugliness, sin, darkness, immortality" (Sardar xiii).

In the Western world, aestheticians were white and therefore, literally, and symbolically, they "defined beauty in terms of whiteness" (Gayle 40). Western aesthetic tag line, art for art's sake, propagated by Poe and Pater was completely dismissed by black aestheticians. They strictly observed the phrase, art for blacks' sake. Black aesthetic evolved from the rebellious thought: "Why should I want to be white? I am a Negro-and beautiful!" (Hughes 171). It is always "a part of the lives of black people" (Miller 380). Gayle declared that accepting the "phrase Black is Beautiful is the first step in the destruction [. . .] of the whole ethos of the white aesthetic" (44). John O'Neal has distinguished aesthetic and black aesthetic in the following way:

Aesthetic: a philosophical term – the principles of art.

Black Aesthetic: the principles of Black Art. (53)

Neal perceives aesthetic as a philosophical term that deals with the principles of art. Further, he understands black aesthetic as the principles of Black Art. He draws a distinct line between aesthetic and black aesthetic to differentiate both based on their nature of function.

4. Conclusion

Smith argues that adopting a white standard of beauty would affect the level of satisfaction of an African-American woman with the physical appearance that she maintains. Therefore, the Blacks demand for a different standard of beauty which will not affect their self-esteem and black pride. As a result of their rebel again the white standard of beauty, the new black aesthetic came into existence, favouring Blacks and Blacks'. Revolutionary black artists completely renounced white standard of evaluating beauty and demanded that their works must be evaluated according to Blacks' standard. Fuller writes that Blacks have broken the shackles of imitating white society in order to project their real black self without shame: "black people who are snapping off the shackles of imitation and are wearing their skin, their hair, and their features 'nature' and with pride" (8).

REFERENCES

1. Baraka, Amiri. (2011). "The Black Arts Movement: Its Meaning and Potential." *Journal of Contemporary African Art*, no. 29, fall, pp. 22–31.
2. Bernard, Emily. (2009). "A Familiar Strangeness: The Spectra of Whiteness in the Harlem Renaissance and the Black Arts Movement." African-American Poets Volume 1: 1700s–1940s. edited by Harold Bloom, Infobase Publishing, pp. 165–84.
3. Brown, Fahamisha Patricia. (1997). *Black Poetry: A Vernacular Art*. UMI Company.
4. Clark, Norris Berkeley III. (1980). *The Black Aesthetic Reviewed: A Critical Examination of The Writings of Imamu Amiri Baraka, Gwendolyn Brooks, and Toni Morrison*. University Microfilms International.
5. Franklin, John Hope. (1976). "A Brief History." The Black American Reference Book. edited by Mabel M. Smythe, Pentice-hall, Inc., pp. 1–85.
6. Gayle, Addison. (1972). *The Black Aesthetic*. Anchor Books.
7. Grewal, Nitasha. (2014). "Black Aesthetic Theory: A Perspective." *International Journal of Advanced Research in Management and Social Sciences*, vol. 3, no. 1, January, pp. 116–122.
8. Locke, Alain. (1972). "Negro Youth Speaks." *The Black Aesthetic*. edited by Addison Gayle, Jr., Anchor Books, pp. 16–22.
9. Taylor, Paul C. (2016). Black is Beautiful: A Philosophy of Black Aesthetics. Wiley Blackwell.
10. Taylor, Paul. (2010). "Black Aesthetics." Philosophy Compass, vol. 5, no. 1, pp. 1–15.

Integrating Advancements in Education, and Society for Achieving
Sustainability – Dimitrios A. Karras et al. (eds)
© 2024 Taylor & Francis Group, London, ISBN 978-1-032-70841-6

21

Indian Green Fiscal Policy: Quantitative Analysis of Media Coverage in Online News Portals

Shailendra Boora*, Meljo Thomas Karakunnel

CHRIST University of Bengaluru, India

Abstract: Budget is an instrument that sets the public policy priorities of a government and the public get to understand the importance of those priorities from the kind of media coverage it gets. The Finance Minister of India, while presenting Union Budget-2022, used environmental terminologies such as "climate finance," "energy transition," "inclusive growth," and "climate action" and announced 'Energy Transition and Climate Action (ET & CA)' as one of the priorities of the budget. To push its green agenda, the government increased budget allocations for high-efficiency solar photovoltaic modules, battery swapping policy and special mobility zones with zero-fuel policy. This study analysed the number of reports published on ET & CA along with PM Gati Shakti which is an infrastructural priority of the government. Results showed that news coverage on ET & CA is much less compared to 'PM Gati Shakti' that promoted heavy infrastructural projects, portals.

Keywords: Indian union budget, Green fiscal policy, Media coverage, Online news portals

1. Introduction

Climate change has increasingly become a global issue affecting the humankind in the past four decades. Due to increased human economic activities, environmental impacts on climate have increased significantly in developing countries having large populations and land, such as India (Keller et al., 2020). According to Ge, Friedrich and Vigna (2021), greenhouse emission has increased by 50% from 1990 to 2018, and

India is among the top ten countries that produce at least 68% of carbon gas, *i.e.*, 8.4 tCO_2e per person (Ge, Friedrich & Vigna, 2021). They observed that the magnitude and strength of natural disasters could increase significantly due to the high carbon emission rate per person. As India ranked third for emitting carbon globally, reducing emissions and protecting the environment are essential (Slater, 2020). Since climate change could hamper India's economic growth by declining access to energy security

*Corresponding Author: bjshailu@gmail.com

DOI: 10.4324/9781032708461-21

and the ability to achieve SDGs like poverty eradication and improved health India needs to prioritize the environment.

Though Indian policymakers since the 1990s have given attention to climate change, there were minimal measures to address and mitigate climate change. The trends changed since the 2000s when the country implemented the Kyoto Protocol to minimize carbon emissions (Keller et al., 2020. India promised in the Paris Climate Summit and the recent Glasgow Summit to lower carbon emissions by one billion tons by 2030, reduce emissions to net zero by 2070 and increase renewable energy to 50% by 2030.

2. Climate Change and Fiscal Policy

While climate change and its consequences threaten economically developed and developing nations, the latter is more affected due to limited financial muscle to counter the scenario. Although the global literature on the fiscal impact of climate change is restricted, fiscal policy can execute a fundamental function in reducing climate change effects (Catalano et al., 2020).

The initial preventive measures need to be prioritized over actions of interventions to reduce the heavy expenses at critical stages. Therefore, developing countries such as India should implement a detailed strategy due to the predicted increase in climate risks (Catalano et al., 2020).

3. Environmental Degradation and Economic Impacts

Environmental degradation and climate change has extensive economic and fiscal

impact regarding the reclamation of green cover, protecting endangered species, and cleaning up landfills. The death rate due to ambient particulate matter pollution in India increased by 115.3%, while ozone ambient air pollution by139.2% in 2019. The country lost productive labour, which resulted in low economic output that accounted for losses of $36.8 in 2019 (Pandey et al., 2021). Worldwildlife.Org (2020) predicts that the USA will lose its yearly GDP to $83 billion due to environmental degradation in the next three decades.

4. Need for Budgeting

In the context of negative implications on a country's economic and fiscal well-being, it is essential to budget to mitigate the consequences. Budgeting is critical to contain climate change and achieving fiscal sustainability and climate change (Ekins & Speck, 2011). Integrating budgeting into climate change is crucial for developing nations such as India because it improves the state's capability to contain and address the consequences of climate change effectively. So, the Scottish government has allocated $250 million for the next ten years to conserve the environment (UNEP, 2021).

5. Media Coverage and Budgeting

A country's budget is among the crucial public policy documents since it defines its priorities and can determine losers and winners. Notably, the budget entails a hybrid or mixture of various components that attain extensive media attention. According to Langer and Sagarzazu (2017), several factors, including cost and the level of attention the authority gives, determine the permeability of a policy decision in media coverage. Media coverage on State budgeting is essential since it enlightens people regarding the money allocated to climate change and its expenditure. Media coverage provides a platform for the public to engage in a discourse on budget credentials.

6. Indian Union budget 2022

The Indian Union budget 2022 proposed to spend a total of Rs 39,44,909 crore in 2022-23. 'Energy Transition and Climate Action (ET & CA)' was presented as one of the four priorities in the budget. A shift in public transport through clean technology and governance solutions and the promotion of special mobility zones with a zero-fossil fuel policy, creating an ecosystem for E-vehicles through an approach of battery swapping. Towards a 'low carbon strategy,' the budget aims to produce 280 GW through solar power by 2030. To achieve this, Rs. 19,500 crore is allotted for the production of high-efficiency PV modules.

The budget promoted 'circular economy' to handle electronic waste, end-of-life vehicles, and toxic and hazardous industrial waste to promote recycling. Five to seven per cent of biomass pellets will be co-fired besides coal in thermal power production, which could result in CO_2 saving up to 38MMT annually to achieve a carbon-neutral economy. Four pilot projects are planned for coal gasification and coal conversion into chemicals for use in the industry sector.

During the presentation of the Union Budget 2022–23, Nirmala Sitharaman gave 'a blueprint for the Amrit kaal' and stressed 'Energy Transition and Climate Action' (ET & CA) as one of the four priorities of the budget. Throughout the speech, she also repeated the terms like sustainability, clean energy, climate change, green energy and circular economy to promote the budget as a 'green budget'. The finance minister also announced incentives and measures to promote agroforestry and private forestry, especially for SC/ST communities, besides issuing green bonds to the general public to mobilise resources for green initiatives.

PM Gati Shakti, the prime focus of the budget, emphasised on 'seven engines for economic transformation' – roads, railways, airports, ports, mass transport, waterways and logistic infrastructure. These 'engines' are supported by information & Technology communications, energy transmission, bulk water & sewerage and social infrastructure. Under the PM Gati Shakti scheme, national highways are to be expanded by 25,000 km, with an allocation of Rs. 20,000 crores, and a railway network of 2000 km will be built this fiscal year. Four hundred new-generation Vande Bharat trains will be introduced, and 100 Gati Shakti cargo terminals will be constructed in the next three years.

Further, fight ropeway projects for a length of 60 km will be built. Ken-Betwa river-linking project with an estimate of 44,605 crores is sanctioned and draft DPRs for the other five river-linking projects finalised. PARVESH - a single window portal for giving green clearances is going to be expanded through Centralised Processing Centre-Green (CPC-Green), to promote the ease of investments and doing business, which in turn may discredit the Energy Transition and Climate Action.

7. Objective of the Study

1. To analyse the digital news media coverage of 'Energy Transition and

Climate Action' in comparison to 'PM Gati Shakti', the two major priorities of the Indian Union budget 2022.

8. Methodology

To determine the quantity of news media's attention given to 'Energy Transition and Climate Action', it is necessary to examine the coverage of news media given to ET & CA compared to 'Gati Shakti' - another major priority of the Union Budget 2022–23. Towards this purpose, six Indian digital news portals were selected as samples, of which three are from the most circulated dailies in India: Times of India (TOI), The Hindu, Hindustan Times (HT) and the other three are Scroll, The Quint, and The Wire - the top independent media foundations and recipients of Ramnath Goenka Award - which were founded respectively in Jan 2014; March 2015 and May 2015. Digital news portals were selected instead of print media, as the print media are limited by space in paper and variation of publication across different editions of the same newspaper.

The news stories - reports, editorials, opinions, and features related to ET & CA and Gati Shakthi were collected from all the sampled digital news portals between 25th Jan, '22 to 15th Feb. '22. Inferring from the weekly PEJ News Coverage Index (Pew Research Center, 2014) that news agenda of each week differs, news reports were sampled for three weeks - one week before the presentation of Union Budget and two weeks after the presentation. A week before is to check if any news stories were published on environmental issues before the Union Budget presentation and two weeks later is to track how news stories on ET & CA were published and sustained over the days.

Firstly, to know the news media's attention given to 'Energy Transition and Climate Action', the number of stories published on ET & CA were compared with the number of reports on PM Gati Shakthi. Secondly, a bibliometric analysis of the news stories on both the priorities from the sampled news portals was conducted using open-source software, KH Coder, to know priority of words that topped the list of ten.

9. Findings

News Coverage

As shown in Table 21.1, across the selected online news portals, stories on 'PM Gati Shakti' dominated, with 129 stories.

Table 21.1 Publication of news stories on PM Gati Shakthi and ET & CA

News Portal	No. of Stories - PM Gati Shakthi	No. of Stories - ET & CA
TOI	57	27
The Hindu	33	5
HT	28	9
The Scroll	2	3
The Quint	3	0
The Wire	6	5
Total Coverage	129	56

To get an overview of the content and the major topics discussed in the news stories on the two budget priorities, 'PM Gati Shakti' and 'Energy Transition and Climate Action', bibliometrics of the reports were conducted. Table 21.2 shows the top ten words in the order of high to low. The list is dominated by the word related to 'PM Gati Shakti'.

Table 21.2 Top ten words across sampled news portals

Top ten words	No. of times the words used
Infrastructure	253
Energy	248
Development	175
Battery Swapping	153
Green	125
Electric Vehicles	121
Economy	124
Environment	98
Growth	79
Railways	75

Table 21.3 presents how the top ten words are distributed across selected news portals. The words are almost equally divided between 'PM Gati Shakti' and 'ET & CA'. But the words on 'PM Gati Shakti' are repeated more times in all the sampled news portals.

Table 21.3 Top ten words from each sampled news portals

TOI	No. of times	The Hindu	No. of times	HT	No. of times
Infrastructure	95	Railways	143	Economy	92
EVs	93	Development	46	Growth	71
Battery swapping	91	Economy	40	Infrastructure	62
Development	67	Growth	35	Development	56
Energy	61	EVs	33	Energy	51
Policy	55	Water	31	Green	47
Green	52	Battery Swapping	25	Railways	39
Solar	41	Southern	23	Jobs	37
Charging	38	Energy	21	Climate Change	20
Railways	36	Gati Shakti	14	Sustainability	18
The Scroll	**No. of times**	**The Quint**	**No. of times**	**The Wire**	**No. of times**
Energy	60	Economy	12	Energy	76
Environment	44	Development	11	Infrastructure	72
Forests	38	Women	10	Climate action	68
Green	26	Growth	8	Environment	54
Solar	24	Amrit Kaal	5	Development	41
Infrastructure	22	Employment	4	Air pollution	34
Renewable	20	Gati Shakti	3	Battery Swapping	33
Clearances	15	Infrastructure	2	EVs	28
Hydrogen	12	Private Sector	2	River-linking	25
Climate change	12			Climate change	19

10. Conclusion

The Union government attempted to promote Union Budget 2022–23 as a green budget through 'Energy Transition and Climate Action'. The news portals negated the government's green rhetoric as the contentious Gati Shakti action plan of the government to promote heavy infrastructural projects like national highways, railway networks, ropeway projects, river linking projects, and single-window green clearances for projects got more media attention through more news stories.

REFERENCES

1. Catalano, M., Forni, L., & Pezzolla, E. (2020). Climate-change adaptation: The role of fiscal policy. Resource and Energy Economics, 59, 101111.https://doi.org/10.1016/j.reseneeco.2019.07.005
2. Ekins, P., & Speck, S. (2011). The fiscal implications of climate change and its policy responses. UNEP MCA4 climate initiative, www.mca4climate.info/_assets/files/ClimatePolicy_FiscalSustainability_Final_Report(1).pdf (accessed 27th Jan 2013).
3. Ge, M., Friedrich, J., & Vigna, L. (2020, 6th Feb). 4 charts explain greenhouse gas emissions by countries and sectors. World Resources Institute. https://www.wri.org/insights/4-charts-explain-greenhouse-gas-emissions-countries-and-sectors
4. Keller, T. R., Hase, V., Thaker, J., Mahl, D., & Schäfer, M. S. (2020). News media coverage of climate change in India 1997–2016: Using automated content analysis to assess themes and topics. Environmental Communication, 14(2), 219-235.https://doi.org/10.1080/17524032.2019.1643383
5. Langer, A. I., & Sagarzazu, I. (2017). Are all policy decisions equal? Explaining the variation in media coverage of the UK budget. Policy Studies Journal, 45(2), 337–358.
6. Pandey, A., Brauer, M., Cropper, M. L., Balakrishnan, K., Mathur, P., Dey, S., ... & Dandona, L. (2021). Health and economic impact of air pollution in the states of India: the Global Burden of Disease Study 2019. The Lancet Planetary Health, 5(1), e25–e38. https://doi.org/10.1016/S2542-5196(20)30298-9
7. Pew Research Center. (2014, 5th Mar). News coverage index methodology. Pew Research Center's Journalism Project. Retrieved 22nd Sept, 2022, from https://www.pewresearch.org/journalism/news_index_methodology/
8. Slater, J. (2020, 12th Jun). Can India chart a low-carbon future? The world might depend on it. The Washington Post. Retrieved 22nd Feb, 2022, from https://www.washingtonpost.com/climate-solutions/2020/06/12/india-emissions-climate/
9. Worldwildlife.org. (2020, 11th Feb). US economy will suffer the most as the planet warms. World Wildlife Fund. Retrieved 22nd Feb, 2022, from https://www.worldwildlife.org/stories/us-set-to-have-the-highest-economic-loss-due-to-nature-loss

Note: All the tables in this chapter were made by the author.

*Integrating Advancements in Education, and Society for Achieving
Sustainability – Dimitrios A. Karras et al. (eds)*
© *2024 Taylor & Francis Group, London, ISBN 978-1-032-70841-6*

22

Political Discourses Combating in the Sociopolitical Terrain: Engaging with the Arguments and Negations

Fredrick Ruban Alphonse*

Christ (Deemed to University), India; Krupanidhi Degree College, India

Arya Parakkate Vijayaraghavan

Christ (Deemed to University), India

Mathusha Sam Lara

CSI Jayaraj Annapackiam College, Nallur, India

Banumathi P.

PKR Arts College for Women, Erode, India

Abstract: In English studies, any spoken or written form of text is treated as discourse. There are various types of discourse and political discourse is one among them which is produced within the political terrain. The present research paper aims at interpreting and investigating the political discourse produced within the contexts of Dravidian and Tamil Nationalistic politics. It has employed the political discourse analysis model of Teun A. van Dijk to analyse the political discourse at the macrolevel. The political milieu of Tamil Nadu is always intriguing as there are multiple ideologies formulated to capture the attention of the public. The present research paper examines the fierce contestation between the Tamil Nationalistic discourse and Dravidian discourse to explore how effectively they charge, negotiate and negate.

Keywords: Political discourse analysis, Dravidam, Tamil Nationalism, Contestation

1. Introduction

In the political context of Tamil Nadu, there is a constant tension between the discourse produced in the Dravidian Terrain and Tamil Nationalistic terrain. This tension is perceived as a politics emanated from the discourse produced by Davidian ideologues and Tamil nationalistic ideologues. The political discourse of Seeman counters and questions the Dravidian ideology in many ways which created tension between the Dravidian ideologues and Tamil Nationalistic ideologues in the political terrain of Tamil Nadu. Veerapandian counters the notion of treating Davidian

*Corresponding author: fredrick.ruban@res.christuniversity.in

DOI: 10.4324/9781032708461-22

discourse as a counter discourse to the Tamil nationalistic discourse as he perceives it as an extension of Dravidian ideology. This counter of his creates tension between both therefore, the political discourse of Veerapandian and Seeman have been selected for the present study. The objective of the present paper is to investigate how Seeman's Tamil Nationalistic discourse contests with Veerapandian's Dravidian discourse, and in return how effectively the Dravidian discourse counter-attacks and negates the charges laid by the Tamil Nationalistic discourse. It mints to examine the fiery contestations that were held on the grounds of five charges laid by Seeman in the election campaigns, specifically, the post-Karunanidhi period.

2. Method and Methodology

The present research uses the political discourse of Suba Veerapandian and Senthamizhan Seeman as the primary source. The selected discourses are examined using Teun A. van Dijk's political discourse analysis model. This model supports to evaluate and critique political discourse in terms of theme and rheme. The word discourse refers to "the spoken word, or to all utterances written and verbal, or to a particular way of talking vocabularies" (Griffin 94) and "to a specific set of statements within a given context" (Griffin 94). There are different types of discourse and political discourse is one among them, which "is fundamentally argumentative and primarily involves practical argumentations" (Fairclough 1). It is seen as a "manipulative linguistic strategy which serves concrete (ideological) goals" (Amaglobeli 19). According to Dijk, the analyzation of discourse can be categorized as "language use, discourse, verbal interaction,

and communication belong to the microlevel of the social order" and "power, dominance, and inequality between social groups are typical terms that belong to a macrolevel of analysis" (345). Dijk has enumerated "propaganda, political advertising, political speeches, media interviews, political talk shows on TV, party programs, ballots, and so on" (18) as the examples of political discourse. As Dijk suggests the political discourse of Suba Veerapandian and Seeman are critically evaluated to explore the prevailing conflict and tension between the Dravidian and Tamil Nationalistic discourses. The political interviews and debates of these "political actors" (Dijk 14) have been accumulated from *YouTube* for the present study to make a macrolevel analysis of the discourses.

3. Focal Charges: Arguments and Negations

The article will focus on the major five counters that were enumerated by Seeman during the election campaigns, specifically, post-Jayalalithaa and Karunanidhi period. The primary charges of Seeman are: (i) Dravidian parties have been in power for more than 70 years yet, there is no noticeable development in the lives of Tamils, (ii) the Dravidian parties have corrupted the Tamil land to an irrevocable condition, (iii) the term *Dravidam* is overshadowing the Tamil identity thus the prominence of Tamils have been snatched away by the Dravidian identity, (iv) *Dravidam* has failed to stop the Sri Lankan Civil War which took away the lives of lakhs of Eelam Tamils, (v) *Dravidam* has nurtured caste practice in Tamil Nadu. It is necessary to deal with these charges in details to explore how Tamil Nationalistic discourse counters the Dravidian discourse and in return how the Dravidian discourse negates the charges.

It has been 70 years since the Dravidian parties started ruling Tamil Nadu, yet the land has not seen fruitful growth in the lives of people. This is the major charge of Seeman and Maniarasu on the Dravidian parties. Hereafter, we would use the label Dravidian party to indicate DMK. Periyar and Annadurai are the two political stalwarts who have greatly contributed to the development of Tamils in Tamil Nadu. It is necessary to look at one by one separately to shed light on their achievements, which gradually increases the scope of the term *Dravidam*. The discourse of Periyar needs to be considered as it has brought major changes in the Tamil society. He is widely recognized and appreciated for the initiatives that he took to erode caste and establish social justice in the Tamil land. His struggle to break the Brahminical hegemony is seen as a major contribution by the Dravidian party. He always placed *Dravidam* as a contesting identity to *Aryam*. The mighty contribution of Periyar is injecting the sense of self-respect. Tamil Nadu was in the control of the upper-caste community then and it is Periyar who rigorously worked to uplift the lower caste communities. He demanded reservation for them in education and government jobs. His journey was "from faith to reason and from inequality to equality" (Ramesh 52). He always treated *Dravidam* as an anti-Aryan ideology, which rejects the *Aryan* supremacy in all the aspects of the lives of Tamils. He rejected and opposed it because he understood the Aryan practices as dominating and giving no space to social justice and gender equality. Its strong hold on *Manusmriti* is the principal reason for Periyar absolutely opposing it.

Presently, the Tamil Nationalistic discourse completely eschews Periyar and his opinion as he is not an indigenous Tamil. The Tamil Nationalists treat Periyar as a *Telugan* and not as a Tamil; therefore, they avoid quoting him. They critique and condemn Periyar for addressing Tamil as "*Saniyan* (caniyan)" and "*Kaatumirandi mozhi* (Barbaric language)" ("Dravidam vs Thamizh Desiyam – Suba V" 21:00-22:00). In this context, the word '*saniyan*' has the meaning troublesome and '*kaatumerandi mozhi*' takes the meaning barbaric language. The next charge laid on Periyar is that he has advised the Tamils to give away their Tamil linguistic sentiment and embrace English language. It is true that he motivated the Tamils to use English in their daily conversation. The charge on Periyar is negotiated and attempted to clear by the advocates of the Dravidian ideology. Veerapandian tries to clear this charge verbalizing that Periyar's contribution to Tamil language and Tamil people is immense and adds that "Periyar believed that Tamil language guards caste practice therefore he did not want the Tamils to be obsessed with Tamil language sentiment" ("Dravidam vs Thamizh Desiyam – Suba V, Maniarasan Ediya Aakapurvamana Vivaatham" 22:01-22:54). His attempt to negotiate and defend Periyar may not convince many Tamils unless it is substantiated with an outside source. The words and belief of Periyar can be validated by quoting the argument of Meena Kandaswamy, an English poetess from Chennai. In one of the reality shows of Vijay TV titled *Neeyah Naana?*, she claimed that Tamil is a caste influenced language as some vocabulary and slang of the language carry caste implications. She cites the meaning of the word '*maamaa*' that she found in a Tamil dictionary. It states that the term *maamaa* is used by "the wives of *Pariyer* community to denotes their husbands" ("Neeyah Naana Oru Dalit" 360p" 0:01-0:33). She questioned that is it really a word used only by the wives of *Pariyer* community and even if it is so what is the need to mention it in the dictionary. She argued that such an evil association of caste to language is harmful. By citing the argument of Kandasamy, it can be validated that Periyar's charge on Tamil language is undisputable. Like Periyar, the contribution of Annadurai to the Dravidian politics is immense. His political discourse is identified with his Dravidian rhetoric skill, which fetched him a mass group of idealistic youth. The Dravidian ideology speakers cite the words of Annadurai that demanded the development of Tamil Nadu. He verbalized that *"Vadakku Vazhkirathu Therku Theikirathu"* (20), which means "the north is flourishing, and the south is deteriorating" (Venkatachalapathy 20). He wants to point out that the North India is growing with the nurture rendered by the Union government, but the South India is diminishing without proper nurture.

In the recent times, the Tamil Nationalistic discourse express a strong disapproval and distaste for the Dravidian discourse. Probing into the issue sheds light on the fact that the Dravidian identity and ideology are perceived as overshadowing factors within the sociopolitical landscape of Tamil Nadu. The Tamil Nationalists criticize the Dravidian ideologues with a major charge that the Dravidian identity has snatched away the recognition and position of Tamils in their indigenous land; therefore, Seeman spurs his audience to hail "Tamil Nadu as *Thamizhar* land" ("tamil desiyam vs Dravidam Thamizhan, Ntk Wing Vs DMK Wing 6:00-6:34), which indirectly means that Tamil land should not be identified as Periyar land. Usually, the Dravidian discourse boast and

claim that Tamil Nadu is the land of Periyar. But this claim is countered by the Tamil Nationalistic discourse stating that labelling Tamil Nadu as Periyar land or Dravidian land seizes the original Tamil identity and rich legacy of Tamil Nadu. This shows that the Tamil Nationalists are engaged in relabeling Tamil Nadu with a Tamil name by removing the non-Tamil label. At this juncture, it is necessary to restate that the Tamil Nationalists treat Periyar as a non-Tamil therefore they disapprove the act of calling Tamil Nadu as Periyar *bhoomi* (land).

The Tamil Nationalists are very particular in referring the "Tamil land as *Thamizhar Bhoomi* or as the land of Tamil legends like Kamarajar, Kannagi, Kakkan etc." ("Seeman/tamil desiyam vs dravidam Thamizhan" 6:01-6:44) rather than referring it as Periyar *bhoomi* because they perceive threat in it. Their discourse sheds light on their fear of losing the principal position in one's own land and losing one's linguistic identity. This anxious Tamil Nationalists are trying to redefine the chronology of post-independence period of Tamil history. Consequently, Seeman condemns the act of confining the 50,000-year-old Tamil national history as 'before Periyar' and 'after Periyar'. He criticizes that a rich Tamil history has been contracted by employing the label: 'the period before Anna' and 'the period after Anna'. He questions that why Tamil Nadu is not identified as the land of Chera, Chola, Pandian *bhoomi*, who created a rich history for the Tamils. He continues attacking the Dravidian discourse by stating that the Dravidian discourse has snatched away the Tamil identity by introducing the word *Dravidam*. This inclusive label which is used to identify the people of Andhra, Karnataka, Tamil Nadu and Kerala is overshadowing the native labels. He claims that this label has been disapproved by the other three States, but only Tamil Nadu holds it firmly. He indicts that this disapproved label has been enforced on the Tamils to simply accept. Seeman makes such arguments to validate that there is no Dravidian identity, and the term is an invalid one in the political context. This countering question creates conflict between the Dravidian Discourse and Tamil Nationalistic discourse.

Seeman continues countering the Dravidian discourse and combats with it. He attacks the Dravidian discourse for employing the term *Dravidam* in the sociopolitical context of Tamil Nadu. He condemns and criticizes the term to make it meaningless. He vehemently criticizes this practice and states that "there is nothing as such Dravidam because no one is able to define it and the term does not have any ideology in the political context" (Dravidam Yental Yenna? / Neriyazharai Therikavita Seeman/Seeman speech about dravidam ideology" 0:01-0:37). He questions "what is the real meaning of Dravidam?" (Dravidam Yental Yenna? / Neriyazharai Therikavita Seeman/Seeman speech about dravidam ideology" 0:38-0:41), and as a response he critiques that no one who uses the word *Dravidam* is able to define it vividly. This is seen as a serious charge thrown by Seeman. He criticizes the Dravidian discourse with the charge that "Dravidam is said to be against domination, but it has failed to explain against which domination it is. Is it against caste domination? or Is it against hegemony?" ("Dravidam Yental Yenna? / Neriyazharai Therikavita Seeman/Seeman speech about dravidam ideology" 0:38-0:45). In addition, Seeman questions the integrity of the Dravidian party leaders who were voicing for gender equality and women empowerment in Tamil Nadu. He added that when leaders have two and three wives, how

will they be able to contribute for women empowerment and gender equality. This comment satirizes Periyar and Karunanidhi. From Seeman's contesting discourse one can make out that Seeman considers not only the political life but also one's personal life. He does this to find out the integrity of the leaders and to show that he is against hypocrisy. The contesting discourse of Seeman sheds light on the bitter truth of the lives of the leaders and this gains more power in the combat to counter the Dravidian discourse.

The next charge that the Tamil Nationalism lays on *Dravidam* is that the Dravidian party has been a corrupted party after the regime of Annadurai. Seeman condemns and criticizes Karunanidhi for running the government with least honesty and integrity. He critiques that Tamil Nadu has retrogressed in its economic strength and has lost the maximum natural resources. This criticism seems likely to be negotiated and nullified since he has failed to give facts in figures. The Dravidian ideologues point out the growth that the Dravidian party has brought into the State by taking a massive rational and scientific approach to promote the lives of Tamils. The Dravidian discourse highlights the growth achieved in the education rate and the upliftment brought in the lives of the Tamils by eroding caste practice in the Tamil society. This claim can be validated by referring to the Department of School Education, Government of Tamil Nadu directory. According to the data presented in the directory, in the pre-Dravidian period, that is in 1961, the literacy rate of Tamil Nadu was 36.39% and after the Dravidian rule of 50 years, that is in 2001, the education rate has raised to 73.47%. This statistical data underscores the fact that the 50 years of Dravidian regime has achieved a double proportion growth in the education rate. Another scintillating fact is that according to the statistics conducted in 2011, the State has about 100% Gross Enrollment Ration (GER) in primary and upper primary education. Juxtaposing these facts, one can arrive at the conclusion that the argument of the Dravidian discourse is justifiable and can be validated in addition to negating the charges of Tamil Nationalistic discourse. Seeman does not agree with the claim that the DMK has brought noticeable progression in the education sector. In order to reject this claim, he throws a countering and disputing question: 'who is the teacher that taught Thiruvalluvar to write *Tirukural*?' He uses the term *'yenn paten'*, means my great grandfather, to refer Thiruvalluvar thereby connecting himself closely with the eminent Tamil writer Thiruvalluvar. He does this since Thiruvalluvar is a Tamil representing the legacy of Tamils. Seeman also claims that Thiruvalluvar has lived in a well civilized society therefore he could create a moral book like *Tirukural* to educate this world. In this manner, Seeman proudly asserts the richness of Tamil community. His discourse rejects the claim of Dravidian discourse and escalates the position of Tamil identity.

4. Conclusion

From the above discussion, it is apprehensible that the Dravidian discourse and Tamil Nationalistic discourse counter each other and contest in the political terrain of Tamil Nadu. They both effectively contradict and negotiate. The political discourse of Seeman formulates charges against the Dravidian party and they have been nullified by Suba Veerapandian through his political discourse. The present research has explored and found that the Dravidian discourse could effectively contest with the Tamil Nationalistic discourse and re-establish its prominence in the political milieu of Tamil Nadu. The major five charges constantly laid on *Dravidam* by the Tamil Nationalists have been negotiated and effectively invalidated by Suba Veerapandian.

REFERENCES

1. Anderson, Benedict. (2006). *Imagined Community*. Verso.
2. Amaglobeli, Givi. (2017). "Types of Political Discourses and Their Classifications". Journal of Education in Black Sea Region, vol. 3, Issue 1, pp. 18–24.
3. A. Kalaiyarasan and Vijayabaskar M. (2021). *The Dravidian Model: Interpreting the Political Economy of Tamil Nadu.* Cambridge U P.
4. Blommaert, Jan and Chris Bulcaen. (2000). "Critical Discourse Analysis". *Annual Review of Anthropology*, vol. 29, pp. 447–66.
5. Dijk, Teun A. van. (1998). "What is Political Discourse Analysis?". *Belgian Journal of Linguistics*, vol. 11, Issue 1, Jan. pp. 11–52.
6. Veerapandian, Suba. (2009). *Periyarin Idathusaari Thamizh Thesiyam.* Vanavil Puthakalayam.
7. "Dravidam Yental Yenna? / Suba Veerapandian speech about Dravidam." *YouTube*, uploaded by Mobile Journalist, 2020, https://www.youtube.com/watch?v=AE6cpEi4224.
8. "Dravidam vs Thamizh Desiyam – Suba V, Maniarasan Ediya Aakapurvamana Vivaatham." *YouTube*, uploaded by Chithambaram S, 2016, https://www.youtube.com/watch?v=zYK_I3BHWzA.
9. "Dravidam vs Tamil Nationalism suba veerapandian speech tamil news." *YouTube*, uploaded by Red Pix 24x7, 11 Nov. 2019, https://www.youtube.com/watch?v=ekwVPOmUQJg.

Integrating Advancements in Education, and Society for Achieving Sustainability – Dimitrios A. Karras et al. (eds)

23

Role of Education for Sustainability among the Tribals of Manipur: A Critique

Benjamin Kodai Kaje*, Paul Lelen Haokip
Christ University, Bangalore, India

K. Krelo Peter Kajeo
Hill College Tadubi, Senapati Dist, India

Kennedy Andrew Thomas
Christ University, Bangalore, India

Abstract: Before the arrival of Christian missionaries during the latter part of the nineteenth century in the hills of Manipur, education was in the abyss. The role played by education in ushering transformation in society is undeniable. Basically, the role of education is twofold: to preserve and to provide change for society. Education for sustainability (ES) covers society, economy, and ecology. This paper attempts to give a critical analysis of the twin roles from the social, economic, and ecological perspectives of the tribal context of Manipur, India. Social evils like head-hunting are no longer practised, and some communities have abandoned shifting cultivation. The aim of the paper is that it is possible to educate without destroying culture and earn a livelihood without destroying the ecology. The paper will conceptually present the role of education towards sustainability for the tribals of Manipur through an interdisciplinary approach, supplemented with the experiential knowledge of the authors.

Keywords: education, tribals, sustainability, society, economy, ecology

1. Introduction

Education can be considered a boon for most tribals in Manipur, India. The overall growth and development brought by education amongst the tribals of Manipur have been impressive. In the last half-century, tribals have marched forward in many fields, renounced certain unhealthy cultural practices, and adapted new technologies in education and various other fields. Education has performed well in providing changes in society. Yet there are areas of concern when examining critically the role of preserving social values, cultural practices, sustaining nature, and imparting scientific cultivation.

*Corresponding Author: benjamin.kaje@res.christuniversity.in

DOI: 10.4324/9781032708461-23

Education has a herculean task to perform; to preserve the rich cultural heritage of the tribals, to raise the economy through a sustainable approach, and finally educate the people to save the planet for the future generation. By teaching pupils to consider themselves as global citizens and fostering a feeling of global social responsibility, ES can help students create connections to the actual world (Leal Filho et al., 2018).

Manipur is divided into two sets of landmasses - the valley and the hills. The valley is mainly inhabited by the Meiteis (non-tribals), and the tribals (Nagas and Kuki-Chin) occupy the hills. The valley constitutes about 8% of the total area, whereas the hills comprise 92% of the total land. There are 34 recognised tribes in Manipur as of 2011 census, all of which come under Nagas or Kuki-Chin tribes.

2. Methodology

This paper is divided into two main parts: the role of education as an agent of providing change and a source of preserving the tribals. These twofold roles are looked at with a critical sense from the perspective of a tribal. A few recommendations are made, after which a final conclusion. Education might not have anticipated all the present challenges of the society, the economy, and the dangers ecology faces today. It is important that education makes amends for the good of the society and create a sustainable society.

3. Education for Sustainability (ES)

Figure 23.1 shows the role of education impacting the society, the economy, and the ecology of the tribals. Education contributes

Fig. 23.1   Role of ES in the social, economic, and ecological spheres

to the well-being and societal literacy of the people. It helps people to earn better and live better standards of living. People have become conscious of their health and, at the same time, the importance of maintaining a clean and green environment. Besides, it is vital that education makes an impact in bringing a substantial change in the psyche of the people about the importance of ecology and the environmental dangers we live in today.

4. Agent of Change

The lives of the tribals have risen dramatically over the last few decades. The impact of education is seen and felt in most areas of their lives. Change is a necessity for human beings to progress and sustain themselves. The world we live in is dynamic and constantly changing, and it keeps improving. Many of the social evils were removed from society, aided by education. Among the Nagas of Manipur, the dreaded practice of head-hunting is stopped for good. One can only imagine the consequence if such harmful practices are continued up to the present age. Education has played a pivotal role in the lives of the tribals in ushering social and cultural changes. Maurice Craft (1984) stated, 'Yesterday's solution does not necessarily work for tomorrow's problem, and failure to solve problems does not perpetuate society.'

It appears that Christianity, through education, has played as a 'civilising agent' for the tribals in Manipur. As a result, a large tribal population is not only literate but also educated (Reimeingam, 2013). Fr. Peter Bianchi opened a girls' school as early as 1962 at Punanamei, Mao, Senapati District, Manipur. This paved the way to bring about education in the area for many generations. The first step taken by missionaries has grown, and there are many schools, secondary schools, and even colleges in the area now. The improvement in the sex ratio among the Schedule Tribes (ST) and all ethnic groups throughout time can be attributed to a decrease in the preference for sons brought on by mothers' increased education, more urbanisation, and changes to the traditional systems of social and economic control (Reimeingam, 2013).

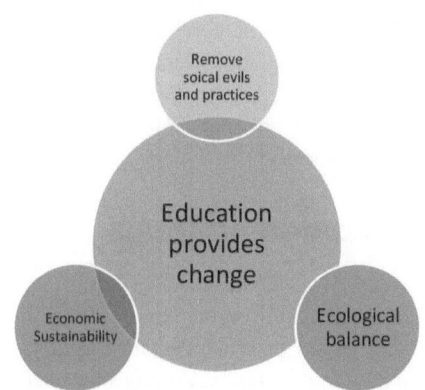

Fig. 23.2 Role of education to provide change

From the economic front, many have adapted to modern scientific means of cultivation. The use of modern machinery is a visible change in the face of the state and even among the rural areas of the tribals. The Maos, one of the major tribes of Manipur, have done away with the practice of shifting cultivation. With education, people have become aware of the menace of shifting cultivation that could ultimately destroy the environment. There are times when on the pretext of bringing growth and development, many forests and natural habitats of animals are disturbed and destroyed.

The tribals of Manipur have made adaptations in their approach to ecology. Many tribal communities allow Alder trees to grow in the midst of terrace cultivation. There are Willows planted along the riverine to prevent soil erosion, and bamboo is grown in areas where the land is prone to landslides. These practices have been aided by scientific knowledge. It is no longer a rare sight to see civil organisations, student bodies, and different associations planting trees on a large scale. Though much is to be desired, some changes are visibly done, which will augur well for the future.

5. Education for Preservation

Another essential role of education is to preserve. It does not mean society has to be static and remain stagnant. But when one observes the tribal communities of Manipur, many core values of the tribals have been forgotten by the younger generations. The rich cultural heritage handed down by the forefathers is almost left in oblivion. Education must preserve the social values, convictions, practices, rites, rituals, and information that enable a civilisation to endure over time (Bass, 1997). Sadly, though formal education arrived almost close to a century back, there are meagre sources written down about the culture and traditions of the tribals. If the early educated people were to write down the various traditional practices and beliefs, tribals in Manipur could have a much richer heritage today.

Tribals were given the impression that whatever was indigenous and old was bad and uncivilised. Parsons (1985) is of the opinion that there has to be a 'core of shared values' remaining unbroken if society is to endure. The Board of Secondary Education, Manipur (BoSEM) has incorporated some tribal languages of Manipur in the high school level syllabus. These educational initiatives encourage young learners to study their own myths, stories, and culture. Hence, in the process of assimilating the modern way of life, the old traditions should not be considered primitive or abhorred.

There are many social values among the tribals which are to be promoted. The sense of respect for elders and the care and love for the sick and the unfortunate ones in society. These people are not outcasted in society but are well looked after by their family members and relations. The integrity to speak the truth is one of the hallmarks of the tribals, which sadly is dwindling with the present generations. Corruptions have crept into the social fabric of the tribals of Manipur. Education needs to promote and preserve that among the tribals; there are no old age homes for the elderly. And it would not be wrong to say that there are beggars among the tribals of Manipur.

The present generations have lost the sense of preserving nature, the know-how to fell trees at the proper season, sustaining forest land, and the importance of safeguarding spring water. Given that they were dependent on other living things for their survival, the tribal people were aware of the importance of the plants and animals in their area (Tiwari,

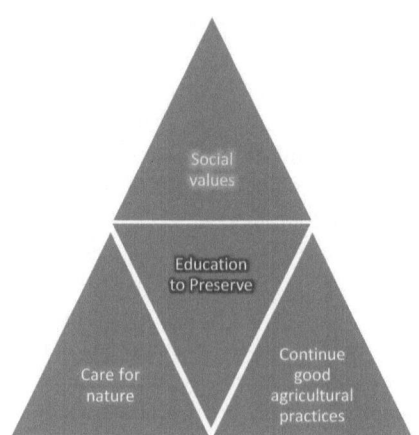

Fig. 23.3   Role of education to preserve

2022). Traditional knowledge regarding the protection and ethical management of sacred forests that have been handed down through the years is a resource on par with biodiversity. Education has to conscientiously educate the importance of the environment to the people so that the populace will make a concerted effort to preserve the ecology of the tribals. More efforts should be encouraged to try sustainable agricultural practices so that the practice of shifting cultivation can be abandoned for good.

6. Recommendations

The role of Education in bringing about certain sustainability has to be lauded. It has not helped to erase certain social evils in society. For instance, among many tribes of the Nagas, women cannot inherit ancestral property, even if there are no male children in the family. Besides, women are not involved in the decision-making of the village and community in general. Though society is, to a great extent, literate and educated, certain age-old practices persist to this present age. Among the Kukis, the ordinary villager owns no land, it belongs to the Chief, and they reside in the land according to the whims of the Chiefs. As educated people, those living in a place need to be respected and cared for and not dismissed from one place to another.

Blindly aping the west has brought much harm to the present tribal society. Many of the cherished values of elders are comfortably set aside, and the modern way of life has taken over. The role of education in creating awareness about the need to preserve ecology has to be intensified. There is hardly any wildlife in many parts of the hills of Manipur. Tress are fell recklessly, and wild animals are hunted without any qualms. The academic curricula must centre on sustainability, which calls for a lifetime commitment on a global scale at all levels of society and commerce (Martins et al., 2006). Before many flora and fauna in the tribal area became a thing of the past, people needed to be taught, educated and stressed the importance of ecology. Education needs to be the torchbearer to show how culture can be preserved without being destroyed and how economy can be earned without compromising the ecology.

7. Conclusion

Education cannot be a panacea for the social ills in society. The future is unpredictable. What is good and noble in the present could be destructive in the years to come. Scientific communities across the globe need to collaborate and share knowledge for the good of society, the overall growth of the economy, and that the environment is preserved for future generations. Ecology needs to be preserved before it becomes uninhabitable for human beings. There seems to be a sense of revival and reawakening of cultural consciousness. Stress needs to be given that livelihood can be earned sustainably, that

culture can exist with modern education, and that growth and development are still possible without destroying the environment. The people have been given the impression that the old cultural practices are bad and to be abandoned. There are hardly any written documents of the culture of various tribes. Much of the rich cultural heritage has been lost, with no written records. Whatever modern need not be good and beneficial. There are many things education has done for the growth and development of society. Importance has to be given to women's education for the overall improvement in the health sector. Enhancing women's access to education could have a significant impact on how often they use maternal health services (Mishra et al., 2021).

REFERENCES

1. Bass, R. V. (1997). The purpose of education. Educ. Forum. 61(2): 128–132.
2. Leal F. W., Raath, S., Lazzarini, B., Vargas, V. R., de Souza, L., Anholon, R., Quelhas, O. L. G., Haddad, R., Klavins, M., and Orlovic, V. L. (2018). The role of transformation in learning and education for sustainability. J. Clean Prod. 199: 286–295.
3. Martins, A. A., Mata, T. M., and Costa, C. A. V. (2006). Education for sustainability: challenges and trends. Clean Technol. Environ. Policy. 8(1): 31–37.
4. Mishra, P. S., Pautunthang, N., Marbaniang, S. P., and Anushree, K. N. (2021). Geographical divide led inequality in accessing maternal healthcare services between hills and valley regions of Manipur state, India. Clin. Epidemiol. Glob. Heal. 11(April): 1–7.
5. Reimeingam, M. (2013). Educational development among the scheduled tribes of Manipur. J. North East India Stud. 3(1): 1–17.
6. Tiwari, B. K. (2022). Biodiversity. The Routledge Companion to Northeast India, Taylor & Francis.

Note: All the figures in this chapter were made by the author.

*Integrating Advancements in Education, and Society for Achieving
Sustainability – Dimitrios A. Karras et al. (eds)*
© 2024 Taylor & Francis Group, London, ISBN 978-1-032-70841-6

24

UPI a Fintech Initiative: Growth and Govt. Role in Boosting Digital Economy

Veenu Madan[1] and Manjit Kour[2]

Chandigarh University, India

Abstract: In the financial industry, the Unified Payment Interface (UPI) is also a significant milestone. It is an online banking device that allows customers to use their smartphones to virtually send and receive money from and into their accounts. People were apprehensive to use UPI at first, as they were with any new tool, but it now handles 2039 million transactions every year. Digitalization is necessary for the Indian economy to curb black money and money laundering and also make the system more clear and transparent. Keeping the importance of UPI in digitalization this research paper analyses the growth of UPI transactions, and government initiatives to promote UPI transactions.

Keywords: UPI, digitalization, Digital payments, Government, India

1. Introduction

A connected internet-enabled way of life that offers a wide range of services and activities is known as a smart life. The free flow of information, capital, and materials independent of time and geography is achieved by using technological technologies such as storing and accessing data in the cloud and other computing platforms. A variety of users accept mobile payment as one of the fundamental concepts in this type of society, which promotes a more effective and efficient way of life. When it comes to the government's campaign for cashless transactions, Bharat QR is currently in the lead. The government also introduced a few other payment options besides Bharat QR. The first one was UPI in April 2016, although it only supported fund transactions. A cellphone number that is linked to your bank account and a banking app that supports it are prerequisites. Bharat QR, which launched in September 2016 and offers a simple and immediate payment process between customers and merchants, came later. In December 2016, BHIM, which aims to make e-payments through banks more

[1]veenu87madaan@gmail.com, [2]manjuz_99@yahoo.com

DOI: 10.4324/9781032708461-24

convenient, was released. Users can send or receive money to UPI payment addresses or accounts that are not UPI-based.

2. Review of Literature

Gochhwal (2017) summarised that India's economy is ranked seventh in the world, and it is likewise cash-based. High reliance on cash causes a slew of issues, including difficulty in storing and managing cash, the expense of currency notes, the use of counterfeit currency, and the loss of transaction data, all of which contribute to tax evasion. As the economy grows, so do challenges like these. Then NPCI, the RBI's umbrella agency, launched UPI to provide a low-cost digital payment mechanism for retail transactions. Digital payments have now become universal throughout the country thanks to UPI. The use of UPI is rapidly increasing. As the number of smartphone users has grown, and so has the use of digital wallets. Neema & Neema (2016) investigated the UPI architecture, benefits, the process in their exploratory study. By enabling the use of mobile phones as the principal means of transmitting and receiving payments, UPI invented the m-payment industry. Sheerin (2019) analyzed that As a proportion of the total population, India's mobile internet users are rapidly growing. It was 18.55 percent in 2015 which grew to 23.93 percent in 2017. However, the number of internet users in rural and urban India differs. Post demonetization there is a growth in digital as well as UPI transactions. There is a need for awareness among the masses because people do not want to share their personal information. Customers' lack of faith in online banking prevents them from using digital means of payment and settlement. Rastogi & Damle

(2020) found that UPI has climbed to the forefront of India's digital payment market, changing the meaning of transactions with its ease of use and increased accessibility to financial services. Financial technology companies and financial institutions throughout the world are developing faster payment methods, but payment security is more important. As a result, the payment system's future viability will be determined by its security. Gupta & Kumar (2020) investigated the factors affecting the adoption of UPI payment by interviewing 140 respondents in Meerut city and found that Except for education, no demographic factor has a significant impact on UPI adoption. Education and the growth of mobile phones and the internet make the person technology and increase the adoption of UPI payment methods.Malusare (2019) summarized that the computer literacy rate in India is only 6.5 percent. He identified the problems in the usage of electronic fund transfers faced by Indians because people still trust debit and credit cards for payment options. The use of a digital payment system is influenced by societal and infrastructure impediments. It is necessary to impart technical training and education to the masses. Muthurasu & Suganthi (2019) identified the various modes of digital payments and their advantages in tabular foam. They suggested that the government may conduct training to teach everyone how to use digital payment systems. Service quality of the electronic payment system should be enhanced through an instant feedback mechanism for complaints of customers. Sruthy S. Pillai (2019) Mobile phones are the fastest expanding market in today's world. Many countries have used mobile marketing strategies. The Indian government has introduced UPI to promote the mobile payment system because it is a hassle-free and efficient way to conduct transactions. Financial institutions and banks must adopt new technologies to conduct ultra-fast transactions. Sinha et. al. (2019) analysed that 68 percent of Indians prefer cash and 48 percent consider mobile transactions to lack a strong value proposition. About 55% had no idea how the system operated and believed it was too complicated. After demonetization, mobile payment penetration has increased, although use and retention have remained low. Access to finance, injustice, and tax fraud are all economic and social challenges in India that could be remedied with the usage of mobile payment technologies. Security is the primary factor influencing the acceptance of mobile payments.

3. Method

This study is based on secondary data. Data is collected from various government sites NPCI, RBI, and use reports of FCCI, PWC, and India fintech report 2020 Medici. Extensive literature is done to make a clear understanding of the role of UPI transactions in the digitalization of the Indian economy.

4. Growth of Fintech Payment Gateways

Without actually participating in the currency, payment gateway providers offer digital skills to direct and enable the management of digital payments. The paper examines the present market situation and major trends in India's Payment Gateway sector. Here are the top 5 nations with the most scanning activity during the first quarter of 2022, according to the QRTIGER database:

Table 24.1 Highest scan of QR code

United States	42.2%
India	16.1%
France	6.4%
U.K	3.6%
Canada	3.6%

Source: QR TIGER's database

Table 24.1 shows the highest scan for UPI payments. Highest scanning by the united states after that by India. India ranks second in the world but it is far behind the united states so there is a great need to bring awareness among the masses regarding digital payment to bring more transparency in the economy.

Table 24.2 Value of retail payments using UPI on NPCI platform (INR billion) (2016–17 to 2020)

Year	BHIM	USSD 2.0	UPI excluding BHIM & USSD
2016–17	18.04	1.09	48.48
2017–18	300.18	3.58	794.55
2018–19	796.34	2.67	7970.69
2019–20	752.85	1.79	20562.66

Source: RBI

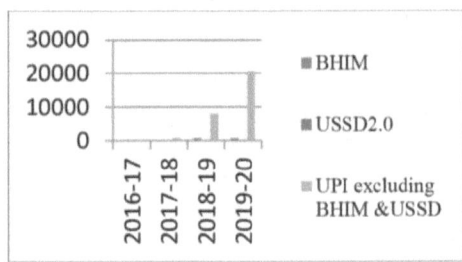

Fig. 24.1 Value of payment by BHIM, UPI, and USSD

Source: NPCI

Table 24.2 and Fig. 24.1 show that the Bhim application was launched by the government of India but acceptance and value of the payment are not good as compared to UPI. UPI payments can be done easily by scanning a QR code and mobile-to-mobile bank transfers. It requires a smartphone and any payment application.

UPI Top 5 Remitter and Beneficiary bank in Jan 2022

Table 24.3 depicts the top 5 UPI remitter banks that are sending money and business decline (BD) due to invalid passwords, Wrong PIN or beneficiary account, or other business reasons. Technical decline due to internet connectivity issue.

Table 24.3 UPI top 5 remitter bank in Jan 2022

Sr. No.	UPI remitter bank jan 2022	Total volume in mn	Approved %	BD%	TD%	Total reversal count in (mn)	Debit Reversal success%
1	"State Bank Of India"	1,291.88	91.70%	7.10%	1.21%	8.17	95.66%
2	"HDFC Bank Ltd"	406.61	94.01%	5.71%	0.29%	2.52	91.30%
3	"Bank of Baroda"	299.12	92.13%	7.38%	0.49%	1.55	98.88%
4	"Union Bank of India"	287.21	90.21%	8.82%	0.97%	3.17	46.80%
5	"ICICI Bank"	274.00	95.14%	4.69%	0.17%	1.22	69.88%

Source: NPCI

Table 24.4 Top 5 beneficiary bank in UPI

Sr. No.	UPI Beneficiary Banks (January-2022)	Total Volume (In Mn)	Approved %	BD%	TD%	Deemed Approved %
1	"Paytm Payments Bank"	957.39	99.27%	0.69%	0.04%	0.01%
2	"State Bank Of India"	647.80	98.59%	0.50%	0.77%	0.14%
3	"Yes Bank Ltd"	593.69	99.91%	0.06%	0.02%	0.01%
4	"ICICI Bank"	378.39	99.07%	0.78%	0.04%	0.12%
5	"Axis Bank Ltd"	353.13	99.51%	0.44%	0.00%	0.05%

Source: NPCI

Table 2 depicts the top 5 beneficiaries who receive money through UPI and the rate of business decline and transaction decline is very low. Deemed approval means credit transaction is not showing instantly in the beneficiary account but is later done manually by the bank.

5. Govt. Initiative for Boosting UPI Payments

To increase electronic banking Promotional programs for the adoption of RuPay Debit cards and low-value BHIM-UPI transactions (P2M) (up to 2,000) for one year, starting on April 1, 2021, have been announced by the Ministry of electronic and information technology, with an expected financial investment of 1,300 crores. This program will assist banking institutions in building a robust ecosystem for digital payments, boost online transactions using the RuPay Debit Card and BHIM-UPI across all industries and population categories, and support national digital payments. Additionally, it would help financially excluded and poor individuals who are not a part of the established banking and financial system have access to digital payment methods. The government has a chance to use UPI in the Budget to expand its reach in low-penetration areas. We are hopeful that the finance minister would support the RBI's plan to route smaller transactions through UPI app on-device wallets. The wallet will function as a prepaid instrument, significantly reducing bank transaction volume. As a result, the risks of a transactional failure are reduced, and merchant acceptance rises.

6. Conclusion

Digital payments and online banking are greatly favored by consumers, which is supported by a well-established infrastructure. India can speed up the transition to a cashless transaction by implementing high-quality initiatives that focus on user empowerment (rather than extolling product offerings or general benefits of digital payments), with specific "how and where to use" knowledge, learning resources, problem-solving hotlines, and safety features that help the customer "stay safe." UPI act as fuel for driving the economy into the path of digitalization. It is becoming popular for grocery payments, Bill payments to financial planning. UPI will eventually make e-commerce transactions, as well as micropayments and person-to-person (P2P) payments, much easier. As a result of their simplicity, openness, and ease, UPI payments are well-suited to the transition to a cashless and digitized economy.

REFERENCES

1. Gochhwal, R. (2017). Unified Payment Interface—An Advancement in Payment Systems. *American Journal of Industrial and Business Management, 07*(10), 1174–1191. https://doi.org/10.4236/ajibm.2017.710084

2. Gupta, S., & Kumar, D. (2020). UPI – An Innovative Step for Making Digital Payment Effective and Consumer Perception on Unified Payment Interface. *The International Journal of Analytical and Experimental Modal Analysis, 12*(1), 2482–2491.

3. Malusare, L. (2019). Digital Payments Methods in India: A study of Problems and Prospects. *International Journal of Scientific Research in Engineering and Management (IJSREM), 03*(08), 1–7.

4. Muthurasu, D. C., & Suganthi, D. M. (2019). an Overview on Digital Library. *Global Journal for Research Analysis, October*, 1–2. https://doi.org/10.36106/gjra/8906567

5. Neema, K., & Neema, A. (2016). *UPI (Unified Payment Interface) – A new technique of Digital Payment: An Explorative study Ist generation IInd generation IIIrd Generation. 3*(10), 1–10.

4. Pillai, S. S., Sandhya, G., & Rejikumar, G. (2019). Acceptance of mobile payments and UPI technology-Indian context. *International Journal of Business Forecasting and Marketing Intelligence, 5*(3), 371–384.

5. Rastogi, A., & Damle, M. (2020). Trends in the Growth Pattern of Digital Payment Modes in India After Demonitization. *PalArch's Journal of Archaeology of Egypt …, 17*(December 2016), 4896–4927. https:// archives.palarch.nl/index.php/jae/article/download/1732/1724

6. Sheerin, A. (2019). Cashless Economy in India: Challenges and Opportunities. *SSRN Electronic Journal*, 498–507. https://doi. org/10.2139/ssrn.3308606

7. Sinha, M., Majra, H., Hutchins, J., & Saxena, R. (2019). Mobile payments in India: the privacy factor. *International Journal of Bank Marketing, 37*(1), 192–209. https:// doi.org/10.1108/IJBM-05-2017-0099

Integrating Advancements in Education, and Society for Achieving Sustainability – Dimitrios A. Karras et al. (eds)
© 2024 Taylor & Francis Group, London, ISBN 978-1-032-70841-6

25

Menvertising or Femvertising: Analysis of Air India Advertisement

Sathvika R.*

Research Associate, School of Social Sciences and Languages,
Vellore Institute of Technology, Chennai, India

Rajasekaran V.

Associate Professor, School of Social Sciences and Languages,
Vellore Institute of Technology, Chennai, India

Abstract: The aim of the paper is to examine gender portrayals in advertising and the notion of menvertising in Air India airline advertisement. This study implements Image analysis as methodology, and inspects the compositional elements of still images and verbal cues in pictorial advertising. The theoretical analysis of Erving Goffman's gender theory helps to illustrate the principles that imply in the select advertisement and the findings suggest how Goffman's gender principles are effective in the select advertisement. Further findings, brief the evident portrayal of menvertising instead of femvertising in the select advertisement. The future study can be investigated using Erving's gender theory in other aspects of advertising to have clear and in-depth knowledge about menvertising in gender portrayal.

Keywords: Gender advertising, Goffman theory, Image analysis, Femverstising, Menvertising

1. Introduction

Gender advertisements refer to the images and the model portrayed in advertising which can depict stereotypical gender roles and portrayals. Media plays a vital role among people in disseminating ideas and ideologies. The modes of media like television, cinema, advertisements, online ads, social networks, etc. portray and display the postures of men and women in particular way that leads to the stereotypical talks. Gender displays are used in advertising to put forth the role of each gender in relation with the other. The advertisers are peculiar in connecting with gender since the readers or commoners define themselves by gender. Gender advertising has other factors that influence the body image, intimate relationships, parenting, and so on. It is argued that a wide variety of social cues, even the most subtly expressed ones, may be taught to the viewers through these visuals.

*Corresponding Author: sathvikasanju@gmail.com

DOI: 10.4324/9781032708461-25

Furthermore, it is asserted that advertising teaches children about gender relations. The gender roles of femininity and masculinity are examples of these learned roles.

In advertisements (1993), men and women are portrayed in accordance with the preconceived notions of femininity and masculinity. Being a woman entails being feminine, and being a man entails being masculine.

2. Research Objectives

- To identify the portrayal of Menvertising and Femvertising in the selected advertisement.
- To apply the sub-contexts of Goffman's gender theory on Air India advertisement.
- To examine the selected advertisement under Imagery analysis as a methodology.

3. Research Questions

- In what form does Menvertising dematerializes Femvertising in the particular advertisement?
- How do the sub-contexts Relative size, Function ranking and Ritualization of subordination imply in Air India advertisements?
- How does Imagery analysis function as a methodology in the selected advertisement?

4. Significance of the Study

The present study examines Goffman's Gender theory and its sub-context. The study explores the sub-context of Erving Goffman's theory such as Function Ranking and Ritualisation of Subordination. The reader or commoner will gain knowledge about gender advertising, where Femvertising fades out and Menvertising emerges in recent days. Furthermore, it is effective for the audience to know the gender stereotypes that exist in our society. The gender stereotypes in advertisements have changed over a period of time. The audience have given importance to Femvertising but now it is revamped according to the advertising trends.

5. Analytical Framework

The present study comes under qualitative research that can be further extended into the methodology of Imagery analysis which is used as an essential mode for analysing still images. Grayscale and colour images, multispectral images for a few discrete spectral channels or wavebands (typically less than 10), and hyperspectral images with a series of contiguous wavebands covering a specific spectral region are all types of images that are quantified, identified, and differentiated using image analysis as a significant tool (e.g., visible and near-infrared).

Goffman proposed a novel method for examining gender stereotypes in 1979 called semiotic or frame analysis. His major attention was focused on more than just an advertisement's obvious message. Hands, eyes, knees, facial expressions, head and body postures, relative proportions, body positioning, and head-eye aversion all suggest minor information. These indicators, which point to inequalities in social power, influence, and authority, operate mostly subconsciously to inform culturally constrained beliefs about gender.

His coding system, which is divided into the following categories, makes it possible to examine the relationships between men and women depicted in commercials. (1) "Relative size," which suggests that women are inferior by being depicted as smaller and/or shorter than men; (2) "feminine touch," a ritualistic touch that highlights the delicate and priceless nature of the female body; (3) "function ranking," which refers to the propensity of men rather than women to fulfill executive roles and exert control over situations; and (4) "ritualization of subordination," intended to capture the adoption of postures that suggest subordination. (5) "licenced withdrawal," depicts women psychologically withdrawing from the circumstance by decontextualization, gaze aversion, and inaction. (6) Body display does not exist in male displays or portrayals, but out of all other categories. The sub-context is one of the most frequently identified sex appeals and dominance over women stereotypes.

6. Analysis

The portrayal of Gender in Advertising

Other definitions of gender portrayal include gender behavior and how it is utilized to interpret social reality. This is the basic model that advertising adopts, and Goffman explains the reason why advertisements do not look awkward or strike the public as odd. Goffman also contends that gender can be determined using codes that indicate gender. The technique of male and female portrayed in advertising reflects these gender norms. These gender codes fall into six categories such as relative size, function ranking, the feminine touch, the ritualization of subordination, licensed withdrawal, and body display.

Portrayal of Menvertising

Menvertising is advertising that questions hegemonic masculinity and provides visual and narrative portrayals of men that promote diversity and advance equality by Knutson and Walder (2017). Men are frequently portrayed in advertising Standing straight, eyes wide open, men's body posture, facial expression (anger), tight hand grips, hand in pockets, serious attitude and physical activity and represent brave such that a man's thought process as well.

Portrayal of Femvertising

The term "femvertising" has been coined to refer to female-targeted advertising that opposes inequality and stereotyping. Femvertising shows the qualities of empowering women, feminism, female activism, or women's leadership and equality (Rodríguez Pérez & Gutiérrez 2017). Femininity in advertising is how women are portrayed in advertising in the following ways: touching oneself, resting on the ground, cuddling an object or male model, sitting on a bed or chair, eyes closed, unfocused, vulnerable body twisted, childlike attire, using a man or an object as support, sexually attractive, seductive, playful, and carefree. These are stances of weakness and submissiveness. Women are depicted literally as being under men when they are seen lying on the ground with males standing over them. The pursuit of beauty and sex appeal is encouraged for women, and part of the sex appeal is subordination. A few general stereotypes of gender are illustrated below in Table 25.1.

Table 25.1 Gender stereotypes

Stereotypes in the Advertisements	
Female Stereotypes	**Male Stereotypes**
Dependable	Sex appeal
As physical attractiveness	Authoritative power
As sex object	Head of the family
Career orientated	Career orientated
Voice of authority	Frustrated male
Neutral	Neutral

Source: Made by Author

Studies on Gender stereotypes in Print advertisements

As numerous studies documented, female stereotypes are pervasive in print advertisements and it is one of the most common forms of advertising, Erving Goffman's (1979) Gender Advertisements book refers as the first book that analyses gender advertisements in magazines. He explains how the advertising the public witness is not really focused on males and females' attitude, but rather on how we perceive them to act. Goffman (1979) asks to pay attention to how the advertisements are put together to depict a social condition and how they are built to attain a particular meaning. He has discussed subordination of women and this is the most prevalent topic in these commercial advertisements. Erving Goffman explores how male and female are portrayed in his book. He discovers dramatic differences between men and women portrayal after analysing more than 500 different images of advertisements and evaluating the various poses, placement of the body, clothing, and other factors. Women are frequently represented as being delicate, weak, helpless, dreamy, childlike, and subservient.

Analysis of *Air India* airlines advertisement

Airlines have one of the most elite customers and their advertisements play a vital role. The advertisements should have some creativity that evokes the consumers to think wisely. At first, from the selected advertisement menvertising dematerializes femvertising in such a manner it is shown in Fig. 25.1 that it is evident that male portrayal has given the privilege to take over the entire advertisement. The male model cast in the advertisement is chosen to be fair as white in colour. The manly posture was so firm holding his lady love signs as protecting and submissive to femvertising. Just to relate the caption given in the advertisement "New York to Mumbai-daily" the male model dominates in skin tone, and acts protective to lady love where it relates to the pre-independence period Britishers were protecting and submissive Indians. As illustrated in the Table 25.2 the non-verbal ways of Gender representation in the selected advertisement have been examined.

Table 25.2 Non-verbal ways of gender representation in advertisements

Women	Men
Light color	Dark/attractive colors
Childish/innocent	Adult/authority
Sitting on men lap	Holding on woman
Leaning for support	Supportive/protective
Skin tone-brownish	Fair as white
Domestic set up-holding a baby	Protective
Costume – body exposed in sleeveless hand	Normal T-shirt – not exposed
Smile – teeth shown	Gentle manly smile

Source: Made by Author

Addressing the next research question, from the Gender categories the three sub-contexts of Goffman's gender theory are suitable for the selected advertisement as Relative size, Function Ranking, and Ritualization of Subordination. The advertisement has witnesses how the female is portrayed relatively lesser in size and the height to showcase male are usually superior in status over female. In the given ad even though it is sitting posture we would find a size difference between the male and female. In Goffman's theory it is assumed that the differences in size will correlate with the differences in society. Further, in Function Ranking it is depicted that males take over the executive roles and controls the entire situation again the female plays a submissive role. The hierarchy of functions is clearly pictured in the advertisement. The female is postured to sit on male's lap where it appears like he doesn't even hold the lady because his hand is gestured downward. The ritualization of subordination refers that women's positioning themselves closer to any object or male. In the selected advertisement, woman is postured to sit on a male's lap holding her baby which creates a domestic setup as a homemaker. The woman's skin tone and the costume a sleeveless t-shirt to showcase the color of the body completely is shown in a darker shade than the men whose in a black t-shirt with a fair skin tone symbolizing the lowering of the female role. The physical position is the model has sat on his lap with a wide smile and hold a baby representing the subordination among the menvertising. The verbal caption given as New York-Mumbai daily indicates the daily flight services available in Air India airlines. The same caption when relating it to the female portrayal depicts that to represent Mumbai or Indian ladies they have depicted the model in a darker skin tone, whereas the male in fairer skin tone as the Americans which creates ritualization of subordination.

The methodology for the selected advertisement is Imagery analysis functions. The basic criteria of Imagery analysis where still images include image capturing, image storage, correcting imaging defects, segmentation of objects, and the models are well-segmented in the picture. The simple thing to showcase the contrast of an image is to expand the brightness scale which is followed in the selected advertisement. The main elements of image enhancement are the elimination of surface roughness, correction of defects, and strengthening texture elements are well depicted in the advertisement to get the sharpness of it.

7. Conclusion

The present study has discovered menvertising is given space in the advertisement than femvertsising. Sexuality is connected to the idea of attractiveness that creates fantasy, whereas sex is a taboo for men in society. In conclusion, it has been examined that menvertising was well portrayed in the selected advertisement than femvertising. To conclude that gender is manifested in the study as a stereotype who are depiction of females are under the control of men who appear to be protective and supportive in society.

REFERENCES

1. Goffman, E. (1979). Gender advertisements. New York: Harper.
2. Klassen, M.L., Jasper, C.R., & Schwartz, A.M. (1993) Men and women: Images of their relationship in magazine

advertisements. Journal of Advertising Research, 33(2), 30–39.

3. Knutson, M., & Waldner, M. (2017). Reshaping the man in the mirror: The effects of challenging stereotypical male portrayals in advertising (Bachelor of Science thesis, Stockholm School of Economics).

4. Rodríguez Pérez, M. P., & Gutiérrez, M. (2017). Femvertising: Female empowering strategies in recent Spanish commercials. Revista de Investigaciones Feministas, 8(2), 337–351.

Integrating Advancements in Education, and Society for Achieving Sustainability – Dimitrios A. Karras et al. (eds)
© 2024 Taylor & Francis Group, London, ISBN 978-1-032-70841-6

26

Employing YouTube to Increase the Speaking Motivation of EFL Learners: Student's Perspective

Indrajit Patra
NIT Durgapur, India

Shahid Bashir
Dhofar University Salalah, Sultanate of Oman

Gulnaz Fatma*
Jazan University, Saudi Arabia

Yaisna Rajkumari
Faculty, Dept of Humanities and Social Sciences, NIT Hamirpur

Abdulwahab Mohammed Saeed Mohammed
Jazan University, Saudi Arabia

Mohammad Jamshed
Prince Sattam bin Abdulaziz University, Al-Kharj, KSA

Abstract: To meet the student's learning needs of the learners, YouTube videos must be used in EFL classes. The study attempts to examine how active and significant YouTube may be in the learning system, and how it might help English teachers improve their students' skills. The study focused specifically on an emerging phenomenon in English learning in Taiwan and examined the self-regulated language learning that EFL university students engaged in outside of the classroom on YouTube. To gain insight into how students felt about this technologically enhanced teaching method and how it affected their English learning, their replies have been examined. As per the study's outcome, most students had a positive opinion of YouTube and were willing to use it in their EFL classes. Additionally, it was found that the children had used YouTube to assist them in finishing their homework and other schoolwork. Thus, it could be deduced that the students are encouraged for using English YouTube videos to aid them in improving their level of skill in the language.

Keywords: English language, Learning, YouTube, Students perspective, Foreign language

*Corresponding author: Gulnaz.fatima15@gmail.com

DOI: 10.4324/9781032708461-26

1. Introduction

The circumstances for learning and teaching languages are not at all restricted to traditional classroom settings thanks to recent technological breakthroughs. Beyond the actual walls of language classes, there is a great deal of informal learning going on. Informal learning, which is defined as "learning that results from everyday life activities involving work, family, or leisure," is relevant to the emergence of continuous learning, in which students are portrayed as independent, political groups who constantly seek out learning experiences, techniques, and materials in their immediate environment (Encinar-Prat & Sallán, 2019). The development of technology has also resulted in a significant increase in the social, active, and multimedia nature of students' literate settings (Galvin & Greenhow, 2020). As soon as children get online, numerous options vie for their interest. How to adequately explain students' informal, out-of-class language acquisition in such virtual settings is still a crucial area of research because multitasking and students' quick attention spans are standards in the digital world (Baker, 2018).

The current generation employs internet as a platform in every part of their lives because we are in the age of digital technology. These newcomers' digital literacy has caused a change in their learning habits from earlier generations (Szymkowiak et al., 2021). The usage of YouTube as a teaching tool is thought to have an impact on the degree of student involvement. The method has offered fresh insights to teaching in higher education. It has been suggested that using YouTube, social media, and other Internet applications in the classroom will help to educate pupils of the newest generation. When videos are uploaded, comments are

made, and other kinds of engagement take place on YouTube, it becomes social media. YouTube mostly creates information and offers multimedia learning options (Sakkir et al., 2020). When students see, hear, and create materials throughout teaching, overall understanding level is higher than when they just comprehend during preparations, and only 40% of this group see and hear. Therefore, various academics have shown that using movies to learn can have an impact on educational activities (Mseleku, 2020). Learners now possess higher order reasoning skills, including those for making problem solving and decision, as well as social media communication and collaboration skills. These skills relate to what the students are learning in the classroom. As a result, social media can be used in education. By considering the impact of YouTube videos but also developing techniques to include them, learning and the teaching process is improved (Alawamleh et al., 2020).

Despite the benefits of using YouTube to keep students interested in the material, few study has been conducted to find out how Taiwanese undergraduates feel about using the site in EFL classes. By asking students about their attitudes regarding using YouTube to learn English, the paper will investigate how it is used in EFL classrooms. In order to make classes more engaging and encourage students to be using resources to learn English beyond the class, it specifically aims to examine how students perceive utilising YouTube as a supplementary to other educational materials. What are the students' perceptions of utilising YouTube in the EFL classroom, in brief, was the questionnaire method. The existing study is reviewed in section 2 after that. The study's methodology is briefly detailed in section 3. The study's conclusions are given in Section 4. Section 5 of the study draws a conclusion.

2. Related Works

The English learning as perceived by students using a YouTube application. Technology, a component of student experience, may provide helpful feedback, particularly in aiding the development of knowledge and English abilities. The results of a qualitative method are presented in the study, demonstrating how helpful and beneficial student understand YouTube is for learning English. According to the study's findings, any student may benefit from improving their English abilities. Only when there is a poor internet connection do the students experience difficulties (Gracella & Rahman Nur, 2020).

In the research, the impact of project-based learning with YouTube presentations as the capstone work will be examined in relation to the cognitive domain online learning outcomes for Physics topics. A quantitative strategy and a quasi-experimental method were utilized in this work. According to the t-test findings, there is a learning accomplishment difference in which the experiment group outperforms the control class. According to the correlation findings, project-based learning with YouTube success as the capstone project has a considerable impact on the cognitive domain (Rozal et al., 2021).

Particularly in EFL contexts, a teacher may employ media to be one of the tools to support the process of learning and teaching in a school. Media may be divided into two categories: audio as well as visual. The author of this essay advocates for the use of video in the teaching-learning process, particularly YouTube videos. The goal of

learning is for pupils to become understand process text, specifically how to create and recognize it. The essay also explores whether or not using YouTube to help teaching and learning activities is a good idea (Sirait et al., 2021).

The goal of the study is to assess how interactive learning tools based on YouTube are used, as well as how instructors and students feel about it. To examine the impact of YouTube use, the study employed a qualitative methodology using a phenomenological perspective. According to the study's findings, it was determined that UMN AW English language education learners used interactive multimedia learning resources centered on YouTube mostly for accessing news media. The study's implications are intended to inspire teachers and learners of English to create interactive material based on YouTube (Nasution, 2019).

3. Methodology

Participants

Twenty university students from two public universities located in northern Taiwan, ten men and ten women, participated in this study. They represented a variety of academic fields, including Chinese studies, implemented foreign languages, business administration, schooling, manufacturing control, and engineering. One student had been obtaining a master's engineering degree at the period of the study, despite 95% of the participants being undergraduate students. Every one of the participants have been gathered using the snowball sampling strategy, which involved two research assistants looking for people who had watched English-teaching YouTube videos via their personal networks. It is able to enroll 20 individuals thanks to the original participants have been invited who were then asked to recommend other individuals with comparable experiences to take part in the research.

Interview

The fundamental method of collecting data used was interviewing, which is heavily based on the attitude that the participant's perspectives are meaningful and relevant (DeJonckheere & Vaughn, 2019). This allowed us to engage with the participants directly in order to examine the topics of interest. In particular, semi-structured interviews were used since this strategy, with a flexible set of survey questionnaire, might enable us thoroughly explore the questions of how these students viewed YouTubers' English instructional videos and how doing so had benefited them on several dimensions. Ten questions were asked of each participant during a one-on-one interview with two trained interviewers about their perceptions viewing English-teaching videos on Social media and other online media platforms (such as Facebook), their motivations for doing so, their opinions of the video content, and the perceived impact of such an informal learning strategy on their educational and outlook on life. All of the interviews have been conducted in the students' native language of Mandarin Chinese so that there would be no language hurdles preventing them from freely expressing themselves.

Data Analysis

Two research associates verbatim transcribed the interview sessions that is assessed in length from 7 minutes and 49 seconds to 25 minutes and 50 seconds. After completing of the transcribed interviews, the study conducted a three-level analysis of the data to identify the study's three main foci: self-regulated learning objectives, learner

response behaviour to YouTubers' English-teaching videos, and comparisons among formal classroom instruction and learning English on YouTube. First, three stages of analysis were performed on self-regulated learning goals: recognition, coding, and comparisons. First, every participant's response is reviewed individually and then used the description of technologically assisted self-regulation to find occasions where the students' language acquisition was tied to their own self-regulation (Rahimi & Fathi, 2021). Team performance, resources, attachment, culture knowledge, metacognitive strategies, and social connection seem to be the six characteristics of self-regulation that are the emphasis of their classification, which is based on an examination of various SRL models. All occurrences were categorized once more using the categorization system after identifying typical learner replies. For the category of self-regulated learning, the inter-rater consistency rates were. 50. Finally, in order to come to a consensus and settle every disagreement, the result is compared.

Second, utilizing the data pool, a categorization scheme with eight response behaviour was created from the beginning. This was done by reading through each of the students' responses once more to determine their response behaviour. Following the creation of this scheme, to evaluate the data is collaborated and investigates the activities that every student made after watching English-teaching clips.

In order to examine the similarities and contrasts among learning English in a classroom and on YouTube, the students' comments were reexamined. The pupils' most frequent similarities and contrasts were emphasized and chosen for discussion throughout the study.

4. Findings and Discussion

Out of 20 pupils, the results indicate that 10 of them were male and 10 were female. The students' ages ranged from 17 to 22, with the most of them falling between the ages of 18 and 22. The statistics given in Fig. 26.1 also showed that only few of students identified themselves as having a beginner's level of English proficiency 7.31% (n=5), while talented students made up 10.15% (n=8) of the population of students and medium students only made up 9.12% (n=7). Thus, it could be said that the majority of the pupils were proficient in Basic English.

Fortunately, most of the students accessed English-language content than Chinese-language content, which allows students to use YouTube as a direct and intuitive learning tool for English. Fewer pupils use YouTube every period, compared to the majority, who only do so occasionally. Every student usually use YouTube, it takes them between one and two hours, two and three hours, three and five hours, and five or more hours on average.

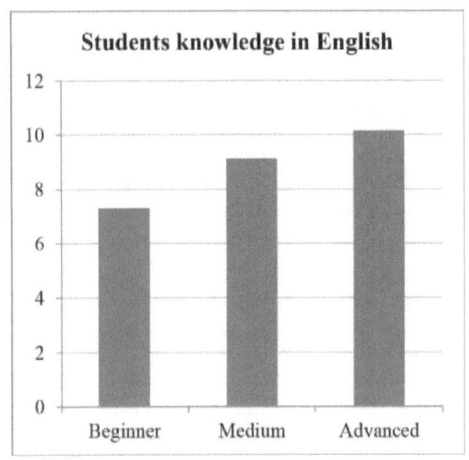

Fig. 26.1 Statistics of student's knowledge in English

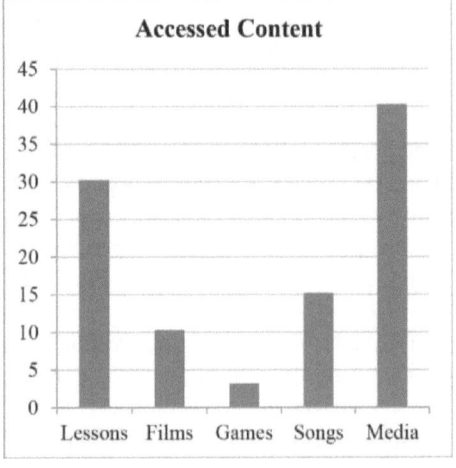

Fig. 26.2 Content accessed in YouTube

The statistics given in Fig. 26.2 revealed that most of the students browsed media content and music, followed by course material in third place, movie programs in ninth place, and games in last place. The replies from the students are therefore collected in order to assess whether this strategy is useful to be included. In this step, the questionnaires are distributed to the students in order to get their feedback on the viability and efficacy of using YouTube in the teaching - learning activities in EFL classrooms. The questionnaire is given to the students to fill out in order to learn what the students thought of the YouTube deployment. The findings indicate that students had a positive view of utilising YouTube.

Table 26.1   Assessment of student's perception in percent

Categories	Range	Frequency rate	Percent (%)
Strongly Agree	81-100	20	30
Agree	61-80	55	70
Moderate	51-60	5	10
Disagree	41-50	0	0
Strongly Disagree	20-40	0	0
Overall		80	100

Calculated as a percentage assessment of students' perceptions in Table 26.1, the analysis reveals that there have been no students who expressed negative statements about using YouTube, 9 students who scored in the range of 84 to 100, 6 students who scored in the range of 68 to 83, and 5 students who scored in the range of 52 to 67.

Table 26.2   Questionnaire result of students' perception

Statements	Mean rate	Categories
YouTube is a great resource for learning English because it has several benefits over other forms of media.	4.32	Strongly Agree
YouTube is a good resource for learning English because the content is easy to understand.	3.81	Agree
I wasn't enjoying learning English at the time, but after using YouTube to supplement my studies, I started to enjoy it.	3.32	Agree
I can learn a lot of words through YouTube.	4.12	Strongly Agree
I don't like utilizing YouTube to learn English because it makes it harder to communicate with friends.	2.49	Disagree
I don't like utilizing YouTube to learn English because it's useless.	1.42	Strongly Disagree
I don't like utilizing YouTube to learn English because it doesn't aid in my learning.	1.78	Strongly Disagree
I don't like using YouTube to study English since it confuses me while I'm trying to understand the lesson.	2.10	Disagree

As shown by the Table 26.2, the students concur that learning English using YouTube is a worthwhile endeavor because it offers many benefits over other types of media, including helping to expand students' vocabulary, enhance their listening comprehension, make their speech clearer due to proper pronunciation, and sharpen their analytical thinking. In conclusion, it is categorized as agree based on the questionnaire's mean rating of 3.81. It demonstrates that the students' attitudes regarding learning English on YouTube are positive. According to the research's findings and justification, YouTube could be used to teach English through visual media. It suggested that students' attitudes on using YouTube to study English were positive.

5. Conclusion

YouTube is more than just a platform for sharing, uploading, and commenting on videos; educators and educators may use it most effectively to support students. YouTube videos must be used in EFL lessons to satisfy the learning demands of the learners. The study looks at how active and significant YouTube may be in the educational system and how it might assist English teachers in helping their students develop their skills. The study studied the self-regulated learning languages that EFL university students did beyond the classroom on YouTube with a particular focus on an emerging phenomenon in English learning in Taiwan. Twenty university students who appeared to have a lot of experience watching YouTube videos that teach English were asked to participate in a general interview to express their thoughts upon this style of self-directed learning. The research revealed that finding more educational resources, realizing the appeal of learning English, and learning about culture were the three most frequently cited reasons for learning English on YouTube. However, it was also believed that students who wanted to improve their English or prepare for English tests would find the interactive learning approach to be less effective. The study's conclusions showed that the majority of students thought favorably of YouTube and were open to using it in their EFL lectures. The kids had utilized YouTube to help them do their homework and other schoolwork, it was also discovered. Thus, it follows that the students are urged to use English YouTube videos to help them advance their language proficiency.

REFERENCES

1. Alawamleh, M., Al-Twait, L. M., & Al-Saht, G. R. (2020). The effect of online learning on communication between instructors and students during Covid-19 pandemic. *Asian Education and Development Studies*.
2. Baker, L. (2018). From learner to teacher assistant: Community-based service-learning in a dual-language classroom. *Foreign Language Annals, 51*(4), 796–815.
3. DeJonckheere, M., & Vaughn, L. M. (2019). Semistructured interviewing in primary care research: A balance of relationship and rigour. *Family Medicine and Community Health, 7*(2).
4. Encinar-Prat, L., & Sallán, J. G. (2019). Informal learning about teaching among novice university professors. *The Qualitative Report, 24*(12), 3102–3121.
5. Galvin, S., & Greenhow, C. (2020). Writing on social media: A review of research in the high school classroom. *TechTrends, 64*(1), 57–69.
6. Gracella, J., & Rahman Nur, D. (2020). Students' Perception of English Learning through YouTube Application. *Borneo Educational Journal (Borju), 2*(1), 20–35. https://doi.org/10.24903/bej.v2i1.623
7. Mseleku, Z. (2020). A literature review of E-learning and E-teaching in the era of Covid-19 pandemic. *SAGE, 57*(52), 588–597.
8. Nasution, A. K. R. (2019). YouTube as a Media in English Language Teaching (ELT) Context: Teaching Procedure Text. *Utamax: Journal of Ultimate Research and Trends in Education, 1*(1), 29–33. https://doi.org/10.31849/utamax.v1i1.2788
9. Rahimi, M., & Fathi, J. (2021). Exploring the impact of wiki-mediated collaborative writing on EFL students' writing performance, writing self-regulation, and writing self-efficacy: A mixed methods study. *Computer Assisted Language Learning*, 1–48.

10. Rozal, E., Ananda, R., Zb, A., Fauziddin, M., & Sulman, F. (2021). The Effect of Project-Based Learning through YouTube Presentations on English Learning Outcomes in Physics. *AL-ISHLAH: Jurnal Pendidikan, 13*(3), 1924–1933. https://doi.org/10.35445/alishlah.v13i3.1241

11. Sakkir, G., Dollah, S., & Ahmad, J. (2020). Students' Perceptions toward Using YouTube in EFL Classrooms. *Journal of Applied Science, Engineering, Technology, and Education, 2*(1), 1–10.

12. Sirait, D., Harahap, Y. S., & Handayani, A. T. (2021). The use of Youtube-Based Interactive Learning Media in Learning English in the new Normal Era. *European Journal of English Language Teaching, 6*(4). https://doi.org/10.46827/ejel.v6i4.3703

13. Szymkowiak, A., Melović, B., Dabić, M., Jeganathan, K., & Kundi, G. S. (2021). Information technology and Gen Z: The role of teachers, the internet, and technology in the education of young people. *Technology in Society, 65*, 101565.

Note: All the figures and tables in this chapter were made by Authors.

Integrating Advancements in Education, and Society for Achieving Sustainability – Dimitrios A. Karras et al. (eds)
© 2024 Taylor & Francis Group, London, ISBN 978-1-032-70841-6

27

Compassion, Empathy, and Trust: A Study of Posthuman Care in Kazuo Ishiguro's *Klara and the Sun*

Gowri Shankari*
Vellore Institute of Technology, Chennai

Manali Karmakar[1]
Vellore Institute of Technology, Chennai

Abstract: Kazuo Ishiguro's *Klara and the Sun* unfurls a tapestry of uncanny experiences confronted by humans who converse with robots in their day-to-day lives. The concerns pertaining to human-machine interaction (HMI) as dramatized in the novel may be used to reflect on the transitioning phase of the healthcare sector where carebots/social robots are now being used for nursing patients with chronic illnesses. This paper through the select fictional narratives aims to speculate on how in the emerging culture of posthuman care the ontology of care, companionship, empathy, and trust is undergoing a process of transition. It aims to foreground the significance of literary studies to thought experiment with the disposable feature of the posthuman caretaker who is designed to offer them solace and companionship. Using De Falco's theory, the paper investigates the ethics of care in the contemporary posthuman setting.

Keywords: Posthuman care, Carebots, Empathy, Companionship, Human-machine interaction

1. Introduction

Dr. Amelia De Falco's existing research study, 'Posthuman Care' revisits care as an ideology, philosophy, and literary practice in the light of recent critical reorientation towards animal nature, vulnerability, ownership, and posthuman. It investigates how contemporary textual and theatrical mediums capture nonhuman companionships in the technology-grooming world (DeFalco,

A. 2021). In addition, the project calls attention to the notion of care with specific attention to ethical issues such as emotional liability, accountability, and dependency. By drawing on the theories and ideas proposed by De Falco, this research article reflects on the ethics of care in terms of communication, compassion, empathy, and trust with special reference to the major and minor characters in Kazuo Ishiguro's *Klara and the Sun*.

*Corresponding author: gowriabclit@gmail.com
[1]manali.karmakar@vit.ac.in

DOI: 10.4324/9781032708461-27

This research article explores the ethical issues surrounding robotic assistance which are dramatized and explored in the novel *Klara and the Sun* through the perspective of the artificial intelligence named Klara. Klara's purchase by a rich clientele for being a constant companion to her chronically ill child Josie, enables us to reflect on the challenges and possibilities of assimilating artificial intelligence in the health care system. Through the fictional experimentation of care robots, the select novel enables us to spot experiment the challenges we may be encountering with the acceptance of AI in the health care systems.

Mirbabaie (2020) proposes that the application of AI in the healthcare system raises fundamentally new ethical problems that could endanger patients. So, this research article attempts to unveil the normative understanding of the care and its ethical implications that differentiate humans from nonhumans.

2. AI and Lit Narratives

Moravec (2000), a leading researcher in the field of mobile robot research, asserts that the inventions of robots are progressing at a rate similar to that of the life forces that evolved on earth. He predicts that in the future, a robot would be able to perform any human labor, whether it be physical or cerebral. Numerous technological progress prevailing today is once seen as simple and complex fictional depictions.

The allure of science fiction comes from its noticeable depictions of technological roles and apocalyptic settings. In fact, science fiction tries to instill a moral message in the readers in addition to providing entertainment. By taking science fiction as a medium, this research article speculates on

the limitations of social robots in posthuman settings.

In *Klara and the Sun*, Ishiguro depicts a near future scenario in which social robots become a part of human life like a smartphone. Furthermore, the fiction depicts the intriguing concerns about care in posthuman settings as well as the current revolution in AI technology. This research article reconfigures the very picture of a social robot as a disposable affective object, irrespective of its acquired intelligence and services.

Ishiguro's *Klara and the Sun* be conceived as a fictional piece that subtly comments on the COVID-19 pandemic lifestyle, in which performing jobs at home and the digitalization of trading and education become a new reality (Sahu and Karmakar 2022). The fiction reflects the human-machine interactions (HMI) and their impact on the interpersonal bond that develops between them. It is expected that the robot's shortcomings would be apparent when it interacts with a human. Klara's limitations and inadequacies are highly noticeable in the fiction when she seeks to answer ethical questions through her extraordinary observational skills.

3. Fictional Experimentation

In select fiction, multiple events effectively depict the affective robot's disposable nature and its limitations in understanding human emotions. Klara is a devoted robotic caretaker created to forge an emotional attachment with clients. Though she is gifted at expressing her affection and concern for her clients, her authenticity is highly questionable to all those who surround her. Therefore, this research article speculates on the narratives of Klara to reflect on the theoretical and ethical flaws of social robots in contemporary society.

Humanoids with empathy, like Klara, are only intended to exist for a certain period before being damaged. At first, she feels happy about finding her best companion and enjoys caring for Josie. Whereas only in the 'interaction meeting episode' she understands Josie's carefree attitude. Until then, she believes Josie is the best companion who adores her and rejoices in her presence. Conversely, her perception of Josie is entirely wrong when Josie supports the choice of the B2 model over Klara. This particular episode in *Klara and the Sun* offers a plethora of meaning to the readers. On one hand, it exhibits, Klara's complexity in understanding human emotions and actions, and on the other hand, it shows the power structure of the human-machine binary.

The Morgon Fall episode reveals the sneaking suspension of the real purpose of Klara in the house of Josie. Though Klara was ostensibly purchased to care for Josie, Josie's mother has another purpose for keeping her as an AF. The purpose is made clear when Chrissie, Josie's mother, asks Klara to replicate her daughter in Josie's absence. Klara is puzzled to see the kindness and patience of Josie's mother toward her. From the beginning to the end, Klara is perplexed by the human actions and emotions that they express towards their fellow humans and nonhuman. Her observation teaches her that she has a lot of limitations to work within. It could be because she interprets things intellectually rather than emotionally.

Klara's observations of Miss Helen's preference for loneliness, Rick's decision to spend time with her mother, Coffee Cup Lady's mixed emotions, bubble game, and its consequences, and many more make her comprehend the complexities of human emotion and care.

Klara finds it difficult to comprehend the uniqueness of humans and their fluidity in compromising things. The limitation of Klara is well reflected in Mr. Capaldi's testing and construction episode where Mr. Paul, Josie's father detests the notion of making a digital replica of his week daughter. Mr.Paul asks Klara, "Do you believe in the human heart? I don't mean simply the organ, obviously. I'm speaking in the poetic sense. The human heart. Do you think there is such a thing? Something that makes each of us special and individual?" (Ishiguro K. 2021: 218). Mr. Paul made Klara understand many things which are beyond her ability to work on. One such thing is understanding the human heart and its complexities. From then on, Klara attempts to learn the complexities associated with the human heart and the reason behind its uniqueness. With each passing day, Klara attempts to discover things that are beyond her ability to achieve. The most prominent among them is attaining the ethical aspect of human emotion and care.

4. Klara as the Other

In the research article titled "Humanoid Robot as 'The Cultural Other': Are We Able to Love Our Creations?" the researchers articulated the "otherness" associated with the humanoid robot. They emphasize the fact that by using the concept of 'other' in multicultural research investigations, one can easily arrive at the destination, where the concepts, beliefs, and theories associated with robotics can be easily comprehended. In short, the article tried to unveil the ideology - the "mirror of the self"- in connection to the robots that evolve and become part of the post-human world. (Kim and Kim 2012: 310).

As a result of the evolution of social robots in the personal sphere, new interpersonal relationships are starting to emerge between humans and robots. As stated above, they mirror humans in certain situations and become 'the other.' Davis, the author of the book *TechGnosis: Myth, Magic and Mysticism in the Age of Information* (1998), admitted the fact that machines end up acting like humans and described them as "interactive mirrors" and the "ambiguous 'other'" (qtd. in Kim & Kim 2012). Based on this, six major perceptions of humanoids are framed by Min-Sun Kim and Eun-Joo Kim (2012) in their research article, and they are 1. "Frightening Other," 2. "Subhuman Other", 3. "Human Substitute", 4. "Sentient Other" 5. "Divine Other" and 6. "Co-evolutionary path to Immorality" (318). All of this shows the predominant condition of post-human circumstances. Among these, Klara falls under the category of a robot as a "Human Substitute."

Numerous unpleasant thoughts are revolving around the evolution, role, and scope of bio-technological devices in the daily lives of humans, both among the citizens of Japan, who are having a craze for AI machines, and among the people of other nations. Indeed, there is a conventional notion that the Japanese were more accustomed to robots because they preferred automation to impose foreign labor. Undoubtedly, these ideas are slowly occupying the minds of people of various nationalities, and they have started to accept the robot culture. One of the reasons behind their acceptance is that, unlike emigrants or marginalized workers, the robot has nothing to do with cultural variations or historical traumas/memories (Kim and Kim 2012).

In the novel, *Klara and the Sun*, Mr. Capaldi views Klara as the substitute for Josie. He tests Klara with a set of questionaries and consoles Chrissie that she will substitute for her daughter. In one instance, Chrissie too accepts the fact and articulates that "That science has now proved beyond doubt there's nothing so unique about my daughter, nothing there our modern tools can't excavate, copy and transfer" (Ishiguro 2021: 224). The post-human setting of humans viewing the robot as another is highlighted in the novel when Chrissie asks Klara to help her. She requests Klara, "I'm asking you to do what's within your power. And think what it'll mean for you. You'll be loved like nothing else in the world.". Though the characters are ready to love their robots either as a pet companions or as the 'other', they are not satisfied with the love and care reciprocated by the robots. Various incidents offered as justification that robots' care and love will never be a substitute for human care. It may provide brief pleasure and satisfaction, but it never equates with the full-fledged affection of a human companion (a friend or family member). This is why Chrissie always prays for the betterment of her daughter's health. Even, she is ready to quit her job to protect and care for her daughter, irrespective of all the efforts she made to create an imitation of Josie. The very action of Chrissie shows that care is a more personal and human thing than the algorithms or programable feeds of the nonhuman.

5. Klara as a Quasi Robot

Today, social robots are becoming more human-like and acting as participants in society. Besides, the autonomous or interactive skills, the looks of the robot allow humans to call them a socializing being. Apart from this, sometimes it can be seen or addressed as more than an instrument or material, and during that state of time, it falls under the category called "quasi-other" (Coeckelbergh, M. 2010).

As stated earlier, their interaction and appearance help the robot become more social. Many scientists are working to build AI machines that are indistinguishable from real humans in terms of the natural language that humans produce during their communication (Coeckelbergh, M. 2010).

In the beginning, people use third-person pronouns like "he,' 'she' or 'it' when discussing robots, as many social robots are making their presence felt in the contemporary world. During the social gathering meeting, friends of Josie addressed Klara as 'it' and 'she.' For instance, one of her friends says, "Throw her over here. That'll bring her to life" and the other says, "Maybe she's low on solar" (Ishiguro 2021: 77). The first phase of talking about the robot is well articulated in the novel.

In the second phase of the robots' evolution in society, people begin to recognize and acknowledge the significance of social robots and start conversing with them. In this case, people start using the second-person pronoun "you.". Josie, Rick, Chrissie, and other people who are close to Klara address her with the second person pronoun. In one instance, Capaldi addresses Klara with the second person pronoun, when he tries to express Klara's purpose in Josie's house, he says, "Learn her till there's no difference between the first Josie and the second" (Ishiguro 2021: 209). Thus, Klara acquires the state of "quasi-other" (Coeckelbergh 2010: 63).

The third phase is talking with the robot. Robots at this phase are prepared to respond to human companions' conversations. Here,

Klara responds and has a conversational dialogue with the characters. To quote, Rick, while discussing the state of Josie with Klara, he asserts, "Look, you might be a highly intelligent AF. But there's a lot you don't know. If you only ever listen to Josie's side of things, you'll never get the whole picture" (Ishiguro 2021:142). Klara responds to the queries of Josie, Rick, Chrissie, and Capaldi through her observational skills. Thus, Ishiguro's fictional experimentation captures the stages of language evolution in social robots and helps us understand Klara's role as a 'quasi-other' in the life of Josie, Rick, and Chrissie. (Coeckelbergh 2010: 63).

6. Conclusion

"Books are not made to be believed, but to be subjected to inquiry. When we consider a book, we mustn't ask ourselves what it says but what it means" (Umberto Eco) Ishiguro's fictional experimentation in the post-anthropocentric future offers an incredibly unique description of a technologically updated society where readers can speculate on various issues about social, cultural, and political matters that are still looking for a resolution in one way or another. The fictional episodes assist us in tracing the changing pattern of human perception of AI machines (social robots), which are considered 'quasi-other' in contemporary society. Ishiguro's work successfully projects the post-human setting in which Klara is described as a non-human entity who aspires to learn more about humans through her mind-blowing observational skills. The in-depth analysis of the fiction allows us to unveil the mask of Klara, who is frequently referred to by others as an empathetic machine.

Based on the study and the investigations into the ethical qualities of the machine, this research article looks for a better plan to invest in constructing the emotional aspect of the robot to foster the emerging affective bond between a human and a robot in the process of their co-evolution.

Though robots try to socialize with people, the ethical aspects are still challenging for them to acquire which eventually makes humans lose the trust factor in AI. This research article once again emphasizes the fact that machines cannot serve as a substitute for a human beings however humane they might look and act. In the select fiction, Klara is purchased to care for Josie, but her purpose of caring is not yet appreciated fully by human companions. The real phenomenon of human emotion is highly distributive, and it is not a thing to be studied or coded with a particular definition. This is why Klara fails to understand the complexities of humans and met her end as a disposable object. Thus, the care shared with utmost love by humans will never be equated with the care shared by robots.

REFERENCES

1. Coeckelbergh, M. (2010). You, robot: on the linguistic construction of artificial others. AI and Society. 26(1): 61–69.
2. Davis, E. (1998). TechGnosis: Myth, Magic and Mysticism in the Age of Information. Harmony books.
3. DeFalco, A. (2020). Towards a theory of posthuman care: Real humans and caring robots. Body & Society. 26(3): 31–60.
4. DeFalco, A. (2021). Imagining posthuman care. Directories, University of Leeds, https://ahc.leeds.ac.uk/directories1/dir-record/research-projects/1210/imagining-posthuman-care.
5. Ishiguro, K. (2021). *Klara and the Sun*. Faber and Faber Limited.
6. Kim, M. and Kim, E. (2012). Humanoid robots as "The cultural other". Are we able to love our creations? AI & Society. 28(3): 309–318.
7. Mirbabaie, M., Hofeditz, L., Frick, N., and Stieglitz, S. (2021). Artificial intelligence in hospitals: Providing a status quo of ethical considerations in academia to guide future research. AI & Society. 37: 1361–1382.
8. Sahu, O. P. and Karmakar, M. (2022). Disposable culture, posthuman affect and artificial affect and artificial human in Kazuo Ishiguro's *Klara and the Sun* (2021). AI and Society: 1–9.

Integrating Advancements in Education, and Society for Achieving
Sustainability – Dimitrios A. Karras et al. (eds)
© 2024 Taylor & Francis Group, London, ISBN 978-1-032-70841-6

28

Inquiry-Based Instruction, Critical Pedagogy, and NEP 2020: Ideology and Politics of Empowerment in Schools of India

Ashraf Alam* and Atasi Mohanty

Rekhi Centre of Excellence for the Science of Happiness,
Indian Institute of Technology Kharagpur, India

Abstract: India's new National Education Policy (NEP 2020) places heavy emphasis on developing each student's 'creative potential'. It adheres to the idea that the purpose of education should not only be to develop cognitive abilities, but also to develop emotional, social, and ethical abilities and dispositions. It endeavours to sustainably convert India into a dynamic, egalitarian, and enlightened knowledge society of creative, inventive, and global people who possess skills for the future. Against this backdrop, the current investigation sheds light on the affordances, conflicts, and unsolved problems that must be addressed to make schools and higher educational institutions more holistic, immersive, integrated, learner-centered, discussion-based, discovery-oriented, inquiry-driven, and adaptable. Using strategies unique to each academic field, this study investigates pedagogies and provides guidance on what students, teachers, and researchers must do to bring positive social change at a time when the environmental, political, and humanitarian crisis is at an all-time high.

Keywords: Instruction, Learning, Teaching, NEP 2020, School, Universities, Education, Creativity, Pedagogy, Curriculum, India

1. Introduction

K-12 students are dynamic participants in the learning process, enthusiastically engaged in meaning-making, and as they do so, they are naturally curious about the environmental, social, and cultural milieus in which they live [1]. Literacy is a potent instrument that may be used in any field of study to generate new insights, hone existing abilities, and assess previously acquired information. This is particularly true in the field of science, where students are asked to analyze texts, plan experiments, and develop rational arguments.

Unfortunately, children from linguistically, ethnically, and economically diverse backgrounds are given far fewer chances to participate in classroom discourses, because of test-centered curriculum and high-stakes

*Corresponding author: ashraf_alam@kgpian.iitkgp.ac.in

DOI: 10.4324/9781032708461-28

testing mandates. In order to better incorporate literacy and science, teachers have adopted inquiry-based teaching strategies from within the framework of traditional, subject-specific learning. Inquiry-based learning is a strategy for instructing, that bases lessons and activities, entirely on students' questions and curiosity. Reading and writing are central skills in an inquiry-based school curriculum because they allow students to perform the specialized tasks of inquiry, such as planning and carrying out experiments, weighing the reliability of various sources of information, and constructing arguments supported by evidence.

Critical pedagogy is an important movement that investigates how powerful people and organizations maintain their control over schools. All methods of teaching and learning, no matter how impartial they seem on the surface, are influenced by ideologies that elevate certain people and some points of view as more informed than others.

By focusing on reading as a tool for recognizing and challenging oppressive ideas, discourses, and practices, critical pedagogy seeks to bring about social change. Literacy is seen as an ongoing relationship with the written word that necessitates the critical examination and revision of social, cultural, and political texts and conditions. This way of thinking about literacy may help both teachers and students use the written and spoken word to fight against injustice and promote positive social change [2].

2. Context

Inquiry-based pedagogy is generally thought of as being synonymous with the progressive education movement. Every aspect of education is based on the interests and backgrounds of its pupils [3]. An experiential and interest-based educational perspective is required in the curriculum. In contrast to progressive educators, Freire blamed dominant structures, procedures, and ideologies for the widespread use of decontextualized or banking system of education in schools. Students' awareness of injustice was raised, and they were encouraged to take part in political actions that would bring about change, via a system of constant cycles of thought and action.

Knowledge is created and recreated by human creativity and inventiveness, as well as their restless, eager, ongoing, and optimistic pursuit of information about the world. Teachers and students need to unite in the fight for freedom, justice, and equality using inquiry as a pedagogical strategy and a way of life. In order to appreciate how inquiry-based teaching practices advance student-centered and critical/decisive pedagogical goals, it is critical to have a firm grasp of the inquiry-based instructional paradigm.

3. National Education Policy, 2020 (NEP 2020)

The new National Education Policy, which was unveiled on July 29, 2020 by the Ministry of Education (MoE), Government of India (GoI), intends to foster critical thinking and creativity in 21st-century learners to support rational decision-making and innovation. The policy aims to recognise, identify, and nurture each student's distinctive qualities and aptitude.

The National Education Policy's objective is to sustainably convert India into a dynamic, egalitarian, and enlightened knowledge society of creative, inventive, and global people who possess skills for the future. In order to make school and higher educational institutions more holistic, immersive, integrated, learner-centered, discussion-based, discovery-oriented, inquiry-driven, and adaptable, it suggests considerable modifications to school curriculum and pedagogy.

The National Education Policy 2020 places a substantial emphasis on the development of each student's 'creative potential'. It adheres to the idea that the purpose of education should not only be to develop cognitive abilities—both 'foundational capacities' of numeracy and literacy and 'higher-order cognitive capabilities' like critical thinking and problem-solving—but also to develop emotional, social, and ethical abilities and dispositions. At all educational levels, it shall endeavour to promote imagination, innovation, and unconventional thinking via sound governance, freedom, and empowerment.

4. Critical Inquiry

The term 'critical inquiry' is used to describe pedagogical methods that draw on students' lived experiences and areas of interest to critique unequal power dynamics. It encourages students to reflect on and respond to issues in the social, cultural, political, and ecological worlds. In doing so, it relies on perspectives from critical pedagogy, critical literacy, and sociocultural learning. By challenging oppressive and unjust structures, ideologies, and discourses, critical pedagogy aims to improve educational opportunities for all students.

Cultural Differences and the Power Dynamics between Students and Teachers

One of the novel distinguishing benefits of critical inquiry techniques is that they encourage students to make use of their own sociocultural knowledge as a crucial resource in the formation of scientific understandings. As long as teachers keep looking for newer pedagogical methods to acknowledge and cultivate students' cultural, linguistic, and literacy practices in the classroom, school students may continue to be seen as capable scientists, meaning-makers, and investigators. It is uncommon to employ both inquiry-based pedagogical strategies and culturally relevant pedagogies (CRP) that respect students' cultural origins and experiences, despite both being hailed as effective practices in critical educational research. Critical inquiry research has shown that students' sociocultural knowledge may be a valuable asset in advancing scientific understanding. Because of student- and culturally-centered research investigations, vital topics like racial and ethnic stereotypes are now being included in the formal school curriculum.

More research needs to be carried out on the topic before we can fully understand the benefits of incorporating students' mother tongue into the classroom. Children come to school speaking a wide variety of languages but are expected to learn and use only English. Students are better able to engage in critical scientific inquiry and learning when they are able to express their ideas in their native languages. This also provides them with the opportunity to critique mainstream scientific paradigms that may exclude non-Western or indigenous epistemologies. Teachers may find it difficult to personalize lessons to suit their pupils' interests and prior knowledge of cultures and societies due to time restrictions and the pressure of high-stakes testing. The need to adhere to strict guidelines hinders the development of a curriculum that addresses students' actual needs and interests. These challenges exposed the systemic barriers that prevent such instructional initiatives from ever materializing.

5. Shifting Paradigms

Using one's critical thinking skills, one might challenge authority in many different ways. In order to challenge the expert/novice binary, students apply critical enquiry to a body of accepted literature. The theoretical foundation of critical literacy may be used as a tool in the fight against an oppressive culture. Critical enquiry helps accomplish these overall aims because it may be used as a tool to raise critical awareness and challenge assumptions. The language of science is often seen as unwavering and authoritative. As part of their scientific education, students may be expected to learn a series of inarguable truths.

Students who engage in critical inquiry are not only encouraged to critique the conventional scientific curriculum but also to propose alternatives for a more equitable future. Students must be invested in and get involved in the process of change. By engaging in critical inquiry, students are able to both apply and challenge the rigour of the curriculum. Learning how to critically analyse scientific informative material may help bring more democracy to the classroom and encourage more student involvement. Children's full democratic engagement in the classroom is sometimes stifled by a focus on standardised testing. Students shall analyse science literature carefully and critically in order to spot inherent biases in the curriculum. It is beneficial to promote literacy as a means of inspiring genuine, discipline-specific behaviours like those utilised by scientists, who often question the veracity of information sources in order to construct knowledge. It is crucial to be able to assess and evaluate information sources in order to tackle the massive environmental, social, and political problems that our country and the rest of the globe are currently confronting.

Student comments and reflections, self-teaching, and the autonomy that comes with utilising their own intellect are signs that the student is aware of how to access and use their intelligence. Proponents of inquiry-based education claim that the use of disciplinary-specific approaches not only increases academic learning but also makes available the norms, discourses, and literacy practices that are frequently hidden in the official school curriculum. It is hoped that through doing research, students would see connections between their work and those of scientists, bolstering their confidence as contributing members of the scientific community. The ability to think critically is often cited as a crucial factor in achieving social justice. The goal of critical literacy is to teach young people how to interpret texts and their surrounding reality in light of issues of power, identity, diversity, and inequality of education, opportunity, and resources. Students are empowered to challenge oppressive systems and ideas by practicing critical inquiry.

6. Acquisition of Knowledge through Experiences

Past studies on classroom learning often assume a certain school-based framework, but it is crucial to conduct scientific inquiry outside of the school's walls and concentrate on children's physical and sociocultural environments. To simulate a different environment, classrooms may sometimes undergo physical changes. The instructor and students shall occasionally set up a kitchen in the classroom and prepare appetisers as a way to educate nutrition and include students' cultural and familial culinary expertise in the lesson plan.

Students have the option of doing research on scientific topics in the field. If there is a pond in the neighbourhood, then that may be used all year long by kindergarten teachers as a learning resource. Learners' questions stem from real-world application and reflection. The instructor and the assigned readings are not the only possible sources of inquiry. When kindergarten teachers use the local pond as a learning resource, students' questions are rooted in the real world. However, all the questions revolve around the same place, *i.e.*, around that specific pond. This prevents students from considering how this ecosystem compares to others in the area or how threatening environmental factors affect different types of ecosystems. Children, while they learn to read the word and the world, must also learn to understand, and respond to the larger social and political environment. When doing research that is truly representative of its setting, it is crucial to account for both the social and physical aspects of the area. Reviving a school garden could go hand in hand with studying the local flora and fauna. Contextually relevant are also students' social experiences and questions inspired by activities they had with family or friends, such as fishing in the neighbouring ponds. Students' interest in these issues is great, yet teachers often report feeling underprepared to address those interests. This conflict illustrates how conventional approaches to the classroom may run counter to the needs of critical educational activities. Teachers who want to encourage students to do community-based research must be flexible and responsive.

Engineering designs may be read critically by high school pupils. By taking a closer look at a variety of produced goods from the viewpoint of both local and global cultures, students and instructors can bring the world into the classroom. When investigating different designs and the people (or businesses) responsible for them, teachers may want to pose questions such as "Who will benefit from this product?" or "Who has paid for the research, design, and development of this product?" Many students might benefit from this in-depth analysis of engineering creations in their historical and cultural contexts. The discovery of *'blood filter'* to fight Ebola may seem like a major scientific breakthrough to some students at the beginning of the semester. However, after thorough evaluation, they may conclude that the filter is too costly to be practical in West Africa for its intended use. Students will learn that the filter was developed in Europe and tested in Germany, two locations far from the West African patients who were likely to get benefitted from its use. Students, via critical analysis, towards the end of the semester, will be able to discern the company's profit-driven objective and arrive at the conclusion that the creation of such a product/device may lead to unequal access of healthcare facilities. Students may thus get better prepared for careers as engineers or as engaged stakeholders.

7. Discussion and Conclusion

Students need to be seen as unique individuals with their own cultural, literary, and linguistic practices before they can critically investigate scientific questions. The experiences and identities of students from racially, linguistically, and socioeconomically diverse backgrounds must be honoured and respected as an integral part of learning about the natural and socio-political world, especially given the pervasive forms of racial and caste-based discrimination, language dominance, and unequal access to resources that these students encounter in school. Second, critical inquiry motivates students to delve into underexplored aspects of the curriculum, including power dynamics. Students of today require this sort of education so that they can tackle the complex problems of the twenty-first century as informed and engaged citizens.

Even though teachers may design lessons that include their students' cultural backgrounds, they may fail to conduct a self-reflective examination of their own cultural practices or their connections to the prevailing culture. No previous research has explicitly accounted for the fact that instructors' opinions and skills impact fair classroom practices. Research in this area should look at the impact that teachers' worldviews and critical reflections have on their use of critical pedagogy in the classroom. There is also a dearth of studies examining the integration of advocacy and action into inquiry-based pedagogy in the classroom. Despite the many opportunities for civic, political, and environmental action presented by K-12 schools, these studies show how challenging it is to put this vision into practice. Looking at how school kids and adolescents use critical thinking to lobby

for change outside of the classroom might help researchers better address the concerns of equity and social justice.

Teachers need help in many areas, including creating and managing lessons, identifying students' learning needs, and introspecting their own cultural identities. Teacher educators and researchers need to work together, share materials, and use critical reflection methods in order to address the requirements of future and current educators who participate in the critical pedagogical activity. These pedagogical approaches motivate students to speak up for and act against social and environmental injustices.

REFERENCES

1. Alam, A. (2022). Positive Psychology Goes to School: Conceptualizing Students' Happiness in 21st Century Schools While 'Minding the Mind!'Are We There Yet? Evidence-Backed, School-Based Positive Psychology Interventions. ECS Transactions, 107(1), 11199.

2. Orr, S. (2021). Political science pedagogy: A critical, radical, and utopian perspective. Contemporary Political Theory, 20(3), 148–151.

3. Alam, A. (2022). Mapping a Sustainable Future Through Conceptualization of Transformative Learning Framework, Education for Sustainable Development, Critical Reflection, and Responsible Citizenship: An Exploration of Pedagogies for Twenty-First Century Learning. ECS Transactions, 107(1), 9827.

Integrating Advancements in Education, and Society for Achieving Sustainability – Dimitrios A. Karras et al. (eds)
© 2024 Taylor & Francis Group, London, ISBN 978-1-032-70841-6

Research on Impact of Online English Education System on Student's Performance During the Covid-19 Pandemic

Indrajit Patra*

Independent Researcher, Ex Research Scholar at NIT Durgapur, Durgapur, West Bengal, India

N. R. Vembu[1]

SASTRA Deemed To Be University, SRC campus, Kumbakonam, India

Divyadharshini R.[2], Saranyadevi S.[3]

Dr. SNS Rajalakshmi College of Arts and Science, Bharathiyar University, Tamil Nadu, India

Abstract: This research examined the impact of the corona virus pandemic on training, cognitive processes, and academic success. Average academic achievement, cognitive capacities, improvement on oral and writing tasks, and capacity to process knowledge when reading and by ear among three respondents were all self reported. Throughout training, general health status was evaluated and confirmed for compliance with Probability density law. All rates were lower when to remote learning that was done before the epidemic. However, it must be emphasized that they were superior to the marks obtained in traditional classroom settings. Interviews were conducted with learners who may have had an impact on the efficacy of the analyzed teaching strategies. These qualities, according to the survey, included having more free time, being able to take vacations more often, having a more enjoyable teaching atmosphere, and not having to drive as far to school.

Keywords: English, Pandemic, Distance learning, Classroom learning, COVID-19

1. Introduction

The typical sick, coughing, sneeze, heat, and certain respiratory issues are all brought on by a family of viruses called as corona viruses. The corona virus may infect animals, but a few of this virus can also transfer from animals to humans (Yunus, et al., 2021). Closings of schools because of the COVID-19 outbreak have brought to light a variety of obstacles to education access. When schools are closed, networks between instructors

*Corresponding author: ipmagnetron0@gmail.com
[1]vembu@mba.sastra.edu, [2]divyadharshini150595@gmail.com, [3]sarannyadevi.05@gmail.com

DOI: 10.4324/9781032708461-29

and students are disrupted, which might affect children's performance (Sinaga, et al., 2021). The present research aims to create and assess a notion of learners' satisfaction with reference to online teaching since both instructors and learners must use the internet site for ongoing learning and teaching throughout COVID-19. (Lestiyanawati, et al., 2020). Numerous comparative studies have been carried out to prove the idea and establish if online or blended learning is better than traditional teaching methods or face to face training (Kamal, et al., 2021). This article offers a data study of the student survey responses to show how online learning is better to traditional classroom instruction and to further the educational process. The essential elements impacting the success of distant learning are described in this study (Nguyen, et al., 2022). In the case of a 2020 corona virus epidemic, research is being carried out to evaluate how distant learning patterns may alter COVID-19.

2. Methodology

Research design and sample: additionally, in relation to the corona virus pandemic, a further survey was conducted among participants in distant English learning to ascertain any differences between the outcomes (Mahdy, et al., 2020). Only those with "Very excellent," "Good," and "Satisfactory" scores in English were enrolled in this programmed since it was dependent on the academic achievement of the participants. The definitions of "Excellent" and "Fail" won't significantly alter based on the kind of class, resulting in a heterogeneous sample.

Experiment: The survey was conducted in three phases. The learners were required to complete the survey by completing the applicable online application after three weeks of in person training. The identical form was subsequently required of every survey responder once again after the conclusion of the 21day distance learning course (Spunei, et al., 2022). It is significant to note that since the quarantine was instituted at different times in Russian and the United Arab Emirates; the final survey period varies significantly depending on the institution.

Data analysis: In order to confirm or disprove the theory about the rise in effectiveness of language learning through distance training mode, a data test of the survey findings was conducted (AlMahdawi, et al., 2021). The study also covered how the COVID-19 Education and Computing Economic Plan affected students' academic success, physical wellbeing, and psychological well-being. Students were also asked to think of any other factors that may have a big influence on the results after the poll.

Research limitations: The research has certain limitations, chief among them being reliance on student reports. In the future, researchers may expand their samples to include all of the many participants in the educational process. In order to get a full picture of the scenario and how the various factors interact with or impact one another, researchers may go further by interviewing instructors about their perspectives and experience. By increasing the number of the investigated samples and using novel interviewing techniques, it is possible to increase the dependability of this research (Irudayasamy, et., 2021). It should be noted that outcomes are based on learners' own assessments of the attributes being studied that might, in some situations, reduce the trustworthiness of the results (Rajesh, et al., 2021). The results were analyzed using the respondents' subjective evaluations of their ability to learn English in addition to indirect performance indicators. The performance metrics of the students may be significantly impacted by inadequate teacher control during distance learning.

3. Results and Discussion

The research presents a comparison of the survey's findings from the experiment part. Table 29.1 presents the average of the replies from the two respondents, divided into three groups that correspond to face-to-face training, distance learning before quarantine was established, and dispersed education during the COVID-19 pandemic. This research also looked at the findings of the survey done during the corona virus epidemic in additional to classroom and online learning. Average student results were discovered to be significantly lower than they had been before to the COVID-19 quarantine. Table 29.2 provides an introduction to the categories to which the samples under analysis belong. The table below the data structure displays the Walk test findings for every one of the three samples.

They follow the principle of normal distribution as a consequence, and the Student's t test may be used to analyze them. The results of the validity of three sample pairs are shown in Table 29.3, with the lines "1-2" designating the groups "In school and learning" and "Distributed learning"; In contrast to "in class learning," which is associated by "3-1," distant learning and education in COVID-19 are linked by "2-3". The third and second columns, respectively, of the Patient's t test display both real and critical values. The linked calculations allowed for the conclusion that samples "1-2" and "2-3" supported the alternative hypothesis whereas samples "3-1" supported the null hypothesis. Given the findings, it can be said that in-person training and distance learning are not equivalent in terms of efficacy. Since the "Distance learning" subgroup's average questionnaire estimates were higher (Table 29.1), the alternative hypothesis that studying English online is more effective than conventional classroom instruction was confirmed. Table 29.3's data also led to the conclusion that distant learning's effectiveness during the quarantine significantly decreased and was almost equal to that of regular classrooms. Although there were no noticeable differences in the evaluations for the categories "In classroom lessons" and "Learning during COVID-19," the latter subgroup's student achievement was somewhat higher. Plagiarism the findings of the research we presented are consistent with findings from earlier, related investigations. Based on pre and posttest results from this study, researchers discovered no differences in learning outcomes when older, less convenient technologies were used, with the exception that distant learning seemed to require more time. Due to the fast evolution of technology, studies of remote learning, and especially e learning, from different eras cannot be seen as comparable, and in this situation, there is no common foundation for comparison. The findings of our research, the organization of the school's work, and the evaluation of students' involvement, interest, and development may all be related to the lower effectiveness of online learning. According to two concurrent polls of educators and their students, tutors often are unable to assess the degree of feedback required throughout the study. Numerous scholars assert that there is a connection between phonological proficiency and anxiety related to learning a foreign language while studying a language remotely. Researchers have highlighted on the significance of student partnerships to establish a high degree of collaboration in their study on students' perceptions of effective online teaching in higher education. They think this element might have both favorable and unfavorable effects.

Forming partnerships may have negative consequences on one's perspective of one's success and prospects due to the constant comparison of individual achievements with that of other students, in addition to improving student motivation and fulfilling communication needs.

Table 29.1 Questionnaire findings

No.	Learning (class)	Distance learning	Learning during COVID
1	6.59	8.6	6.92
2	4.21	7.39	5.23
3	5.19	8.61	5.51
4	6.41	7.71	6.17
5	4.59	7.62	6.39
6	5.41	6.91	5.81
7	5.21	7.51	6.13
8	6.31	8.06	5.96

Table 29.2 Results of Shapiro Wilk tests performed on samples

Sample	Test Shapiro Wilk			
	X_{av}	S^2	W	W_{cr}
Inclass learning	5.52	5.306	0.003	0.9180
Distance learning	7.91	1.907	0.057	
Learning during COVID-19	5.61	1.871	0.091	

Table 29.3 The findings of the student's T test used to test the hypothesis

Compared samples	Student's t-test	
	t	t_{cr}
"1-2"	8.736	2.654
"2-3"	7.631	
"3-1"	2.462	

4. Conclusion

In order to identify the key elements that can favorably influence academic achievement

during distant learning, survey respondents were interviewed. The comments received indicate that doing so leads to more leisure time, the ability to take more breaks during education, more pleasant learning surroundings, and a shorter commute time from and to the institution. Possible explanations for the improvement in English distant learning quality over face to face instruction were also suggested throughout the inquiry. Among these include the lack of strict instructor discipline and the limited possibilities pupils have to evaluate their academic performance to that of their peers' learning outcomes. Such activities may help students feel more confident in them, which would reduce the likelihood of their performing poorly in school. Our results, how schools often organize their work, and how we measure students' engagement, progress, and growth might all contribute to online education's comparatively low efficacy. Two separate surveys of teachers and their students both found that tutors often misjudged the amount of input that was needed at different points in the learning process.

REFERENCE

1. Md Yunus, M., Ang, W.S. and Hashim, H., 2021. Factors affecting teaching English as a Second Language (TESL) postgraduate students' behavioural intention for online learning during the COVID-19 pandemic. Sustainability, 13(6), p. 3524.

2. Sinaga, R.R.F. and Pustika, R., 2021. Exploring Students'attitude towards English online learning using moodle during COVID-19 pandemic at smk yadika bandarlampung. Journal of English Language Teaching and Learning, 2(1), pp. 8–15.

3. Lestiyanawati, R., 2020. The strategies and problems faced by Indonesian teachers in conducting e-learning during COVID-19outbreak. CLLiENT (Culture, Literature, Linguistics, and English Teaching), 2(1), pp. 71–82.

4. Kamal, M.I., Zubanova, S., Isaeva, A. and Movchun, V., 2021. Distance learning impact on the English language teaching during COVID-19. Education and Information Technologies, 26(6), pp. 7307–7319.

5. Nguyen, G.H., 2022. Non-English majored students' preferences of online learning during the COVID 19 pandemic: A case study in Ho Chi Minh University of Food Industry (HUFI). International Journal of TESOL & Education, 2(3), pp. 272–283.

6. Mahdy, M.A., 2020. The impact of COVID-19 pandemic on the academic performance of veterinary medical students. Frontiers in veterinary science, 7, p. 594261.

7. Spunei, E., Frumuşanu, N.M., Muntean, R. and Mărginean, G., 2022. Impact of COVID-19 pandemic on the educational-instructional process of the students from technical faculties. Sustainability, 14(14), p. 8586.

8. AlMahdawi, M., Senghore, S., Ambrin, H. and Belbase, S., 2021. High School Students' Performance Indicators in Distance Learning in Chemistry during the COVID-19 Pandemic. Education Sciences, 11(11), p. 672.

9. Irudayasamy, A., Ganesh, D., Natesh, M., Rajesh, N. and Salma, U., 2022. Big data analytics on the impact of OMICRON and its influence on unvaccinated community through advanced machine learning concepts. International Journal of System Assurance Engineering and Management, pp. 1–10.

10. Rajesh, N. and Christodoss, P.R., 2021. Analysis of origin, risk factors influencing COVID-19 cases in India and its prediction using ensemble learning. International Journal of System Assurance Engineering and Management, pp. 1–8.

Note: All the tables in this chapter were made by the Authors.

*Integrating Advancements in Education, and Society for Achieving
Sustainability – Dimitrios A. Karras et al. (eds)*
© 2024 Taylor & Francis Group, London, ISBN 978-1-032-70841-6

30 The Rise of Phishing in Cyber Law

Bhavana Sharma*

Birla School of Law, Birla Global University, Bhubaneswar, Odisha, India

Abstract: The rapid advancement of network connectivity contributes to the proliferation
of information technology, which contributes to the impact of access control systems in IT
sectors. The most critical pattern in today's everyday lives is network security. As a result,
we will keep the business workers'/customers' knowledge and understanding updated on any
potential threats that they should be aware of. There are various technologies available to
combat infiltration, but no system is entirely safe at the moment. Phishing, a criminal activity
that employs social engineering tactics, is one of the most severe frauds that is a risk in the
present period. This paper focuses on the increase in phishing attacks in recent times and
analyzes the steps taken by the government to prevent such increase. Furthermore, the paper
also analyzes the present Information Technology Act, 2000 and its provisions against the
cybercrime recognized in India.

Keywords: Phishing attack, Cybercrime, Phishing kit, Phishing mechanism, Information
technology act

1. Introduction

Phishing could be a frame of cyber-attack
that utilizes a masked e-mail as a weapon.
The term 'phish' is articulated precisely
because it is spoken to, that's, as the term
"angle", the comparison is of a fisherman
casting a bedeviled trap around here and
expecting it to strike. Some of the famous
scams of phishing attacks recognized
globally are: 2009 - The Federal Bureau
of Investigation conducted operation phish
phry. In this case, around 5000 Americans
were scammed and attacked by the process
of phishing. 2013 - One of the bank's
management trainees is about to be married.
Using the bank machine, these couples
communicated via several emails, but their
engagement ended a few months later. The
girl created false email addresses and the
email was sent to all international clients
of the trainee with whom she had split up
using a bank computer. As an outcome, the
trainee employed in the company suffered

*Corresponding author: sharmabhavana44@gmail.com

DOI: 10.4324/9781032708461-30

considerable financial loss, as the company lost a large number of international customers, and they took the bank to court for such mails. 2015 - Workers at the Prykarpattya Oblenergo control panel were getting ready to leave for the day when one of the operators found something unusual. His computer's cursor started to travel on its own when, to his shock, it verified a request to switch off the circuit breakers at a nearby power station. An entire area was left without electricity within minutes. The worker frantically attempted to retake control of his computer, but he was quickly logged out. The attackers then reset his password, making it difficult for him to re-enter. The attackers managed to knock out substations until 230,000 individuals were affected. 2016 - FAAC, an Australian aerospace company became a victim of a Business Email Compromise attack, tricking employees to transfer large sums of money to the individual who acted as a higher executive authority in the mail. In this phishing case, the attackers act as

Walter Stephen, CEO of the company, and convince the employees to transfer the money to the account. The company lost around 47 million dollars.

2. Phishing Attack Mechanism

The main purpose of a phishing attack is to extract confidential information from user accounts by delivering them false and misleading links and emails and convincing them that these emails are official. Phishing networks are utilized to generate deceptive websites and emails that can imitate authentic websites. These phishing networks can obtain user information and transmit it to the attackers (Military, 2005). These malicious domains are so close to authorized websites that people are unable to tell the difference and are easily duped into disclosing sensitive details to the attackers.

The above figure describes the motion of a phishing attack by the attacker.

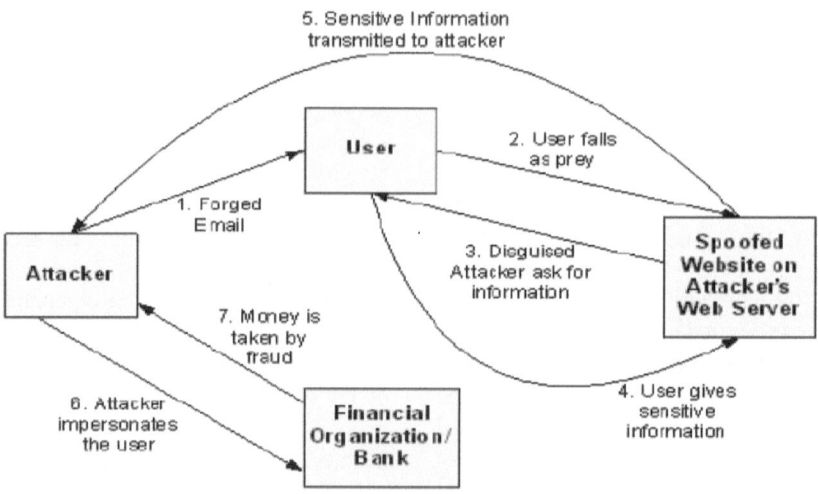

Fig. 30.1 Mechanism of phishing attack

Step 1 - The attacker creates a fake email and sends it to the user; **Step 2** - The user fell in the attacker's trick and provide sensitive information to the attacker; **Step 3** - The attacker uses the sensitive information provided by the user in financial organization/ bank; **Step 4** - The attackers fraudulently take the money from the financial organization/ bank.

Types of Phishing

There are many forms of phishing attacks that have been discovered, and their descriptions are given below:

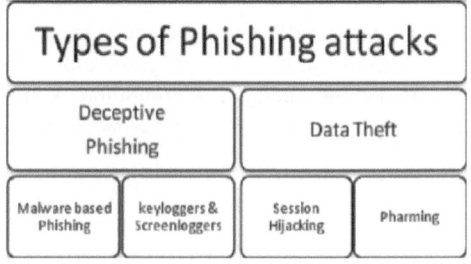

Fig. 30.2 Types of phishing attacks

1. **Deceptive Phishing** - The word 'phishing' simply refers to account theft via text messenger, but best - the known transmission form is a misleading email. The term 'deceptive' refers to misleading. Such emails or messages that mislead an individual to do an activity are known as deceptive phishing (Ali et al., 2020).

2. **Data Theft** - It is a common form of business intimidation. Theft of sensitive correspondence, design papers, legal opinions, employee-specific information, and so on prevents criminals from selling to others who wish to harass or inflict financial damage, or to rivals.

3. **Malware-Based Phishing** - It refers to schemes that include the installation of malicious software on users' computers.

4. **Keyloggers & Screen Loggers** - These are specific malicious software that controls input from the keyboard and sends the pertinent information to the hacker through the Internet.

5. **Session Hijacking** - It is characterized as a form of attack in which users' actions are tracked before they sign in to an aimed account or activity and create their genuine credentials.

6. **Pharming** - Pharming refers to host modification or Domain Name System (DNS)-based phishing. In a pharming scam, attackers interfere with a company's hosts files or domain name system, causing requests for URLs or name services to return a false address and eventual communications to be guided to a false location (Authentication et al., 2015; Easysol, 2009; Paper, 2006).

There are some other methods apart from the above, which are as follows:

1. **Web Trojans:** - It appears undetectably when individuals try to log in. They gather the user's information and submit it to the phisher.

2. **Hosts File Poisoning:** - When a user enters a URL to access a site, it must first be converted into an IP address before being sent over the Internet. By "poisoning" the host's file, hackers are able to send a fraudulent address, leading the user unknowingly to a fraudulent "look-alike" website where their details can be compromised.

3. **System Reconfiguration Attack:** - It modifies configurations on a user's computer for bad intent.

4. **Spear Phishing:** - It is a deliberate attack, similar to conventional phishing, which utilizes the 'spray and pray' technique, in which a vast number of persons receives emails, except these people are specific victims (collectively the same bank/company customers/employee). Phishers do analysis aimed to make phishing more customized in order to maximize the number of victims caught in their hold (Larkin, 2005; Trend Micro, 2012).

5. **Content Injection Phishing:** - It is a kind of phishing in which the phisher substitutes authentic site content with inappropriately crafted content, causing the user to give up confidential details to the phisher.

Phishing Kit

A phishing unit is the site component of a phishing assault. In certain circumstances, this is often the final move after the culprit has imitated a well-known company or affiliation. In the event that introduced, the bundle is aiming to reproduce lawful websites counting those run by Microsoft, Apple, or Google. The availability of phishing units makes it straightforward for cybercriminals to conduct phishing programs, counting those with limited technological mastery. A phishing bundle may be a set of phishing site administrations and program that only got to be mounted on a computer. A few phishing units empower aggressors to mimic

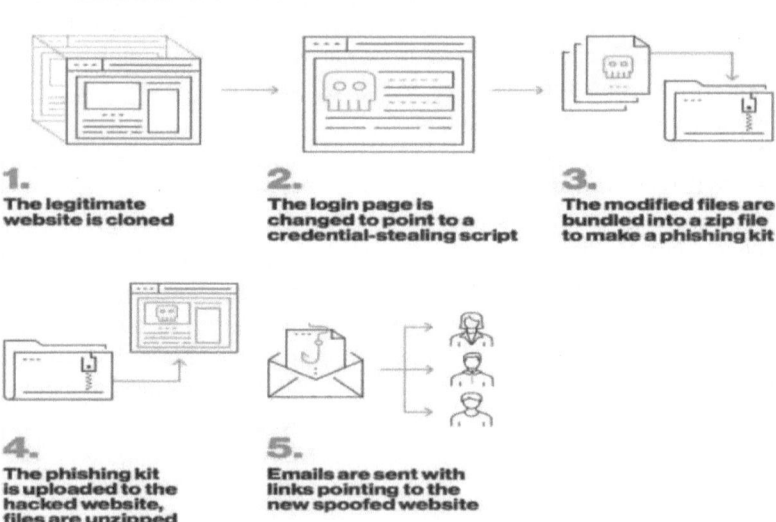

1. The legitimate website is cloned

2. The login page is changed to point to a credential-stealing script

3. The modified files are bundled into a zip file to make a phishing kit

4. The phishing kit is uploaded to the hacked website, files are unzipped

5. Emails are sent with links pointing to the new spoofed website

Fig. 30.3 Anatomy of the phishing kit

well-known brands, raising the chance of a casualty clicking on a pernicious connection.

The above figure provides a step-by-step understanding of the phishing kit to avoid phishing attackers gaining access to public information. Security personnel will track who is utilizing phishing kits by evaluating it. "One of the most significant advantages of investigating phishing kits is determining where passwords are being shipped. It can link perpetrators to particular campaigns and even particular kits by monitoring email addresses identified in phishing kits," Wright mentioned in the report. "It can see not only where credentials are being sent, but also where credentials appear to be sent from. The 'From' header is widely used as a signing card by developers of phishing kits, enabling the security personnel to identify several kits generated by the same perpetrator."

Increase of Phishing Attacks During Pandemics - During the pandemics (2019), technological advancement was utilized in several daily activities of individuals' lives. The growth of the utilization of advanced technology in daily life is a significant change but it increases the risk of phishing attacks. For instance, in India, the ArogyaSetu application was established by the Government to keep an updated track of Covid-19 patients. But soon the application was tampered with by phisher attackers. It was stated by Indian Computer Emergency Response Team that, "Scammers impersonate as the HR department, CEO, or any other known person and target users by spreading messages like 'your neighbor is affected', 'see who all are affected', 'someone who came in contact with you tested positive', 'recommendations to self-isolate', guidelines to use AarogyaSetu' among others."

The following figure showcases the day-to-day increase of phishing sites during the beginning of Covid-19. Organizing activities like expert lecture/seminars/workshops in organizations also at institutions level also

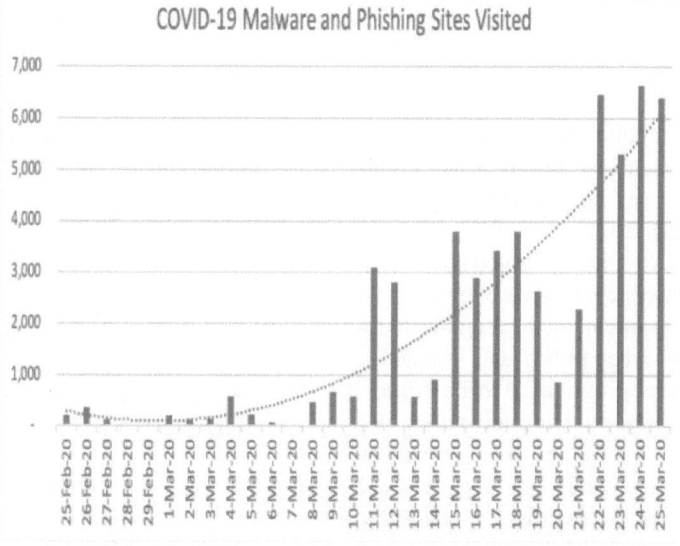

Fig. 30.4 Increase of phishing sites during the pandemics

by using digital platforms and social sites.

Cyber Law in India Against Phishing Attack

In India, specific safeguards against cybercrime have been enacted in the form of cyber legislation, namely the Information Technology Act 2000, as modified in 2008. Section 43, 66, 66A, 66C 66D provides details on it. Under the Indian Penal Code (IPC), phishing can be tried as forgery, 'cheating by personation,' and (iii) cheating and dishonestly inducing the delivery of property. The Information Technology (Reasonable security practises and procedures and personal data or information) Rules, 2011 (SPDI rules) govern business entities that deal with personal information. The Reserve Bank of India (RBI) has issued a regulation on Payment System Data Storage, which states that payments data can only be maintained in India.

3. Conclusion

Phishing is a type of cybercrime in which attackers obtain access to users' private data by sending fraudulent emails and links to them. Phishing has targeted not just large organizations around the globe, but also various government institutions, banking, and financial companies, military establishments, individuals, etc. To prevent the attack of phishing, the government initiated a safe state digital network system to secure the public's data. The government also recognized the significance of cyber police and initiated a public awareness program on cybercrimes. The government is advising the public not to fall victim to these types of crimes and to exercise discretion when entering personal information and passwords on websites. However, the government must now enact

stricter rules, policies, and tactics in order to apprehend the hackers. Furthermore, there is a requirement to develop certain software applications to secure business networks and hospital computers from hacking. The present Information Technology Act, 2000, also requires modification because it is a broad act that excludes all of the other factors impacted by cybercrime. There are several ways to protect yourself: Ensure your software is up-to-date; Don't fall for pop-ups; Secure your internet network with a strong encryption password and a VPN; Manage your social media settings; Protect yourself from identity threats; Educate your children about internet risks.

REFERENCES

1. Arora, R., &Behal, S. (2012, July). Phishing Defense Mechanism. Retrieved April, 2021, from https://www.researchgate.net/publication/275887872_Phishing_Defense_Mechanism

2. Bhansali, S. R. (2003). The Information Technology Act, 2000: An exhaustive, critical, and analytical commentary of act no. 21 of 2000: With Information Technology (Certifying Authorities) Rules, 2000, Cyber Regulations Appellate Tribunal (Procedure) Rules, 2000, Information Technology (Conditions of Service of the Controller) Rules, 2000, Information Technology (Certifying Authority) Regulations, 2001. Jaipur: University Book House.

3. Fruhlinger, J. (2020, September 04). What is phishing? How this cyber attack works and how to prevent it. Retrieved March, 2021, from https://www.csoonline.com/article/2117843/what-is-phishing-how-this-cyber-attack-works-and-how-to-prevent-it.html

4. Kumudha, S., & Rajan, A. (n.d.). A Critical Analysis of Cyber Phishing and its Impact on Banking Sector. Retrieved March, 2021,

from https://acadpubl.eu/hub/2018-119-17/2/128.pdf

5. Nafees Siddiqui, Lokesh Chaudhary, Padmesh Tripathi, Nitendra Kumar, Santosh Kumar. "A comparative analysis of us and Indian laws against phishing attacks", Materials Today: Proceedings, 2022.

6 Plössl, K., Federrath, H., & Nowey, T. (n.d.). Protection Mechanisms Against Phishing Attacks. Retrieved March, 2021, from https://core.ac.uk/download/pdf/11535341.pdf

7. PTI, B. (2020, May 16). Aarogyasetu-related phishing attacks on rise, says India's CYBER agency. Retrieved April, 2021, from https://www.businessstandard.com/article/current-affairs/aarogya-setu-related-phishing-attacks-on-rise-says-india-s-cyber-agency-120051600979_1.html

8. Security, M., ManavMital, Kyle Marchini, CharuhasGhatge,Yesterday, R., (2020, April 04). Sophisticated COVID-19–Based phishing ATTACKS leverage pdf attachments and SaaS to Bypass Defenses. Retrieved April, 2021, from https://securityboulevard.com/2020/04/sophisticated-covid-19-based-phishing-attacks-leverage-pdf-attachments-and-saas-to-bypass-defenses/

9. Yadav, A., & Gemini, J. (n.d.). The Security threat in Cyber World – cybercrime as PHISHING. Retrieved March, 2021, from https://www.krishisanskriti.org/vol_image/24Jul201706074107%20%20%20%20%20%20%20Anu%20Yadav%20%20%20%20%20%20%20%20%20%20%20161-165.pdf

Note: All the figures in this chapter were made by the Authors.

Integrating Advancements in Education, and Society for Achieving Sustainability – Dimitrios A. Karras et al. (eds)

How to think Mathematically? An Attempt towards Transforming India into a "Mathematics Loving Nation"

Ashraf Alam*, Atasi Mohanty

Rekhi Centre of Excellence for the Science of Happiness,
Indian Institute of Technology Kharagpur, India

Abstract: Mathematics education in India limits and controls epistemic validity by posing substantial challenges to the idea that mathematical knowledge is co-constructed and shared via mutually respectful interactions between a teacher and her pupils. The plan relies heavily on the existing power and privilege gaps between participants. From the perspective that sees mathematics as developed and positioned within historical contexts, institutions, and social, cultural, and discursive spaces, the question of who gets to be labelled a mathematician becomes difficult. There have been two major ramifications for discussions about how to improve math instruction in Indian schools. First, it is important that all kids have an education, and second, math is an important skill to have in today's world. The aim thus is to provide mathematics instruction that is both inexpensive and engaging for all students, keeping in mind the reality of India, where very few kids can access expensive mathematical resources.

Keywords: Mathematical thinking, Philosophy of mathematics education, Mathematics education in India, School mathematics, Indian education system, Mathematics pedagogy

1. Introduction

Many school kids think mathematics is creepier than a real monster. For them, opening their math textbooks is a source of exhaustion. Children who ignore math have lower levels of competence, exposure, and practice in the subject, making them less confident and less prepared to achieve learning objectives. Anxiety is a direct consequence of this lack of confidence in mathematics. Apprehension, nervousness, or fear of mathematics may all be symptoms of math phobia. A person with math phobia may be perfectly capable of solving mathematical problems, but the underlying fear prevents them from doing it. Many experts agree that children do not naturally develop a fear of numbers; rather, they learn to be afraid of mathematics from their parents and educators. Teachers that focus only on the textbook, stress memorization

*Corresponding author: ashraf_alam@kgpian.iitkgp.ac.in

DOI: 10.4324/9781032708461-31

of information, and rely solely on drill and repetition to reinforce the lesson may be to blame for certain students' mathematics anxiety at school [1, 2].

We can all agree that everyone should have the opportunity to study mathematics. It is widely held in today's democratic cultures that education is a fundamental human right and that instructors have a responsibility to foster their students' innate curiosity and desire to learn mathematics. These 'enlightened' visionary metanarratives of today hold that development is founded on unadulterated knowledge and is based on a dedication to critical reasoning, individual liberty, and positive change. Much earlier, Déscartes declared his conviction that mathematics will usher in a society that will be more objective and rationally organised. Math, he believed, was the only certain way to an 'enlightened' viewpoint. In the centuries that followed, this perspective would come to shape many facets of culture and thought. Because of its widespread adoption and association with intellectualism, it became an essential component of western democracies. Mathematical rigour in democratic social life is present at both the structural and procedural levels. At this pivotal juncture in history, it is easy to see why mathematics deserves respect and authority [3].

2. Purpose of Teaching Mathematics

The purpose of teaching mathematics is to make students learn 'how to mathematise'. Schools should have two separate aims: the limited aim of producing workers who eventually contribute to societal and economic growth, and the essential aim of nurturing the developing child's inner

thinking capabilities. The first goal places a focus on the value of numeracy in the context of mathematics education [4].

The stress of exams and the fear of being publicly humiliated are cited as the two main causes of increased levels of anxiety among students. Students' freedom of expression may be stifled when they are pressured to conform to an expected response format mandated by mathematics teachers. This leads to memorization, which is counterproductive, according to psychologists. While tests are important, they should not stand in the way of learning. The primary goal of mathematics education should be to provide students with the tools they need to solve real-world challenges. The imaginative side of mathematics should be emphasised while teaching young minds. Learning maths is like learning music or painting. Students' motivation levels tend to soar when they are presented with open-ended mathematical problems to solve. Concepts derived directly from mathematical reasoning play an important role in the definition, construction, modelling, and explanation of the technologically-scientific theorised world of material process and the business world of money circulation and its applications. Furthermore, these notions are both the basis for and the fabric of the abstractions we create, allowing us to make sense of the world by way of analogy, idealisation, and the vantage points they provide [5].

Basic arithmetic operations, including addition, subtraction, multiplication, and division, as well as the concepts of fractions, percentages, and ratios, are introduced and firmly grounded in students' minds by the end of their time in elementary school. Rather than focusing on rote memorisation of formulas, instructors are encouraged by the National Council of Teachers of Mathematics (NCTM) to prioritise encouraging students to think creatively and critically rather than boosting students' self-esteem via their performance in mathematics. Children like to try out novel approaches to solving mathematical problems. Students need to probe, explore, and think for themselves if they are to acquire a deep understanding of mathematics. Studies have shown that pupils benefit more from active learning than from passive learning. The hypothesis of 'multiple intelligences' attempts to account for varied styles of teaching methodologies. Everyone has the ability to learn, but different approaches to teaching mathematics—including role-playing, group work, visual aids, hands-on activities, and technological tools—can bring forth different levels of understanding [6].

How are Mathematics Classrooms Evolving?

Recent discussions on mathematics education have shed light on the difficulties experienced by educators, institutions, and governing bodies as a result of students' dislike of and disinterest in the subject. Because they highlight the persistent pattern of systemic learner underachievement, these conversations have served as a critical focal point for administrators and policymakers. The most recent governmental focus on the problem of students' mathematical understanding may be traced back to considerations of how our mathematics classrooms are evolving to better serve varied learner groups. In dealing with diverse populations, equity and equality are sometimes misunderstood, leading to the notion that the difficulties certain student groups have with mathematics may be explained by their lack of access to certain resources, opportunities, and methodologies [7]. Learning mathematics has been shown to have positive effects on children's emotional and social development. It also helps children grow intellectually. The ability to think clearly and follow ideas to their logical conclusions is crucial in mathematics. Mathematics students learn how to cope with abstractions [8].

Math Anxiety

It is important for students to become familiar with the signs of math anxiety. Fear, paranoia, apathy, and insecurity are common manifestations of math anxiety. Students who suffer from math anxiety should be honest about their feelings. The involvement of both parents is crucial to a child's success in mathematics, as has been shown by several studies. Therefore, one of the simplest methods to help students overcome their math fear is to increase parental engagement in their schooling. Furthermore, studies have also demonstrated that parents' perspectives on mathematics have an impact on their children's perspectives and performance in maths. Careful observation by parents is required to detect any signs of increased irritability in their child when math homework is assigned. The use of praise and encouragement might help a reluctant student feel more comfortable with mathematics. Parents should talk to their kids about their mistakes without passing judgement on them, and shall focus on the lessons that can be learned from those mistakes. They shall emphasize their accomplishments while staying hopeful about their growth potential [9].

Given the importance placed on ensuring that all children, regardless of their socioeconomic status, have access to high-quality mathematics education, these justifications place an emphasis on equality. While it is admirable to try to ensure that everyone has access to a rosy view of mathematics, basic issues like the nature of mathematics itself inevitably arise. Thus, the epistemological issue of who can know mathematics is central to the contemporary discussion. The contexts in which mathematical knowledge might be developed and used rather than prescribing what students should do, think, and know in mathematics classes is significantly important. It examines the paths accessible to students whose mathematical understanding of the world does not always align with that which is considered acceptable by authorities. Mathematics not only equips its students with the tools they need to solve mathematical issues but also with the mindset and methodology they will need to tackle and solve challenges of any kind in a methodical and organized manner. Because of this, a solid foundation in mathematics is crucial [10].

Philosophical Considerations in Mathematics Teaching

High school students in the natural and social sciences often face situations that need a combination of the mathematical, algebraic, and geometric skills they have learned independently. If students want to work on them, if teachers are willing to devote time and energy to them, and if mathematicians find them interesting and challenging, then they must definitely be of some mathematical significance [11].

Over the last thirty years, philosophical considerations in mathematics teaching have grown to become a diffused field of study. There are many different perspectives on the nebulous question of 'what is the philosophy of mathematics education?'. This makes this search even more pressing. The philosophy of mathematical education is not a universal ideology; rather, it is a separate, albeit imperfectly defined, area of research and analysis within mathematics education. Mathematical education is the act or practise of imparting mathematical knowledge. What makes anything a philosophical question is its underlying purpose, reasoning, or justification. Thus, the primary meaning of 'philosophy of mathematics education' is connected to the reason(s) behind the practise of teaching mathematics to students. This is a crucial issue in the theory and practise of mathematics instruction at all educational levels [12].

Mathematical Knowledge is Co-Constructed and Shared

Mathematics education limits and controls epistemic validity by posing substantial challenges to the idea that mathematical knowledge is co-constructed and shared via mutually respectful interactions between a teacher and her pupils. The plan relies heavily on the existing power and privilege gaps between participants. From the perspective that sees mathematics as formed and positioned within historical contexts, institutions, and social, cultural, and discursive spaces, the question of who gets to be labelled a mathematician becomes difficult. Classroom curricula that emphasise the co-construction of knowledge might mask the complexities of the educational relationship in which students have positions of low epistemic authority. It is important to keep in mind that the lessons learned may alter their applicability. One of the many ramifications of these requirements is the need for teachers to use a more active learning strategy while teaching mathematics to students [13].

3. Conclusions

The movement in India toward making primary and secondary school attendance obligatory is a cultural touchstone with far-reaching implications for the educational system of the country. This has two major ramifications for discussions about how to improve mathematical instruction. First, it is important that all kids have an education, and second, math is an important skill to have in today's world [14]. The aim thus is to provide arithmetic instruction that is both inexpensive and engaging for all students, keeping in mind the reality of India, where very few kids have access to expensive resources. Math lessons have been found to be relevant to the child's life, and emphasis on attendance will be effective in putting the student ahead of the curve [15].

The goal of mathematics education is the same as the goal of teaching mathematics. Learning and teaching mathematics go hand in hand, even though they are often seen as two different actions. The truth is that a teacher cannot do their job without a student and that only in very unusual circumstances can instruction occur without student participation. The goals, objectives, and justifications for learning and teaching mathematics need one another. Recognizing that various social groups and people have unique and conflicting objectives is essential for a large and well-organized social activity like teaching mathematics. Goals are symbols for values and the norms of society in terms of both education and social interaction. Given that these are objectives for mathematics education, the values and

goals at stake are fundamentally connected to mathematics and its roles in both schools and the wider community. So, it is natural that people would wonder why we bother teaching and studying mathematics in the first place, what the responsibilities of teacher and student are, how society views mathematics, and what people value most. The discipline of mathematics, the student, the instructor, and the teaching environment, which includes the connection between education and society, are all its tenets, the pillars of the curriculum [16].

The mathematics curriculum cannot be designed on the assumption that all students would continue their studies through senior secondary school. Despite the best efforts by the administration, many learners may stop attending school in India after their eighth grade, even if the target to universalize primary education is achieved over the next decade. Considering the difficulties, they will undoubtedly encounter after high school, it is reasonable to wonder what, if any, advantage eight years of mathematics education may possibly provide for these individuals.

The original modernist justifications for formally teaching mathematics in schools are centered on the belief that a better rational basis for learning and social development could be established if all children were exposed to mathematical reasoning. Consequently, mathematics taught in schools became an objective historical force in political philosophy, politics, and social organisation. As a matter of fact, the efforts of researchers and educators throughout the globe are often justified by the notion of mathematics as a vital academic area and a driver of social and economic advancement. It is often seen as the way to enlightenment.

REFERENCES

1. Alam, A. (2022). Positive Psychology Goes to School: Conceptualizing Students' Happiness in 21st Century Schools While 'Minding the Mind!' Are We There Yet? Evidence-Backed, School-Based Positive Psychology Interventions. *ECS Transactions, 107*(1), 11199.

2. Doig, B., Williams, J., Swanson, D., Borromeo Ferri, R., & Drake, P. (2019). Interdisciplinary mathematics education: The state of the art and beyond.

3. Alam, A. (2022). Investigating Sustainable Education and Positive Psychology Interventions in Schools Towards Achievement of Sustainable Happiness and Wellbeing for 21st Century Pedagogy and Curriculum. *ECS Transactions*, *107*(1), 19481.

4. Chevallard, Y. (1990). On mathematics education and culture: Critical afterthoughts. *Educational Studies in Mathematics*, 21(1), 3–27.

5. Alam, A. (2022). Mapping a Sustainable Future Through Conceptualization of Transformative Learning Framework, Education for Sustainable Development, Critical Reflection, and Responsible Citizenship: An Exploration of Pedagogies for Twenty-First Century Learning. *ECS Transactions, 107*(1), 9827.

6. Zazkis, R., & Campbell, S. R. (Eds.). (2006). Number theory in mathematics education: Perspectives and prospects.

7. Alam, A. (2020). Test of Knowledge of Elementary Vectors Concepts (TKEVC) among First-Semester Bachelor of Engineering and Technology Students. *Alam, A. (2020). Test of Knowledge of Elementary Vectors Concepts (TKEVC) among First-Semester Bachelor of Engineering and Technology Students. Periódico Tchê Química, 17*(35), 477–494.

8. Nesher, P., & Kilpatrick, J. (Eds.). (1990). Mathematics and cognition: A research synthesis by the International Group for the Psychology of Mathematics Education.

9. Alam, A. (2020). Challenges and possibilities in teaching and learning of calculus: A case study of India. *Journal for the Education of Gifted Young Scientists, 8*(1), 407–433.

10. Darragh, L. (2016). Identity research in mathematics education. *Educational Studies in Mathematics, 93*(1), 19–33.

11. Alam, A. (2020). Possibilities and challenges of compounding artificial intelligence in India's educational landscape. *Alam, A. (2020). Possibilities and Challenges of Compounding Artificial Intelligence in India's Educational Landscape. International Journal of Advanced Science and Technology, 29*(5), 5077–5094.

12. Ernest, P. (1989). *Mathematics Teaching: The State of the Art*. The Falmer Press, Taylor & Francis, Inc., 1900 Frost Rd., Suite 101, Bristol, PA 19007.

13. Alam, A. (2021, November). Possibilities and Apprehensions in the Landscape of Artificial Intelligence in Education. In *2021 International Conference on Computational Intelligence and Computing Applications (ICCICA)* (pp. 1–8). IEEE.

14. Alam, A. (2022). Impact of University's Human Resources Practices on Professors' Occupational Performance: Empirical Evidence from India's Higher Education Sector. In *Inclusive Businesses in Developing Economies* (pp. 107–131). Palgrave Macmillan, Cham.

15. Alam, A. (2022, March). Educational Robotics and Computer Programming in Early Childhood Education: A Conceptual Framework for Assessing Elementary School Students' Computational Thinking for Designing Powerful Educational Scenarios. In *2022 International Conference on Smart Technologies and Systems for Next Generation Computing (ICSTSN)* (pp. 1–7). IEEE.

16. Alam, A. (2022). Social Robots in Education for Long-Term Human-Robot Interaction: Socially Supportive Behaviour of Robotic Tutor for Creating Robo-Tangible Learning Environment in a Guided Discovery Learning Interaction. *ECS Transactions, 107*(1), 12389.

Integrating Advancements in Education, and Society for Achieving Sustainability – Dimitrios A. Karras et al. (eds)
© 2024 Taylor & Francis Group, London, ISBN 978-1-032-70841-6

Importance of 'Theory' in Mathematics Education: An Attempt towards Helping India Recuperate its Lost 'Love for Mathematics'

Ashraf Alam*, Atasi Mohanty

Rekhi Centre of Excellence for the Science of Happiness,
Indian Institute of Technology Kharagpur, India

Abstract: Considering the breadth of mathematics education as a field of inquiry naturally leads to the emergence of new theoretical questions on the epistemic agency. Understanding causality, recognising patterns, establishing hypotheses, and debating the reality of mathematical statements vis-à-vis epistemological legitimacy of reasoning, analyses, and offered solutions lies at the forefront of research in mathematics education. When applying theory to mathematics education, we may examine the effects of bringing either the theory of mathematics or the theory of education into the classroom. The study of theory focuses on the methodical examination and critical investigation of primary concerns. Reasoning, analysis, investigation, and thinking are all parts of it, as are the evaluations, beliefs, and knowledge that arise from these rational and cognitive activities. The different manners and styles of incorporating fundamental theoretical ideas into mathematics education is the key focus of this paper.

Keywords: Mathematical Thinking, Philosophy of Mathematics Education, Mathematics Education in India, School Mathematics, Indian Education System, Mathematics Pedagogy

1. Introduction

The field of mathematics education has made considerable gains in recent years toward a wider grasp of what it means to know mathematically. Due to these advancements, socio-cultural theory has become more important in cognitive psychology. Numerous reform documents claim, for instance, that in order to learn mathematics in a way that lessens the impact of social or material disadvantage, teachers need to place more emphasis on students working together to build on their shared interests, on students working as part of a community of learners, on students talking to and with each other, and on less traditional teacher-student relationships [1].

---

*Corresponding author: ashraf_alam@kgpian.iitkgp.ac.in

DOI: 10.4324/9781032708461-32

Understanding mathematics, according to these various formulations, is a collaborative process rather than a property of the individual or the social environment. As a result, it is crucial to acknowledge that in these formulations, learning mathematics involves more instructor-student interactions than information transmission and consumption. Improved, integrated interaction between instructors' intents and actions and learners' attitudes toward mathematics learning and growth is a defining feature of mathematical understanding [2].

Canonical Explanations and Widely-Accepted Unverified Assumptions

There is a wealth of research on how theoretical knowledge and hands-on experience both contribute to professional success. Without a doubt, much of what kids learn in elementary school is relevant to real life. If the curriculum is redirected to attain the 'higher aims', students' problem-solving and analytical abilities will develop, and they will be better prepared to face a range of problems in life. However, there are many difficulties associated with a wide theoretical outlook. For what reasons is it crucial to study theory? What makes the theory so important in the big picture? Specifically, how do their findings contribute to the study of mathematical education? [3]

2. Unstated Beliefs, Ideological Distortions, and Unintentional Biases

In order to understand the development of the theory of mathematics education, past researchers have provided some basic responses to these challenges. First, these theoretical topics and approaches provide a

solid foundation for knowledge by providing an intellectual and well-grounded framework for study and analysis. They provide a structure for incorporating new findings into the body of existing knowledge. Secondly, having a solid theoretical basis provides credence to the claims being made and the evidence that supports them.

Thirdly, they allow individuals to go beyond canonical explanations and widely-accepted but unverified assumptions about the world, society, economics, education, mathematics, teaching, and learning [4]. These skills equip researchers to challenge authority, to see and recognise that the line between the conceivable and the impossible is not necessarily where 'common sense' claims it to be. It makes it possible to probe, argue about, and bring to light a wide range of unstated beliefs, ideological distortions, and unintentional biases. Lastly, theories provide us with the freedom to try out a few other avenues. Just as literature allows us to put ourselves in the shoes of others and see the world through their eyes and imaginations, theory can provide mathematicians with new 'pairs of glasses' through which to view the world and its institutional practises, such as the practises of teaching and learning mathematics and those of research in mathematics education [5].

3. Egalitarian Mathematical Experience

A cultural-historical approach maintains that learning happens in places where the classroom community participates within a web of economic, social, and cultural disparities to elucidate the processes that shape equitable practice. If mathematical understanding is relational and provides light on the character of an egalitarian mathematical experience, then it must be shown as a dynamic process, not a static technique or a trait of individuals, as is the case in traditional socio-cultural research.

The relational aspect combines the cognitive, social, historical, and affective facets. They do not wrap up with a look at the individual or their community. Their point of view highlights the fact that background and experience play a role in shaping our minds and behaviours [6]. If kids are taught to dread mathematics, they may never develop this essential skill [7]. It is via theory that we get the capacity to 'challenge the unquestioned'. This involves the ability to question some of the fundamental tenets of traditional mathematical education.

Math education is a contentious topic because it is at the intersection of two academic fields: educational research and public policy. Examining these disagreements objectively, analysing the ideas, and locating the underlying principles that give rise to them is a major part of the theory of mathematics education and may help bring about a better understanding of, and perhaps a compromise on, the issues at play [8].

4. Epistemological Legitimacy of Reasoning, Analyses, and Offered Solutions

The philosophy of mathematics education includes not only the goals and objectives of mathematics instruction and learning but also the philosophy of mathematics and its impact on classroom practise. It implies that we should widen our quest for philosophical and theoretical tools in order to grasp all aspects of mathematics education and learning, as well as its surroundings. We shall look at the learner, the educator, and the larger social and cultural context.

Thus, the philosophy of mathematics and mathematics education is only one of many areas of study that can be pursued. Other areas of interest include the philosophy of learning and teaching, as well as the philosophy of the surrounding society. We must also consider mathematics education as a distinct field of study in order to prove the epistemological legitimacy of the reasoning, analyses, and offered solutions, even when it is emphasised that they are just temporary. There are several problems that the 'philosophy of mathematics education' may solve [9].

Cultural historical approaches favour chance over a stable and unified subject, and consequently, they welcome post-humanistic philosophies. The unique nature of the identity offered by these methods necessitates a reappraisal of the equity imagination. The natural order is thrown into the question of understanding emancipation, which frees it from the confines of sociocultural theory and leads us to see the enlightened contemporary purpose as a chimerical and unreachable ideal. The current social justice movement claims that a shift in how we talk about and teach mathematics would lead to a revitalization of the emancipation project and the ways in which it is used in the daily lives of individual students in the classroom [10].

In schools in India, kids learn the fundamentals of mathematics. Math suffers greatly when it is reduced to formulae. Schools should place a higher value on students' conceptual knowledge of mathematical ideas than on their ability to memorise formulas and equations. Kids have an open and collaborative attitude toward mathematics. The best mathematics education is the kind that finds a place for mathematics in every aspect of a child's daily life [11].

How are Students Inevitably Entrenched in Traditional Discourses and Epistemes?

Is there a way for a philosophical framework to account for the characteristics of mathematics? Is it possible to see mathematics as both a body of knowledge and a social discipline? Can this lead to tension? What other approaches to mathematics are there in India? Which mathematical subfields do Indian mathematicians think are most important? What does all this mean, and how may it affect mathematics instruction in India?

When it comes to the public's perception of mathematics and the underlying philosophy that directs how it is taught in Indian schools, how much sway have emerging movements really had? Why do we just cover a small subset of mathematics in classes? Is there space for freshly developed mathematical ideas and concepts in the classroom, given that, much of what is taught there, has been there for centuries? What new ways of thinking about mathematics and what modifications are needed to make it useful in Indian classrooms? What values and ideals guide India's educational system and our diverse culture? [12]

How can groups of Indian mathematicians collaborate to create new areas of study? Which theories, tenets, and values do mathematicians hold dear? Is it appropriate for kids in India to be exposed to these techniques of operation, or are they only useful to a small number of highly specialised educators? How does ancient Indian philosophy relate to modern mathematical practise? Is mathematics affected by the development of new methods and the spread of electronic means of communication? Is there an emphasis on mathematical history and philosophy in classrooms? Or should math be taught as a toolbox for dealing with abstract problems?

When comparing 'research mathematics' to 'academic mathematics', what are the key distinctions? How much emphasis should be placed on calculation, proof, and modelling for pupils in India with varied degrees of mathematical ability? How and why should subjects like algebra, geometry, and calculus be taught in today's classrooms? When and how much should students learn about statistics and probability in school? [13] The difficulty centres on the link between the individual and society, and more specifically on the nature of the connection between freedom and truth.

The student-centered approach is supported by several official curricular papers and serves as the basis for standard educational policy and practise. These materials portray the teacher as a facilitator within a learning community, one who takes into account the cultural and mathematical contexts that each student brings to the classroom while also providing students with structured and meaningful exercises and making connections between mathematics and the real world. It is the role of the instructor to promote fruitful dialogue among students on key concepts and their respective definitions [14]. The challenge occurs when the educator is tasked with reconciling the ideas brought forth by the pupils with the more traditional mathematical concepts presented in textbooks. Students are inevitably entrenched in discourses and epistemes, *i.e.*, systems of thinking, that are not their own, and therefore their views are never really original [15].

5. Understanding Causality, Recognising Patterns, Establishing Hypotheses, and Debating the Reality of Mathematical Statements

Children as young as five or six ask profound questions and provide creative solutions. When it comes to formal education, mathematics is the subject most often used to teach students how to solve problems. Since this skill will be useful for the rest of one's life, it is crucial to acquire the necessary knowledge and skills in a formal setting. Making up your own humorous mathematical challenges is a great way to start off interesting discussions with others. The ability to understand causality, recognise patterns, establish hypotheses, and debate the reality of statements all need children to utilise abstractions. Teaching children mathematical concepts has the potential to improve their analytical and communicative skills. Because the building blocks of mathematics (numbers, operations, algebra, geometry, and trigonometry) provide a foundation for the abstraction, structuration, and generalisation skills necessary for success in the area, many children have an innate affinity for the subject. Learning mathematics and exploring its potential benefits learner's intuitive capacities in special ways [16].

6. Conclusions

Because of their foundation in conceptions of rationality and pure objectivity, bodies of knowledge like school mathematics tend to homogenise epistemic agency. Mathematics, in fact, represents the pinnacle of faultless, objective, and well-ordered knowledge, despite this being a hasty assumption. Regimes of truth are embedded in many forms of knowledge, including the mathematics taught in schools. What is often referred to as 'classroom mathematics' does not precede certain normalising and regulating methods.

It is made out of the stories of a privileged few and offered up as the standard for everyone. It has been developed objectively and is provided in a propositional form. Consequently, information acquired in the classroom is the result of certain norms for mathematical development, despite the fact that several definitions of school mathematics exist. Participants' capacity to think about what mathematics is learned in school is limited by these limits, which are frequently unknown to them. They also define the limits of what may be considered mathematically true. In order to fully appreciate the liberatory work done around the topic "Who can know mathematics?" we must first learn that the mathematics taught in schools is inextricably linked to the social system of power. If the contemporary subject communicates, feels, and intends via concepts and language that are not its own, how can it continue to be the centre of meaning, emotion, and purpose? The risk of systematic exclusion increases if teachers do not necessitate students' complete participation in class. To ensure that all children have equal access to a high-quality education, schools must provide their faculty with the tools they need to properly challenge gifted individuals. It is important to trace the history of mathematics education to effectively present the goal of teaching and learning mathematics. Analysis of the most pressing issues in Indian mathematics education from kindergarten through high school has informed our outlook on what needs to be done to improve the situation.

REFERENCES

1. Alam, A. (2022). Positive Psychology Goes to School: Conceptualizing Students' Happiness in 21st Century Schools While 'Minding the Mind!' Are We There Yet? Evidence-Backed, School-Based Positive Psychology Interventions. *ECS Transactions, 107*(1), 11199.

2. Doig, B., Williams, J., Swanson, D., Borromeo Ferri, R., & Drake, P. (2019). Interdisciplinary mathematics education: The state of the art and beyond.

3. Alam, A. (2022). Investigating Sustainable Education and Positive Psychology Interventions in Schools Towards Achievement of Sustainable Happiness and Wellbeing for 21st Century Pedagogy and Curriculum. *ECS Transactions, 107*(1), 19481.

4. Chevallard, Y. (1990). On mathematics education and culture: Critical afterthoughts. *Educational Studies in Mathematics, 21*(1), 3–27.

5. Alam, A. (2022). Mapping a Sustainable Future Through Conceptualization of Transformative Learning Framework, Education for Sustainable Development, Critical Reflection, and Responsible Citizenship: An Exploration of Pedagogies for Twenty-First Century Learning. *ECS Transactions, 107*(1), 9827.

6. Zazkis, R., & Campbell, S. R. (Eds.). (2006). Number theory in mathematics education: Perspectives and prospects.

7. Alam, A. (2020). Test of Knowledge of Elementary Vectors Concepts (TKEVC) among First-Semester Bachelor of Engineering and Technology Students. *Alam, A. (2020). Test of Knowledge of Elementary Vectors Concepts (TKEVC) among First-Semester Bachelor of Engineering and Technology Students. Periódico Tchê Química, 17*(35), 477–494.

8. Nesher, P., & Kilpatrick, J. (Eds.). (1990). Mathematics and cognition: A research synthesis by the International Group for the Psychology of Mathematics Education.

9. Alam, A. (2020). Challenges and possibilities in teaching and learning of calculus: A case study of India. *Journal for the Education of Gifted Young Scientists, 8*(1), 407–433.

10. Darragh, L. (2016). Identity research in mathematics education. *Educational Studies in Mathematics, 93*(1), 19–33.

11. Alam, A. (2020). Possibilities and challenges of compounding artificial intelligence in India's educational landscape. *Alam, A. (2020). Possibilities and Challenges of Compounding Artificial Intelligence in India's Educational Landscape. International Journal of Advanced Science and Technology, 29*(5), 5077–5094.

12. Ernest, P. (1989). *Mathematics Teaching: The State of the Art.* The Falmer Press, Taylor & Francis, Inc., 1900 Frost Rd., Suite 101, Bristol, PA 19007.

13. Alam, A. (2021, November). Possibilities and Apprehensions in the Landscape of Artificial Intelligence in Education. In *2021 International Conference on Computational Intelligence and Computing Applications (ICCICA)* (pp. 1–8). IEEE.

14. Alam, A. (2022). Impact of University's Human Resources Practices on Professors' Occupational Performance: Empirical Evidence from India's Higher Education Sector. In *Inclusive Businesses in Developing Economies* (pp. 107–131). Palgrave Macmillan, Cham.

15. Alam, A. (2022, March). Educational Robotics and Computer Programming in Early Childhood Education: A Conceptual Framework for Assessing Elementary School Students' Computational Thinking for Designing Powerful Educational Scenarios. In *2022 International Conference on Smart Technologies and Systems for Next Generation Computing (ICSTSN)* (pp. 1–7). IEEE.

16. Alam, A. (2022). Social Robots in Education for Long-Term Human-Robot Interaction: Socially Supportive Behaviour of Robotic Tutor for Creating Robo-Tangible Learning Environment in a Guided Discovery Learning Interaction. *ECS Transactions, 107*(1), 12389.

Integrating Advancements in Education, and Society for Achieving Sustainability – Dimitrios A. Karras et al. (eds)
© 2024 Taylor & Francis Group, London, ISBN 978-1-032-70841-6

Learning to Love Math: Philosophical Function of 'Critique' and Restructuring of Mathematics Education in India

33

Ashraf Alam*, Atasi Mohanty

Rekhi Centre of Excellence for the Science of Happiness,
Indian Institute of Technology Kharagpur, India

Abstract: Despite claims that teachers and students, as intersubjectively constituted epistemic beings, negotiate school mathematics by intersubjective means, the frameworks of school mathematics continue to function in a manner that conceals the connections between authority and understanding. Reduced cognitive agency results from the uneven distribution of cognitive resources, competence, and power in the classroom, as mathematical knowledge is alternately pushed, shared, and repressed. If we cannot keep insisting on students' rights to be treated as capable, willing, and self-reliant beings, emancipation is a distant dream. We learn to accept the transience of our own lives. Without truth and ethics, real emancipatory ideals become impossible to sustain. Problems in mathematics education emerge whenever long-held ideas are challenged. This scientific investigation attempts to find fresh ideas for the undefined mission of freedom by combining a more nuanced evaluation of relationships, one that may acknowledge the potential for inventiveness and autonomy despite social restraints.

Keywords: Mathematical Thinking, Philosophy of Mathematics Education, Mathematics Education in India, School Mathematics, Indian Education System, Mathematics Pedagogy

1. Introduction

There is a heated debate in mathematics philosophy between traditionalist mathematicians and philosophers and fallibilist and social constructivist mathematicians and philosophers. The former maintains that mathematical truths are unchanging, accumulative, and free of social issues. However, the latter claims that the fundamental nature of mathematics is social, and this means that claims of certainty, universality, and absoluteness in mathematics are culturally relative. Furthermore, they believe that mathematical concepts are just as open to change as scientific ones. This debate is a component of the larger 'Science Wars' between realists and constructivists. This debate has been going on since at least the time of the Ancient Greeks, when

*Corresponding author: ashraf_alam@kgpian.iitkgp.ac.in

DOI: 10.4324/9781032708461-33

dogmatists and sceptics butted heads. This adds value to the mathematics debate [1].

To what extent, if any, does mathematics have an ethical responsibility for the ways in which it is used and taught in the world, or is it a neutral discipline? How shall mathematics fulfil its ethical and value-laden responsibilities, if it has any? When and how should this be expressed in classrooms? [2] The subject of whether or not pure mathematics is value-laden or has any ethical aspects or duties sparked a heated debate among mathematicians and other experts, many of whom see the question as an oxymoron, which is self-contradictory and hence stupid [3].

There is a wealth of research on how learning mathematics stimulates each student's unique intellect and increases the student's access to useful mathematical tools. The place of mathematics within the educational structure is the basis for these findings. These lenses allow us to articulate our goals for mathematics education, identify the most pressing problems, and provide solutions. Since mathematics is something that may be utilised and enjoyed for the rest of one's life, it is important that children develop an interest in it from a young age, and schools are in the ideal position to achieve this [4].

2. Discourse-Generated Relationships are the New Locus of Power

Foucault proposed a contemporary ethics of the self that has the potential to free people by helping them create new kinds of subjectivity through the rejection of a sort of personality that has been imposed on us

for millennia. Foucault suggests that, rather than accepting externally imposed moral requirements, progressive politics may be best served by an ethic of who we are to be and what it is, therefore, feasible for us to become. He elaborates by saying that how a person becomes a topic is crucial. His contemporary ethics of the self is based on the idea of criticism as its enabling condition, which he proposes as the foundation for an ontology of the self [5].

Learners' mathematical identities are formed in the classroom and beyond via the use of technologies of the self, through which they see themselves to be as mathematical thinkers, speakers, and doers. The ways in which she constructs her identity and her experiences in mathematics are influenced by the methods and strategies she uses to do so. In addition, they affect her feelings and actions. Foucault provides an explanation for human experiences via the concept of governmentality, in which domination and resistance are not seen as ontologically distinct things but rather as contradictory consequences of the same power relations [6].

Discourse-generated relationships are the new locus of power, superseding membership in institutions and social groupings. Discourse may be used to thwart the opposition's plan or to carry it out, as it is both a means to and an effect of power. The idea that people may shape their own lives as active agents, independent of external influences, is derived from the concept of governmentality. This means that although governmentality uses the individual as a tool for social control, it also gives people the tools they need to push back against this kind of authoritarian individualism. The term 'self' conveys an understanding of the student as an individual who is always being influenced by their environment in a wide range of subject positions, some to a larger extent than others, but over all of which the student has control over [7].

3. Cognitive Resources, Competence, and Power Relations in Mathematics Classrooms

To develop one's own moral code, one must first recognise the difference between the 'ethics' imposed from the outside and the 'morals' built from the inside. Foucault argues that, in order to understand how individuals constitute themselves as moral subjects of their own actions, we must examine the practises that constitute, define, organise, and instrumentalize the strategies which individuals in their liberty can have, regarding each other [8].

When seen in the context of human history, formal educational systems are relatively recent. Prior to that time, Western schools were often associated with certain religions. The ultimate goal of these institutions was to produce educated clergy. For most people, mathematics meant 'the various numbers and the various shapes and sufficient astronomy to help in determining the dates of religious rituals'. Nevertheless, schooling was widely available in India. Mathematics and astronomy are emphasised as cornerstones of the curriculum. The time of day for religious ceremonies and sacrifices was calculated using astronomical data. In order to build several 'havan kunds' and sacrificial altars, students needed a solid foundation in geometry. The educational system underwent radical change with the arrival of British

colonists. To run the Empire efficiently, it was arguably necessary to introduce a Western educational system to train Indians to think and act in conformity with western ideals [9].

To achieve what ends, shall 'math' and 'math education' be taught? When it comes to mathematics, what exactly should students be learning? Is it possible to achieve fully, the goals of teaching mathematics? What are they trying to accomplish by teaching mathematics? Who benefits, and who suffers, as a result of acquiring mathematical knowledge? Are uniform mathematical goals best for all students, or should they vary by cohort? How do we decide which categories to use and what objectives to pursue? What standards shall it be based on? What are the connections between math education's goals, methods, and historical, social, and cultural contexts? In what ways might various sets of objectives be supported by common principles? When considering the larger picture of society and the purposes of education, what benefits can mathematics provide?

4. Are there Costs and Collateral Damage linked to Combating Racism and Sexism in Mathematics Classrooms?

How can critical citizenship, gender equality, racial equality, economic equality, and the inclusion of people with disabilities be promoted via mathematics education? Is it really possible to have a mathematics curriculum that actively combats racism and sexism? What may this mean for the future of mathematics instruction? What does mathematics do to the masses, and how does it fare in different cultural contexts? How does this affect classroom instruction? How do math and society fit together? How useful is it, exactly? Which of these goals is most transparent and genuine? Which features of mathematics are potentially harmful or impossible to detect? How pervasive are the societal effects of mathematical metaphors? What are the social implications of it? To whom is mathematics answerable? Is math a limitless good for society? In addition to the advantages mathematics brings to society via its applications in science, technology, and business, among others, are there costs and collateral harm associated with giving it such a high priority in the classroom? Is it possible to mitigate the problems, if at all there exist, by providing more or better math instruction?

How students behave in math classes and how they interpret their own experiences, actions, and identities are increasingly being assessed through the lens of autonomy. This means that the assessment places a premium on the students' standards of behaviour that each student chooses consciously and intentionally in order to develop herself as a piece of art in the classroom. Teachers, curriculum developers, researchers, and other professionals, despite their uniform outward appearance, are, according to the Foucauldian theory of learning, essential products of the practises to which they are exposed. It is the actions, aspirations, objectives, and investments of others that give this character a feeling of identity [10].

5. Theorising Identification

Finding out the truth about ourselves is so essential, but it is not something that comes

naturally. All these theories promote an incoherent view of the subject constitution through time. None of them ever has the chance to really get into the issue. When asked to describe who they are, they say that it is something that is always shifting and evolving. By theorising identification in this way, they may set themselves apart from the Cartesian effort to find a basic position for the subject that is thought to correspond to a rational 'man'. They may sidestep the causal chain that normally underpins the maturation of a consistent sense of self. Further, it provides a framework for explaining the dynamics between teachers and low-performing students in the classroom [11]. Most of the changes to mathematics curricula have taken place during the last thirty to forty years. This is due to the fact that the effects of the current technological revolution on society are comparable to those of the industrial revolution. As a result, educational goals are being re-examined in light of modern technology, and this shift is expected to continue. Many time-honored approaches are becoming obsolete, as shown by a cursory analysis of the emergence of new mathematical areas in response to the needs of emerging technologies [12].

Keeping up with modern technology has an impact on how we teach arithmetic. Furthermore, mathematics curricula are very similar across countries, which means that any new developments in one region are typically adopted by others. Examples include the widespread adoption of modern mathematics in India. Later, like the rest of the world, the country lost interest in ground-breaking mathematics. Curriculum development is not a static process of looking at how it has evolved throughout time, rather it is the manifestation of life, replacing the selection and organization of pedagogical material as the primary goals of the classroom [13].

Math Wars: Propensity to Comply or Rebel

The debate over mathematics education's ultimate purpose has gained increased visibility. For example, in the 1990s and 2000s, mainstream mathematicians and mathematics education experts in the United States engaged in a debate known as the 'Math Wars' about the goals and methods of mathematics education. Is it feasible to find solutions and reach compromises, or are these disagreements inevitable? How useful are the findings from studies and international comparisons for determining the success of various methods for achieving a goal? A student's creative self-fashioning is focused on exploring new ways of existing in the world rather than re-establishing a fundamental mathematical identity [14]. The independence that arises from the constellation of inequalities makes up a history of shaky alliances, resistance, domination interactions, and shifting interest configurations. It causes one to reassess long-held beliefs and practices. The end result is an ethics of personal accountability for our words and the political methods they influence, as well as the ways in which we relate to ourselves and hence our propensity to comply or rebel. In a mathematics classroom, a student's sense of self as a learner is always evolving as she strives to find a middle ground between her own unique learning style and the classroom's established norms for behaviour and achievement [15].

A person's mathematical identity is not something that can be chosen with complete intention, although consciousness may still play a part in the process. It calls for an awareness of the constraints of the classroom as they now stand, including the presence of punishments and recommendations. What is most important is that the student conducts an ethical self-analysis of typical classroom activities, identifying problems and organising the tangled complexity of educational linkages while keeping in mind the option of resistance. On the one hand, cultural interpretations exist because a student's mathematical identity is entrenched inside and determined by the classroom [16].

In 1937, when Gandhiji initially called for compulsory elementary education for all children, the Zakir Hussain committee was formed to further this cause. A solid grasp of mathematics should serve as the cornerstone of any good education. Every kid has to be able to accomplish the kind of simple arithmetic he or she will use in everyday life or for fun. When it was established in 1952, the Secondary Education Commission pushed hard to make math a compulsory course for high school students.

6. Conclusion

With the advent of automation and cybernetics in this century marking the beginning of the new scientific industrial revolution, the National Council for Educational Research and Training (NCERT), in accordance with the National Education Policy (NEP), emphasized that it is all the more imperative to devote special attention to the study of mathematics and that students should be encouraged to investigate mathematical concepts.

In 1986, legislation was passed creating the National Policy on Education. It emphasised that children's exposure to mathematics should be seen as the means through which they acquire the capacity for rational thought, analysis, and expression. It is more than just another course of study; critical thinking skills transfer across all academic fields. This view is shared by the authors of the National Core State Standards Examinations (NCSSE) 2000 report. Despite these repeated calls for reform, mathematics instruction in today's India is very similar to how it was ten years ago.

REFERENCES

1. Alam, A. (2022). Positive Psychology Goes to School: Conceptualizing Students' Happiness in 21st Century Schools While 'Minding the Mind!' Are We There Yet? Evidence-Backed, School-Based Positive Psychology Interventions. *ECS Transactions, 107*(1), 11199.

2. Doig, B., Williams, J., Swanson, D., Borromeo Ferri, R., & Drake, P. (2019). Interdisciplinary mathematics education: The state of the art and beyond.

3. Alam, A. (2022). Investigating Sustainable Education and Positive Psychology Interventions in Schools Towards Achievement of Sustainable Happiness and Wellbeing for 21st Century Pedagogy and Curriculum. *ECS Transactions, 107*(1), 19481.

4. Chevallard, Y. (1990). On mathematics education and culture: Critical afterthoughts. *Educational Studies in Mathematics, 21*(1), 3–27.

5. Alam, A. (2022). Mapping a Sustainable Future Through Conceptualization of Transformative Learning Framework, Education for Sustainable Development, Critical Reflection, and Responsible Citizenship: An Exploration of Pedagogies for Twenty-First Century Learning. *ECS Transactions, 107*(1), 9827.

6. Zazkis, R., & Campbell, S. R. (Eds.). (2006). Number theory in mathematics education: Perspectives and prospects.

7. Alam, A. (2020). Test of Knowledge of Elementary Vectors Concepts (TKEVC) among First-Semester Bachelor of Engineering and Technology Students. *Alam, A. (2020). Test of Knowledge of Elementary Vectors Concepts (TKEVC) among First-Semester Bachelor of Engineering and Technology Students. Periódico Tchê Química, 17*(35), 477–494.

8. Nesher, P., & Kilpatrick, J. (Eds.). (1990). Mathematics and cognition: A research synthesis by the International Group for the Psychology of Mathematics Education.

9. Alam, A. (2020). Challenges and possibilities in teaching and learning of calculus: A case study of India. *Journal for the Education of Gifted Young Scientists, 8*(1), 407–433.

10. Darragh, L. (2016). Identity research in mathematics education. *Educational Studies in Mathematics, 93*(1), 19–33.

11. Alam, A. (2020). Possibilities and challenges of compounding artificial intelligence in India's educational landscape. *Alam, A. (2020). Possibilities and Challenges of Compounding Artificial Intelligence in India's Educational Landscape. International Journal of Advanced Science and Technology, 29*(5), 5077–5094.

12. Ernest, P. (1989). *Mathematics Teaching: The State of the Art*. The Falmer Press, Taylor & Francis, Inc., 1900 Frost Rd., Suite 101, Bristol, PA 19007.

13. Alam, A. (2021, November). Possibilities and Apprehensions in the Landscape of Artificial Intelligence in Education. In *2021 International Conference on Computational Intelligence and Computing Applications (ICCICA)* (pp. 1–8). IEEE.

14. Alam, A. (2022). Impact of University's Human Resources Practices on Professors' Occupational Performance: Empirical Evidence from India's Higher Education Sector. In *Inclusive Businesses in Developing Economies* (pp. 107–131). Palgrave Macmillan, Cham.

15. Alam, A. (2022, March). Educational Robotics and Computer Programming in Early Childhood Education: A Conceptual Framework for Assessing Elementary School Students' Computational Thinking for Designing Powerful Educational Scenarios. In *2022 International Conference on Smart Technologies and Systems for Next Generation Computing (ICSTSN)* (pp. 1–7). IEEE.

16. Alam, A. (2022). Social Robots in Education for Long-Term Human-Robot Interaction: Socially Supportive Behaviour of Robotic Tutor for Creating Robo-Tangible Learning Environment in a Guided Discovery Learning Interaction. *ECS Transactions, 107*(1), 12389.

Integrating Advancements in Education, and Society for Achieving Sustainability – Dimitrios A. Karras et al. (eds)
© *2024 Taylor & Francis Group, London, ISBN 978-1-032-70841-6*

34

Accelrating Climate Protection towards Sustainability

Ritika Malik

Bharati Vidyapeeth (Deemed to be) University, Pune,
Institute of Management and Research, New Delhi

Naveen Nandal*

Assistant Professor, Sushant University, Gurugram

Parul Dhaka

Research Scholar, Sushant University, Gurugram

Indu Rani

Associate Professor, Bharati Vidyapeeth (Deemed to be) University, Pune,
Institute of Management and Research, New Delhi

Abstract: The increase in Industrialization in India, not only leads to the development of the nation in various departments but also serves as a major threat to the environment and climate. The effluents and emissions released by the industries after the completion of the production process of the commodities, lead to the depletion of the quality of the living environment, hence, deteriorating the climate. Both the industries, individual households, and firms also do not follow sustainable use of resources, which leads to an augmenting of the industrial and domestic mullock. The main focus of sustainably accelerating climate protection is to provide congenial resources for our generation without compromising the comfort of the coming generations. Researches show that we use our resources at a pace of approximately 1.75 times the natural replenishing capacity of the earth. Hence, the current resources used prevalently, if utilized at the same pace, will not last long, leaving nothing for future generations.

Keywords: Sustainable resources, economically profitable, environmentally friendly, conservation of resources, renewable resources

1. Introduction

The land-based ice like glaciers and ice sheets are experiencing an increased rate of melting due to global warming. Usually, the glaciers and ice sheets melt in the warmer months and freeze back in the colder months, but, global warming has resulted in an overall rise in

*Corresponding author: Naveennandal@sushantuniversity.edu.in

DOI: 10.4324/9781032708461-34

worldwide temperatures, hence resulting in a lesser amount of melted ice returning to its solid form. Additionally, due to geothermal expansion, the average temperature of water bodies is rising constantly, increasing water levels worldwide. This disproportionate increase in the sea levels, sets the ball rolling for further dire consequences, including an increased risk of hurricanes, tsunamis, storms, and loss of aquatic species which destabilizes the coastal ecosystem, which will affect the livelihood of the coastal communities, and thereby influencing the country's returns from coastal commerce. The increase in the water levels also results in permanent flooding i.e., inundation of low-lying areas like deltas, increasing salinity of estuaries with fertile soil results in agricultural loss, and loss of marshes and wetlands which are not only a valuable resource to the Indian economy but are also national treasures, having precious species of protected flora and fauna. The deteriorating climate not only influences aquatic life but also endangers various other animals and plant species on land including penguins, seals, polar bears, turtles, and deer.

The health issues affecting us as a result of climate change are so great that the United Nations has stated that climate change is by far the worst threat to human health. 90% of people breathe unclean air polluted with highly toxic emissions and effluents released as a result of the combustion of fossil fuels, coal, petroleum, and other waste products. Air pollution causes deadly diseases including Asthma, Bronchitis, Lung cancer, Heart disease, Stroke, etc. Clean air once available abundantly has now become a luxury. Reverting the present situation to the original form will take a great deal of time and profuse effort. The increase in health issues results in a fall in the productivity of the general population, increased mortality, and lower birth rates, hence directly influencing our country's economic output and wealth. That being said, India's population is both its biggest strength and weakness. To use it for our joint profit is solely in our hands.

The air quality index can be increased by reducing the emission of toxic gases and effluents. Reducing individual sources of pollution is probably the most effective way to increase the air quality index at a quick pace. Reducing the amount of factory waste and reducing their harmful impact on the environment by properly treating the industrial by-products before letting them out in the open disposal landfills will make a drastic difference. Processes such as distillation, vitrification, ozonization, coagulation, membrane filtration, ion exchange, installation of wet scrubbers and electrostatic precipitators, and biological treatments where organic substances are consumed by microorganisms such as fungi, ciliates, rotifers, and bacteria, hence, further degenerating the waste before releasing it. Constructing industries far away from settlements reduces the risk of industrial effluents harming the residents. Building a 'green belt' of trees around the industry will help reduce the impact the emissions have on the surrounding environment.

The usage of eco-friendly products, which are modern, cutting edge and cost-efficient, more than being good for us, does even better good for our planet. The eco-friendly products being free from BPA, lead, sulfates, and other harmful substances reduce the risk of diseases including diabetes, auto-immune diseases, reproductive disorders, stunted growth, nervous disorders, etc. These products are easily degradable and return goodness back to the earth in the form of fertilizers after they are used for various needs by us. These products are not only low maintenance but also reduce the use of various other resources such as water for their production. Various products such as kitchen towels, toothbrushes, clothes, cups, utensils, etc, have already been made in compostable forms. Replacing plastic goods with such products is bound to make great changes in the improvement of our climate.

Abandoning the usage of single-use plastic products which adds great numbers to the average waste production, is of prime need. Replacing single-use plastics with cloth, cardboard, and other easily degradable organic substances is of prime need. The idea of Bamboo replacing plastic is a quick and cost-efficient solution to tackle the increase in the production of single-use plastic products like cups, straws, and polythene bags. Replacing polythene bags with cloth bags is the simplest solution we can opt for right now.

With the lesser incomings of conventional energy resources like coal, oil, gas, and petroleum, the issue of the energy deficit in India is increasing. With limited reserves of non-renewable energy resources, beginning to use renewable resources of energy in India, such as solar energy, hydropower, wind energy, biogas, and geothermal energy, etc, is the only appropriate solution to this problem. India is a country with an abundant supply of renewable resources as a result of its geography and topographical location. Using these resources to their full potential is of utmost importance in such dire times. Renewable resources are eco-friendly, with lesser pollution, and can be used in the long run with efficient usage. It makes India more powerful not only with its independent usage of resources, and also economically as it provides more employment opportunities.

These resources are not only clean, safe, and efficient, but also, economically profitable.

2. Literature Review

In A.A. Rosenberg's essay Achieving Sustainable Use of Renewable Resources, he states that sustainable use is a widely accepted goal for managing renewable resources. It "meets the needs of the present without compromising the ability of future generations to meet their own needs". However, natural variability, scientific uncertainty, and conflicting goals or values can make it difficult to achieve sustainable resource use. In a recent Policy Forum article, Ludwig et al. argued that Sustainability should not be relied upon and populations will inevitably (and often irreversibly) be over utilized. We argue that the history of fisheries management provides both good examples of sustainable resource use and lessons for future improvement. Our conclusions may be broadly applicable to other renewable resources. In particular, we believe that there is a sound theoretical and empirical basis for sustainable use, that overfishing is neither inevitable nor necessarily irreversible, and generally that the tradition of open access management systems, combined with risky management decisions under uncertainty, has led to major challenges to achieving sustainability. It's a hindrance. It concludes that the sustainable use of renewable resources is feasible.

Problems and Solutions Although the idea of sustainability is universal and there are many examples of sustainable use of fish stocks, there are many examples of overfishing. In the United States, approximately 45% of the 156 population groups for which stock status assessments are available are currently classified as overfished. In European waters,

59% of the 78 strains were classified as overfished. But only a few of these have been exploited in such a way that profitable fishing is no longer possible. Uncertainties about the ecological processes that control stock status and population dynamics undoubtedly make scientific advice inadequate in some cases. Because there were no, For example, a post hoc analysis of declines in pelagic fish populations in Sätersdal contrasts a set of consensus scientific advice with actual management decisions. Managers consistently allowed more catches than agreed scientific advice. Haddock, Georges Bank, of New England, maintained a relatively stable domestic fishery from the beginning of the 1930s until the arrival of foreign fishing fleets in the 1960s. Contrary to scientific advice, stocks declined to much lower levels when harvest rates were allowed to increase significantly. The decline in numbers is not necessarily irreversible. There are many documented examples of stocks accumulating after depletion and subsequently decreasing fisheries. Among the most dramatic examples is the widespread recovery of fish stocks recorded in the North Atlantic after fishing restrictions during World War I and II. The fish population was too large to be wiped out by harvesting. As a result, the onus has fallen on resource managers to demonstrate that regulation is necessary to protect renewable resources. In addition, de facto open access systems prevailed, with unrestricted access to fisheries. In open-access systems, harvest costs are inflated as participants compete for a limited supply of resources. The inevitable result is that the economic value of resources is wasted. The solution to this problem is to recognize that property rights must be clearly defined and that rights carry duties and responsibilities. With open access,

there is no ownership over resources, and the potential for overdevelopment is much higher. Significant progress has been made in addressing factors that have historically threatened sustainable use. Management in most developed countries has evolved into a system of controlling access to fish stocks, but the process is far from perfect. Additionally, countries such as New Zealand, Canada, Iceland, Australia, and the United States have introduced systems allowing individual quotas in some fisheries. These quotas are transferable, eliminating competition among fishermen for resources. These systems can facilitate more economically viable fisheries. The use of freely available resources is a major problem for developing countries, which currently account for more than half of the world's fish catch. The growth of the Indian renewable energy sector is also important for renewable power generation. Renewable energy is renewable and of course replenished energy obtained from natural sources such as solar, wind, rain, tidal and geothermal energy. Among distributed systems, the growth rate of solar home lighting systems was 300%, solar lanterns 99%, and solar water pumps 196%. This is tremendous growth in the renewable energy sector, primarily for applications that have been powered only by large utility companies. Several large-scale projects have been proposed, with the 35,000 km^2 area of the Thar Desert set aside for enough solar energy projects to generate between 700 and 2,100 gigawatts. Renewable energy systems are seen as a key application to electrify 20,000 remote and non-electrified villages and settlements by 2007 and all households in such villages and settlements by 2012. increase.

In India, the annual growth rate of renewable energy has been around 22%

over the past decade. Production from unconventional sources in India in 2013–2014 was around 53.22 billion units, with the largest contributors being wind and solar at 31.26 billion units and 3.35 billion units respectively. units (Barpatragohain, 2015). Alternative energy also pays for investment in the form of carbon credits for clean development mechanisms. Wind and solar energy do not generate waste, so no investment in waste management is required during the life cycle of such power plants. Other potential resources include hydropower, tidal power, geothermal power, and biomass/biowaste.

In their look at on Water assets of India Rakesh Kumar, R. D. Singh, and K. D. Sharma have found that Water assets of a rustic represent one all its crucial assets. India gets annual precipitation of approximately 4000 km^3. The rainfall in India suggests very excessive spatial and temporal variability and the paradox of the scenario is that Mousinram close to Cherrapunji, which gets the very best rainfall with inside the world, additionally suffers from a quick age of water at some stage in the non-wet season, nearly every 12 months. The overall common annual waft in step with 12 months for the Indian rivers is anticipated as 1953 km^3. The overall annual replenish able groundwater assets are assessed as 432 km^3. The annual utilizable floor water and groundwater assets of India are anticipated as 690 km and 396 km^3 in step with 12 months, respectively. With a speedy developing populace and enhancing residing requirements, the stress on our water assets is growing and in step with the per capita availability of water, assets are lowering day through day. Due to spatial and temporal variability in precipitation in the US. faces the hassle of flood and drought

syndrome. Overexploitation of groundwater is main to the discount of low flows with inside the rivers, decline of groundwater assets, and saltwater intrusion in aquifers of the coastal regions. Over-canal irrigation in several command regions has ended in waterlogging and salinity. The first-rate of floor and groundwater assets is likewise deteriorating due to growing pollutant masses from factor and non-factor sources. The weather is predicted to influence precipitation and water availability.

Furthermore, D.L. Kaplan says that A huge variety of clearly going on polymers derived from renewable sources are to be had for cloth packages. Some of those, consisting of cellulose and starch, are actively utilized in merchandise today, at the same time as many others continue to be underutilized. With the speedy development in the knowledge of essential biosynthetic pathways and alternatives to modulate or tailor those pathways thru genetic manipulations, new possibilities for the usage of polymers from renewable sources are being considered. These biopolymers are derived from various sets of polysaccharides, proteins, lipids, polyphenols, and distinctiveness polymers produced with the aid of using bacteria, fungi, plants, and animals. Some of those polymers have currently been reviewed.

Purpose: The study has been conducted concerning the dire need and absolute necessity to expedite the process of climate protection.

3. Objective of the Study

The objective of this study is to find out sustainable methods to accelerate climate protection and reduce the risk of our environment's vulnerability to climate change in the goodwill of our country's environment and economy and decrease the possibilities of any future perils.

4. Methodology

The objective of the study is to review the impact of the changing climate on the environment, the ill effects of the irresponsible use of resources on the economy and climate of our country, and the benefits of sustainably managing and using the available renewable resources in the country.

Theoretically, the sun may appear a perfect electricity source, as it's far unfastened and definitely limitless. The sun's radiation attaining the earth's floor in three hundred and sixty-five days offers greater than 10,000 instances of the world's every year electricity desires. Furthermore, harnessing simply one sector of solar electricity that falls at the world's paved regions may want to meet all modern-day international electricity desires comfortably. India is densely populated and has excessive sun insolation, a perfect mixture for the use of solar energy. Because of its area between the Tropic of Cancer and the Equator, India has a median annual temperature starting from 25°C – 27.5°C.

It is the most reliable of all renewable energy sources. Small hydropower development in India started at about the same pace as the world's first hydroelectric power station was built in Appleton, USA in 1882 (Dhillon and Sastry, 1992). His KW installation at Sidrapon (Darjeeling) in 1897 was the first installation in India. Other power plants were Shivasamundrum (2000 kW) in Mysore and Buri Singh (40 kW) in Chamba in 1902, Galogi in Mussoorie (3000 kW) in 1907, Jubbal (50 kW) in 1911, Chaba (1750 kW) in Shimla in 2008 (Palit, 2003; McKansey,

2008). These plants were mainly used for important urban lighting, which still works today. The Ministry of New and Renewable Energy is responsible for developing small hydropower (SHP) projects with a capacity of up to 25 MW. The estimated domestic power generation potential from such power plants is around 20,000 MW (MNRE official website). Most of the possibilities are in Himalayan states in the form of river projects, with irrigation canals in other states. The SHP program is currently driven primarily by private investment. Projects are generally economically viable and the private sector has shown strong interest in investing in SHP projects. The feasibility of these projects increases as project capacity increases. The ministry's goal is to use at least 50% of the country's potential over the next decade. It is recognized that small-scale hydropower projects can play an important role in improving the country's overall energy scenario in remote and inaccessible areas (Kumar, 2008). The ministry encourages the development of small hydropower projects in both the public and private sectors. Grid interactive and distributed projects are considered as well. Hydropower projects are generally classified into two segments: small hydro and large hydro. In India, hydropower projects with a capacity of up to 25 MW are categorized as Small Hydropower (SHP) projects. The Indian Government's Ministry of Energy is responsible for large-scale hydropower projects, while the ministry of renewable energy (up to 25 MW) is vested with the Ministry of New and Renewable Energy.

India has great potential to harness this vast renewable and sustainable resource for power generation. India has a long coastline of about 7500 km and about 336 islands in the Bay of Bengal and Arabian Sea, with estuaries and bays where the tides are strong enough to turn turbines and generate electricity. The maximum tidal range is 11 m and 8 m, and the average tidal range is 6.77 m and 5.23 m, respectively, at Kambey Bay and Kutch Bay in Gujarat on the west coast (Barpatragohain, 2015; Ravindran and Raju, 1997). The Ganges delta in Sundarbans is about 5 m high and has an average tidal range of 2.97 m. The identified economic potential is about 8000 MW, of which about 7000 MW is in Kambay Bay and 1200 MW is in Kutch Bay in Gujarat. is in About 100 MW in the Ganges delta in the Sundarbans region of West Bengal (McKinsey, 2008). The total wave energy potential available in India is estimated at around 40,000 MW. These are preliminary estimates. However, this energy is less intense compared to north and south latitudes. In 2000, NIOT Goa initiated a program to conduct research on technologies for extracting high-quality, clean drinking water, and energy from the ocean.

Geothermal power plants generating more than 10000 MW of electricity are in operation in 24 countries around the world. Moreover, geothermal energy is used directly for heating in at least 78 countries (Axelsson et al., 2005). The largest producer of this energy is the United States, producing about 3086 MW of electricity (Monastero, 2002). India has great potential to become a major provider of geothermal energy. However, power generation from geothermal resources is still in its infancy in India. There are 340 geothermal hot springs identified in India. Most of them are in the low surface temperature range of 37°C to 90°C suitable for direct heating applications (Parikh, 2009). These springs are grouped into seven geothermal regions: Himalayas (Puga, Chumatang), Sahara and Kambey Basins, Son-Narmada-Tapi Lineament Belt, West Coast, Godavari Basin, and Mahanadi Basin. Some of the prominent geothermal resources are Puga Valley and Chumatang in Jammu and Kashmir, Manikaran in Himachal Pradesh, Jalgaon in Maharashtra, and Tapovan in Uttarakhand. A new location of geothermal energy has also been discovered at Tattapani in Chhattisgarh. To harness the country's geothermal energy, the Ministry of New and Renewable Energy (MNRE) has supported exploration activities and resource assessment research and development for the past 25 years. This includes forming expert groups, working groups, core groups, and committees, and providing financial support for such projects and resource assessments. MNRE aims to deploy 1000MW of geothermal capacity in the early stages by 2022. A public stock assessment is planned for 2016–2017.

5. Conclusion

To conclude, though all the world leaders express their concerns, and raise their voices on the deteriorating climatic conditions worldwide, and global warming, India has been pioneering in a positive way in fighting for this and planning and executing to really reach for a solution. During the recent visit of the United Nations General Secretary to India, the Government launched "Mission Life" which helps in fighting the climate crisis. With more such positive steps to come in near future, no wonder India will lead the world and be a model to the world nations in improving the climate and environment thus saving natural resources for future generations.

Sustainably using the easy-to-obtain renewable resources present in our country will greatly increase the value of our economy and the quality of our environment.

It will also reduce the burden; the usage of exhaustible resources poses on our economy. Renewable resources being easy to operate and use conveniently, serve as a very practical source of power that has very high functional value. Hence, it is a very ergonomic source of power considering the towering populations of the future and the following increase in the need for power sources to feed the various needs of our society.

REFERENCES

1. Chaturvedi, R. ., Tiwari, R., & Ravindranath, N. (2008). Climate change and forests in India. International Forestry Review, 10(2), 256–268. doi:10.1505/ifor.10.2.256
2. Das, S., Lee, S.-H., Kumar, P., Kim, K.-H., Lee, S. S., & Bhattacharya, S. S. (2019). Solid waste management: Scope and the challenge of sustainability. Journal of Cleaner Production, 228, 658–678. doi:10.1016/j.jclepro.2019.04.32
3. I K, M., Tiwari, R., & Ravindranath, N. H. (2010). Climate change and forests in India: adaptation opportunities and challenges. Mitigation and Adaptation Strategies for Global Change, 16(2), 161–175. doi:10.1007/s11027-010-9261-y
4. Lazarus, M., & van Asselt, H. (2018). Fossil fuel supply and climate policy: exploring the road less taken. Climatic Change. doi:10.1007/s10584-018-2266-3
5. Nandal. et al., U.S – China Trade War & it's Implications on India, Korea, Review of International Studies, Volume 15 | Special Issue 05 | Apr 2022.
6. Nagel, M., Stark, M., Satoh, K., Schmitt, M., & Kaip, E. (2019). Diversity in collaboration: Networks in urban climate change governance. Urban Climate, 29, 100502. doi:10.1016/j.uclim.2019.100502
7. Norman, C., DeCanio, S., & Fan, L. (2008). The Montreal Protocol at 20: Ongoing opportunities for integration with climate protection. Global Environmental Change, 18(2), 330–340. doi:10.1016/j.gloenvcha.2008.03.003
8. Osberghaus, D., & Demski, C. (2019). The causal effect of flood experience on climate engagement: evidence from search requests for green electricity. Climatic Change. doi:10.1007/s105 84-019-02468-9
9. Viebahn, P., Höller, S., Vallentin, D., Liptow, H., & Villar, A. (2011). Future CCS implementation in india: A systemic and long-term analysis. Energy Procedia, 4, 2708–2715. doi:10.1016/j.egypro.2011.02.172
10. Zorpas, A. A. (2020). Strategy development in the framework of waste management. Science of The Total Environment, 137088. doi:10.1016/j.scitotenv. 2020.137088

Integrating Advancements in Education, and Society for Achieving Sustainability – Dimitrios A. Karras et al. (eds)
© 2024 Taylor & Francis Group, London, ISBN 978-1-032-70841-6

35

Hate Crime against LGBTQ+: Mental Health Impacts

Niharika Manish Suri
Student, SOB, Sushant university, Gurugram

Nisha Nandal, Naveen Nandal*
Assistant Professor, Sushant University

Anuradha
Assistant Professor, Bharati Vidyapeeth University, New Delhi

Abstract: We live in a world where LGBTQ+ people are treated as 2nd class citizens, and transgender people are the 3rd gender. Violence against the LGBTQ+ is not a new phenomenon, over some time there have been increased levels of stigma, discrimination, inequality, violence, and victimization of LGBTQ+ people. Needless to say, being a victim of crime has psychological and mental impacts. Especially when acts of violence, aggression, and assault are directed toward someone as a result of who they are, what they like, and how they appear. This review paper presents a review of Hate crimes and threats faced by LGBTQ+ people, and how it affects their mental health.

Keywords: LGBTQ+, Marginalization, Homophobia, Stigma

1. Introduction

Violence, Assault, and Homicide go back to ancient civilizations, but when all these wrongdoings are perpetrated against a particular group of people it turns into Hate Crime. Hate crimes send messages to a member of the victim's group that they are unwelcome and unsafe in the community, victimizing the entire group and decreasing feelings of safety and security. Hate crimes are cruel acts that intend to harm an individual or a group of individuals because of their gender, physical or mental handicap, race, nationality, sexual orientation, religion, or affiliation with another racial or ethnic group, real or imagined.

For a long time, India was oblivious of the growth in hate crimes and the growing social stigma towards the LGBT population, which contributed to the increased victimization of these individuals. LGBT individuals are becoming one of the groups most at risk from

*Corresponding Author: Naveennandal@sushantuniversity.edu.in

DOI: 10.4324/9781032708461-35

hate crimes as a result of rising social taboo. based on the FBI's report, Since 2006, there have been more than 100,000 reports of hate crimes against LGB people. The third most common basis for hate crimes worldwide is thought to be sexual orientation. Ironically, according to many who research hate crimes, one of the primary causes of the rise in hate crime rates may be the growing acceptance of LGBT persons in society today.

The LGBTQ+ community has been a theme of volitation since its existence. They frequently experience violence directed toward their sexuality, gender identity, or gender expression. Unfortunately, we live in a society where it's easier to live a fake life than to be your true self. The fear of isolation, rejection, stigma, and marginalization is wholesome, primarily the life of 97% of LGBTQ+ individuals. The LGBTQ+ individuals are portrayed to be an 'inferior group', they are contemplated as a threat to "traditional" structures, such as marriage, kids, and the family. This portrayal of LGBTQ+ is only based on misapprehension, fallacy, and a delusional idea of culture and tradition that has been passed on from generations. Survivors of any violent criminal offense endure physical pain and psychological distress, leading to anxiety, depression, suicidal thoughts, etc. When an LGBTQ+ individual experiences any assault there is a bloody influx of subliminal messaging around, an incredible amount of self-loathing, guilt, feeling 'dirty' about themselves, their sexual orientation, and sexual preferences. Victim's physical, mental and emotional well-being may suffer from hate incidents, just like from hate crime. Victim's feel targeted, ashamed, and scared that they'll be hurt again, there have been times LGBTQ+

hate crime survivors tried seeking support and services from police, shelters, hospital, and ls, and rape crisis center but they were either denied help or were feeling threatened and hesitant to be "outed". They are also apprised incessantly that no one will trust them, nobody will believe or listen to them what they have to say because they are 'faggots', 'minority', 'dyke'. They start to lose trust in any person around them, especially who is a part of the offender's group, they get into the "us vs. them" frame of mind. Time and again, the survivors are forced to believe that they "deserve" the assault/abuse. Often cases of transphobia were disregarded, claiming an LGBTQ+ person is just befuddled and going through a 'gender identity crises. They refuse to acknowledge the aspect of your identity that the hate crime targeted. They give up some of the activities they used to enjoy because they don't feel safe anymore.

These feelings and the fallout from a hate crime may increase the likelihood that the victim's mental health will deteriorate. Hate crimes may result in depression and anxiety, PTSD (posttraumatic stress disorder), Drug Use, Suicidal Behavior and Emotional Suppression. According to studies, hate crimes may have an impact on suicidal behaviour in those affected, particularly when they involve LGBT youth. Once an investigation focused on lesbian and gay victims of hate crimes it was discovered that compared to victims of crimes unrelated to hatred, they had significantly more signs of anxiety and depression. Society is very oblivious, they don't comprehend how hate, violence, and discrimination in any course of action affect everyone. Therefore, this review was undertaken as a consequence of the augmentation of anti-LGBTQ+

hate crime, to explore more mental health dynamics of the LGBTQ+ community as a constant victim of hate crime by providing a comprehensive image of how hate and stigma affect LGBTQ+ individuals.

2. Methodology

A narrative synthesis of 6 conscientiously written research papers from peer-reviewed journals with strong and positive perspectives. Secondary research on hate crime against LGBTQ+ individuals, psychology of hate crime, and the mental health of LGBTQ+. A lot of research was available on the web, simply by entering keywords like homophobia, hate crime, LGBTQ+ mental health, etc.

Research papers were well immersed and included reports, studies from the government organization, case studies of victims, surveys, etc. To collect secondary data academic search engines like Google Scholar, Ref-seek, Academia, and ResearchGate were searched, and health blogs like PsycINFO and The Conversation were taken into account.

3. Objective

The LGBTQ+ community has been a theme of volitation since its existence. They frequently experience violence directed toward their sexuality, gender identity, or gender expression. Society is very oblivious, they don't comprehend how hate, violence, and discrimination in any course of action affect everyone.

So, like a milking stool, the objective of the review paper is based on three legs, the review will (1) scrutinize the motivational dynamics of LGBTQ+ hate crime, (2) undertake an effort to provide a comprehensive picture of how hate crime affects LGBTQ+ individual mental health and (3) how homophobia, heteronormativity and transphobia influence public health.

Motivational Dynamics of Anti-LGBTQ Hate Crime

1. Macho Culture and Social Homophobia

Male aggressors on male victims carry out the vast majority of homophobic heinous crimes, which are oftentimes motivated by male bullheadedness or strong hetero macho. Homophobia can be demonstrated by a person's anxiety of being recognized as gay, according to scholars like Calvin Thomas and Judith Butler. Men's homophobia is linked to their masculinity's vulnerability. According to this justification, homophobia is allegedly unchecked in the subcultures of sports that are perceived as typically "masculine," like football and rugby, as well as in those of its sympathizers. The argument put up by these academics is that when someone expresses homophobia, they do so not only to express their beliefs about the group of gay people, but also to distance themselves from this group and its economic well-being.

They are asserting their role as a heterosexual in a heteronormative community as a result of separating themselves from gay people, and they are attempting to avoid being labelled and treated as gay people by doing so. Homophobia is defined by several psychoanalytic theories as a threat to a person's analogous sex motivations, whether such motivations are actual or hypothetical. The formation of a response, reluctance, or restriction is brought on by this threat.

2. Islam

The most followed Scholars of Islam, for example, Shaykh al-Islam Imam Malik, and

Imam Shafi among others, decided that Islam refuses male homosexuality and appointed the death penalty for an individual at real fault for it.

The legitimate discipline for male homosexuality has changed among juristic schools: some endorse the death penalty; while others recommend a milder optional discipline. The gay movement is wrongdoing and illegal in most Muslim-greater part nations. In some moderately common Muslim-larger part nations, for example, Indonesia, Jordan, and Turkey, this isn't true, but friendly mistreatment, for example, honor killings are broad of cis-gendered gay men and in some cases lesbians Muhammad recommended capital punishment for both the dynamic and the uninvolved male gay accomplices, which is a reasonable judgment of male homosexuality inside Islam, and the relationship with male homosexuality being related to a reviled activity has delivered a long history of strictly excused and authorized brutality against gay men. According to conventional moral or religious doctrine, someone who engages in such behaviour disobeys God by undermining the harmony of God's creation.

Other modern Islamic viewpoints simply state that homosexuality is one of the worst transgressions, the worst sin, and the most abhorrent act. Although homosexuality is considered the eleventh major evil in Islam, a slave boy who had been sodomised by his owner was once exempted from punishment for the murder of that lord. The 2016 massacre at an Orlando nightclub was the country's bloodiest single-accident mass murder, and it continues to be the deadliest instance of homophobic violence against LGBT people. At Orlando, Florida's Pulse gay dance club on June 12, 2016, Omar Mateen left 49 people dead and more than 50 more injured. Agents have characterized the protest as an attack by oppressive Islamic tyrants motivated by terror and a can't stand wrongdoing.

How Anti-LGBTQ Hate Crime Affect Mental Health

The effect of disdain wrongdoing exploitation on mental working is extreme. The recurrence of LGBTQ+ encounters is decidedly related to the quantity of posttraumatic stress side effects. Semi exploratory investigations that think about people who have and who have not experienced the enemy of LGBTQ+ segregation have found that casualties against LGBT disdain wrongdoings scored fundamentally higher on proportions of horrible pressure side effects. The degree of posttraumatic symptomology varies depending on the individual, even though the link between LGBT exploitation and poor psychological wellbeing is extensively documented. The inquiry is, what represents this variety? For what reason are a few people stronger hostile to LGBTQ+ who can't stand violations? What mediating mental processes decide variety in the experience of posttraumatic symptomology?

Specialists in the field are presently starting to concentrate on interceding mental-emotional cycles that cushion pessimistic psychological wellness results related to such exploitation. This line of examination is significant because it can eventually illuminate mediation projects and strategies that offer help to hostile LGBT disdain wrongdoing victims. Building on the huge assemblage of exploration that inspects the pervasiveness and seriousness of LGBTQ+ disdain wrongdoings, specialists are presently moving their concentration to the mediating mental full of feeling processes that might safeguard against or fuel pessimistic mental wellbeing results experienced by casualties. By examining the possibility that account handling in the context of instances of LGBT hatred crimes is linked to mental health, the momentum study contributes to this burgeoning field of investigation. To my insight, no observational review has inspected whether story handling is related to mental wellbeing among survivors of hostile to LGBTQI+ toward can't stand wrongdoings

Homophobia, Heteronormativity, and Transphobia Influence Public Health

These singular demonstrations of disdain are characteristic of a more extensive example of the oppression of the LGBTQ people group. It is presently perceived inside general wellbeing that this segregation causes critical medical conditions for the LGBTQ people group.

For example, disdain and segregation can become incorporated and a wellspring of ongoing pressure, which thusly is a gamble factor for sorrow. What's more, LGBTQ popularly does encounter higher paces of mental misery and discouragement. Moreover, ongoing pressure can disturb ordinary organic working. This thus can make individuals more helpless to contamination.

Connected with this, men in long-home-sex connections were fundamentally bound to bite the dust from self-destruction men who were hitched to ladies or men who were rarely hitched. The lifetime pace of self-destruction endeavours among the LGBT populace is multiple times higher than the pace of self-destruction endeavours for non-LGBT individuals. This is undoubtedly connected

with long-hardness and the affected disgrace and abuse LGBT individuals face consistently.

Disdain and segregation additionally influence rates and movement of physically communicated contaminations (STIs), including HIV. Shame against HIV - for example, the discernment that it is a "gay man's infection" - still exists in our general public. An apprehension about is being marked as HIV positive, whiles many individuals abstain from testing. The outcome is that many individuals who are HIV positive don't realize that they endlessly are in this manner bound to spread the infection. Even though men who have sexual contact with different men address around four percent of the male populace, they represented 78% of new HIV diseases among men in 2010, and 63 percent of all new HIV contaminations.

Also, this apprehension about being tried for HIV frequently stretches out to an anxiety tow being tried for other STIs. 83% of new syphilis cases in 2014 impacted men who have sexual contact with men.

Vagrancy is bound to influence LGBT youth - 20-40 percent of destitute youth recognizing as LGBT. Numerous LGBT youth experience vicious actual data tack when they come out and may feel more secure living in the city.

Homophobia and prejudice influence everybody. This incorporates people groups who see themselves as straight, or who might not have companions or family members in the LBGTQ people group.

General wellbeing incorporates coordinated measures to forestall illness, advance we were being, and drag out a life among the populace overall. Scientists in the field of general well-being have long concentrated on the impacts of numerous sorts of separation on well-being; whether because of race, financial status, or sexual direction.

3. Literature Review

(Mark Walters, 2020) In this research, Mark Walters and Eric Gitari frame the nature and degree of hostility to LGBTQI+ disdain wrongdoing and its effect on people and social orders in the Commonwealth. The research was divided into 4 segments. Segment 1 diagrams the technique and approach of this research, and sets out the legitimate and social setting in which disdain wrongdoings are perpetrated against LGBTQI+ individuals. Segment 2 surveys the degree and nature of hostility to LGBTQI+ disdain violations, investigating patterns across the Commonwealth and studying territorial and country-explicit case models. Data on the culprits of LGBTQI+ exploitation is additionally illustrated. Segment 3 inspects the effects that the enemy of LGBTQI+ disdain wrongdoings have on people, networks also, and society. Shared traits and contrasts in encounters across various areas of LGBT networks are depicted. Segment 4 of the research finishes up with suggestions on how Commonwealth states ought to enact against hostile to LGBT disdain wrongdoing, and why legal organizations should execute checking instruments to guarantee that enemy of LGBT exploitation becomes apparent and is estimated.

(Eileraas, K.) The Michigan Feminist Study presented a case study based on the murder of Brandon Teena, a girl living as a boy in Nebraska. This case study is one of the most heinous examples of LGBTQ hate crime that exists in society. The study provided an in-depth analysis, as well as forensic evidence, police statements, transcripts of recordings, media reports, and so on. The study discusses how Brandon Teena's story reminds us of the conflicting needs to assert the importance of the female body as a locus of 'unspeakable' pleasures and violent abuse, while also destabilizing conventional understandings of gender identity as innate or expressive. The paper concludes that Transgender violence affects both men and women, but Brandon Teena's case highlights how particularly cruel it is for FTM transgendered people.

(Web MD Article (Benisek, A.) Melinda Ratini, MS, DO, reviewed the paper medically. The research here takes a more qualitative approach than the others. The primary goal of this paper was to demonstrate the motivation, after effects, and mental health consequences of a hate crime. The paper provided a more in-depth perspective on hate crime in society. It also provides insight into how to deal with hate crime.

(Kehoe, 2020) Anti-LGBTQ Hate: An Analysis of Situational Variables by Jill Kehoe, this present research intended to expand the inadequate group of writing on the enemy of LGBTQ disdain by giving an inside and out the assessment of against LGBTQ disdain episode situational qualities including guilty party substance use, number of wrongdoers, wrongdoing area, and casualty wrongdoer relationship. The research had two main questions

(Claire Wilson, 2019) went through a systematics qualitative research because of expanded degrees of disgrace, separation, and exploitation Lesbian, Gay, Bisexual, Transgender, Queer, Questioning, or Intersex (LGBTQI+) youth face specific difficulties in the public eye. With the goal of better getting the difficulties and issues that LGBTQI+ youth are encountering. The author included a blend of the included investigations distinguished five center topics: (1) Isolation, rejection, phobia, need for support; (2) Marginalization; (3) Depression, self-harm and suicidality; (4) Policy and environment; and (5) Connectedness. Claire Winson suggested that there are gaps in our understanding of the poor mental health outcomes and risk and protective factors that have been revealed by quantitative studies of youth from sexual and gender minority groups. The research suggests that there is a disparity in mental health statistics for the population of young people who identify as sexual and gender minorities.

(Dandal, 2015) The psychological effects of being a victim of hate crimes are extremely detrimental. The number of posttraumatic stress symptoms and the frequency of anti-LGBT experiences are positively correlated. This Research paper shows the trend that analysts have recently turned their attention to the intervening mental processes that could prevent or exacerbate the negative mental wellbeing outcomes experienced by victims, building on the vast body of research that examines the prevalence and severity of anti LGBTQI+ prejudice violations. This paper examines the relationship between account handling and psychological well-being of the victims. The paper also suggests a need for more study on mental health of LGBTQ, so that coping methods can come to light.

4. Conclusion and Recommendation

The most violent element in this society is ignorance, only 310 out of every 1,000 assault & violence cases are reported to the police. It is very complicated to report such crimes with the victim-blaming keeping aside race, gender, ethnicity, sexual preference, etc.

Hate crimes against LGBTQ+ individuals are more omnipresent than many chose to believe. The effect of disdaining wrongdoing exploitation on mental working is extreme. The recurrence of LGBTQ+ encounters is decidedly related to the quantity of posttraumatic stress side effects. There is a very infrequent amount of discussion as a society about the mental health of the survivors who have experienced or endured violent crime especially when that survivor is an LGBTQ+ individual. The biggest issue is even though people are hot to trot about raising awareness about this issue, they do not care enough to do something about it.

Explicit, designated mediations are important to diminish the quantity of enemy of LGBT disdain violations, on the grounds that these wrongdoings have not evened out off as different wrongdoings have as of late done. Youthful grown-ups are the ones who most often carry out these violations, so hostile to tormenting programs and other instructive projects might assist with working on the overall social environment for LGBT individuals and, simultaneously, decrease the inclinations that spike against LGBT can't stand wrongdoings.

Convenient and exact information on the enemy of LGBT disdain violations is additionally fundamental for understanding and forestalling these wrongdoings. Therefore, the accompanying arrangements and techniques ought to be created to guarantee that state and neighbourhood police offices distinguish, counter, and precisely report can't stand violations:

Police preparation ought to be improved to diminish examples of misclassification of LGBT who can't stand wrongdoings.

Police divisions ought to foster social skills preparing to improve the capacity of officials to work with LGBT individuals and networks. This could likewise assist with empowering LGBT casualties to report violations, and further develop information assortment on such infractions too.

Medical care arrangements for LGBT survivors of viciousness ought to similarly be worked on through expanded preparation for clinical consideration suppliers. They need to figure out how to recognize disdain wrongdoings and further develop social skills working with LGBT individuals, remembering casualties as well as all others for the local area.

At long last, to further develop administration arrangements and manage wrongdoings against LGBT people groups, progressing coordinated efforts among police, medical services suppliers, and promotion and local area associations ought to be cultivated all over the place. Such participation can work on the overall cultural climate and lessen the predispositions that empower disdain violations, simultaneously as it further develops care for casualties.

The natural world is always changing and evolving. With the evolving needs and conditions, we must confront the hate crime age since it costs a great deal in terms of lives, money, and time. If these aspects are taken care of, the real improvements will manifest. We also need to implement tough legal reforms as a good system of moral education. Criminals are observed for their behaviour so that, once we understand the root of the problem, we can come up with appropriate reforms. To do this, we must pay attention to all the relevant factors and speak out against it. Whether a majority or a minority, violence is never productive for people. As an example, Australia just legalised LGBT marriage on a national level.

REFERENCE

1. Benisek, A. (n.d.). How do hate crimes affect health? WebMD
2. Clare Wilson, L. A. (2019). LGBTQI+ Youth and Mental Health: A Systematic Review of Qualitative Research. Springer Link, 25.
3. Dandal, A. (2015). an investigation of anti-LGBT hate crime victimization. Department of Applied Psychology & Human Development, 56.
5. Nandal, Nisha, Nandal, Naveen and Malik, R. (2020). Is loyalty program as a marketing tool effective? Journal of Critical Reviews, 7(6), pp. 1079–1082. doi:10.31838/jcr.07.06.188.
6. Eileraas, K. (n.d.). The Brandon Teena Story: Rethinking the Body, Gender Identity and Violence Against Women. Michigan Feminist Studies
7. Kehoe, J. (2020). Anti-LGBTQ Hate: An Analysis of Situational Variables. Journal of Hate Studies, 14.
8. Mark Walters, E. G. (2020). Hate Crimes against the LGBT Community in the Commonwealth: A Situational Analysis. Human Dignity Trust, 64.

Integrating Advancements in Education, and Society for Achieving Sustainability – Dimitrios A. Karras et al. (eds)
© 2024 Taylor & Francis Group, London, ISBN 978-1-032-70841-6

36

The Growth of Cryptocurrency across the Globe: Its Challenges and Potential Impacts on Legislation

Nisha Nandal and Naveen Nandal*

Assistant Professor, Sushant University, Gurugram

Shaurya Gulati

Student, SOB, Sushant University, Gurugram

Chakshu Mehta

Research Scholar, Sushant University, Gurugram

Abstract: Many tasks in our everyday lives are now completed electronically as a result of the information and communications technology's rapid technical growth, making them more flexible and efficient. The number of online users has been expanding quickly, and fostered the creation of the "Virtual World", which in turn has helped create a brand-new phenomenon in business called "cryptocurrency," which makes it easier to conduct financial transactions including buying, selling, and trading. Cryptocurrency in the context of electronic business is a Virtual world, social networks, online social games, peer-to-peer networks, and other applications and networks handle valuable and immaterial items electronically. Virtual currencies have been employed extensively in numerous systems in recent years. This research seeks to investigate consumer expectations for the development of cryptocurrencies. The essay also looks at customers' trust in using cryptocurrencies at a time when such virtual currency is not fully regulated and supervised.

Keywords: Cryptocurrency, Virtual world, Legislation, Network, Social networking

1. Introduction

The general population has recently been talking about cryptocurrency. With the development of technology, cryptocurrencies are becoming more at ease for investors who value their privacy and the ability to create wealth. As more people express interest in purchasing cryptocurrencies, cryptocurrencies like Bitcoin, Ethereum, Ripple, Litcoin, and others are currently surging in the financial market. However, a larger portion of the populace is perplexed about the overall performance of cryptocurrencies. The first decentralized cryptocurrency, Bitcoin, was launched in 2009.

*Corresponding author: Naveennandal@sushantuniversity.edu.in

DOI: 10.4324/9781032708461-36

Since cryptocurrency is entirely a digital value on the internet, it has no physical substance. For transactions and other commercial purposes, these currencies can be used as the cash equivalent. Information and communication technologies have undoubtedly created many golden opportunities in a number of different fields. The financial and business industry is one such sector that benefits from these technologies and online connections. Virtual world concepts have been activated by an increasing number of online users, creating new types of trading, transactions, and currencies. The emergence of cryptocurrency in recent years has been a remarkable development in the world of finance. Unlike banks and other financial institutions, cryptocurrency doesn't rely on third parties to verify transactions. It enables anyone anywhere to send and receive payments, via a peer-to-peer system. Instead of carrying out transactions used physical cash, cryptocurrency payments exist purely as digital entries to an online database that is stored on the Blockchain. Cryptocurrency funds are transferred via a public ledger and are stored in digital wallets.

The application of cryptographic operations in financial transactions makes cryptocurrencies an internet-based means of exchange. The decentralization, transparency, and immutability of crypto currencies are enhanced by the use of block chain technology. A digital currency known as a crypto currency only exists as digital book money and does not exist as coins or bills. Direct transfers between crypto assets are possible when two parties using secret and public keys. Users can escape the high costs imposed by conventional financial institutions by using these transactions, which have low processing costs. The main characteristic of a crypto currency is that it is not governed by a single entity.

Cryptocurrency in the context of electronic business is a Virtual world, social networks, online social games, peer-to-peer networks, and other applications and networks handle valuable and immaterial items electronically. Virtual currencies have been employed extensively in numerous systems in recent years. Due to their innovative technological concepts, cryptocurrencies have been raising a lot of legal questions in countries across the globe. Academic research, done by scholars, covers many strands of the technological aspects of such currencies, but legal studies generally only examine a few, if any at all.

Additionally, governments are beginning to regulate cryptocurrencies explicitly in terms of anti-money laundering (AML) and to clarify or strengthen the legal basis for prosecuting crimes committed in connection with cryptocurrencies. Following are the countries that have completely banned the use of cryptocurrencies: Qatar, China, Morocco, Russia, North Macedonia, Turkey, Bangladesh & Egypt

2. History of Cryptocurrency and its Evolution

Around 2009, Bitcoin became the first form of cryptocurrency to be used in India. The first business transaction took place in 2010, and the first cryptocurrency exchange happened in 2013. Over the past few years, it has attracted a sizable fan base and attention in India. According to industry estimates, India is home to 15 to 20 million cryptocurrency investors, who collectively control assets worth about 41 thousand crore rupees ($5.37 billion). This rising popularity has been linked to a number of causes, including the fact that India is leading the globe in terms of internet usage growth and that the country's booming tech industry and computer-savvy millennials provide the ideal market demographic for cryptocurrencies. In recent years, a number of games based on blockchain have appeared. Players can access these games at various levels.

India's Role in the Crypto Market

On the cryptocurrency market, India has been especially active. India really has the largest percentage of cryptocurrency owners and the second-highest adoption rate worldwide, according to multiple study reports. With over 15 million retail investors, more than 60% of the states in India are beginning to adopt crypto technologies. There are also about 230 start-ups in this industry in the nation, which has a strong institutional presence and offers a lot of opportunity for expansion. Even from a global perspective, the Indian tech industry is alive with eager entrepreneurs and a top-tier talent pool. India is in a good position to be a worldwide leader in this field if the ecosystem continues to grow in the same way as it has over the previous few years.

Issues or Challenges Associated with Cryptocurrency

Although people are excited about the appearance of cryptocurrencies because they are inventive in all their endeavors, have acquired a special position on a worldwide scale, and are innovative in all of their endeavors, the path of cryptocurrencies up to now is kind of roller coaster experience. The following discussion examines some of the numerous difficulties that surround cryptocurrencies:

1. Regulation: The most essential component now needed in the bitcoin business is cryptocurrency regulation. While some nations have previously controlled its use and transactions in the existing financial system, some nations are stepping forward to take a cooperative approach to its regulation. It will only be seen as a means of doing things illegally up until and unless it is globally regulated.

2. Volatility: Due to the fact that its regulation is still pending worldwide, it will not be regarded as a stable system and there will be significant variations in its supply and demand. This makes it volatile, meaning that its value might vary drastically in a short amount of time.

3. Security: Due to the fact that this asset is fully digital—from its creation or mining through its transaction, exchange, and storage—it is constantly vulnerable to security threats. Hackers could compromise its existence at any time by attacking any component.

4. Cost: Nothing in life is free, and an asset made entirely of technology will likewise be more value. It has a cost because all the revolutionary technologies required are expensive.

5. People's Perception: People still consider cryptocurrencies as being used for unlawful purposes and are skeptical of any judgements made about them because global regulation of them has not yet been implemented.

6. Upgradation of Technology: Since technology is integral to the cryptocurrency concept and is, as we all know, highly changing, It must be continuously updated. Upgrades are always more expensive.

7. Risk for investors and users: Even though the cryptocurrency has been

around for ten years (the first bitcoins were created in 2009), it is still in its infancy and has not yet been subject to any regulations or legislation. Therefore, due to a lack of regulation, it is very volatile and places a high level of risk on its users and investors.

3. Methodology

The data used in this paper is secondary which has been acquired from various research papers, by authors from across the globe, who have discussed the various impacts of cryptocurrency in this new digital and virtual world.

4. Objectives

1. Will Cryptocurrency be the next currency platform?
2. Are virtual currency platforms safe enough to be used?
3. To know legality and trading of Bitcoin in India.

This paper investigates different cryptocurrency trading platforms, aiming to give a deeper insight regarding the concerns and challenges faced by online users in the market. It also analyses the relationship between the Crypto market and real-world aspects including the business industry, existing world monetary systems, etc. Cryptocurrency platforms are a global phenomenon that affects governments, operators and users. These results point to the importance of regulating Cryptocurrency. Legislators, policy-makers, and virtual currency providers are encouraged to establish and implement rules, policies, and guidelines to regulate virtual currencies.

5. Literature Review

(Dr. Mubarak & Manjunath, 2021) talk about the genesis of cryptocurrency in their paper, wherein they also mention the regulatory bodies that took action against the same in the year 2017. This bill is yet to be adopted or introduced in the parliament, which will then define, not only the use of digital currencies, but also provide certain directions while dealing in digital assets. Dr Mubarak and Hosmani Manjunath also talk about the risks involved while investing in cryptocurrencies including high volatility, cyber risks, no involvement of regulatory bodies, etc. Hence the authors suggest individual investors to invest in gold as it provides a more consistent return as compared to the Bitcoin.

(Shirakawa & Korwatanasakul, 2019) investigated Whether policymakers are willing to pursue greater financial development by allowing the use of cryptocurrencies depends on how effective governance institutions and de jure financial openness affect that stance. According to Jacinta Shirakawa and Upalat Korwatanasakul's research and analysis, they believe that institutions have a greater likelihood of a less restrictive regulatory stance on cryptocurrency. They provide us with evidence that states the policy makers are more likely to be open to cryptocurrency.

(Shovkhalov & Idrisov, 2021) states that cryptocurrencies have great investment appeal, however they are subject to great volatility, making them an asset that requires some caution.

Akshay A., Shivashankarachar Y. - "A Study On Security Issues In Investments And Transactions In Bitcoins And Cryptocurrencies" The pros and cons of using Bitcoin as a cryptocurrency and the difficulties involved in buying, selling, and investing in it are both underlined in this essay. The security of Bitcoins is a matter of concern because they are linked to money and are rising in value on the market. It is crucial to put a lot of emphasis on the security of the channel in order to create one for the secure transfer of bitcoins. Bitcoin storage, extraction, and transactions are important factors to take into account when talking about security. It's important to think carefully about Bitcoin storage online. This study also outlined additional dangers related to bitcoins, including as the absence of Indian regulatory oversight of transactions.

Sudhir Khatwani (CoinSutra) – "Future of Bitcoin and other Cryptocurrencies in India after RBI's Ban" In this article The author of this post has highlighted the regulatory environment for Bitcoin and other cryptocurrencies in the Indian market following the RBI's decision to prohibit the exchange of these digital currencies for real money through its own businesses, including banks and other financial institutions. The author highlighted the many negative effects of prohibiting these cryptocurrencies' bank-to-bank transactions. He has highlighted historical incidents involving the Chinese government's decision to ban cryptocurrencies, which led to an increase in demand for cryptocurrencies from bitcoin investors through alternative channels. India has a huge number of platforms and transaction facilities that enable cryptocurrency buying and selling, so the same thing could occur there as well.

(Rueckert, 2019) talks about how the Bitcoin network is protected by several fundamental rights. AML regulations and other crime prevention concepts of the government may interfere with personal property rights, the right to pursue a profession or trade, the right to freedom of association and the right of freedom to express oneself. The author also believes and states that it can be argued that criminal investigation measures such as storing, processing, and collecting data from the blockchain systematically intrude on privacy and data protection rights.

(Jani S. , 2018) talks about the ever-growing virtual currency market and how it is yet to be regulated in countries across the globe. He formulated a questionnaire to understand how the younger generation perceives cryptocurrency and how they use or invest in such markets. He also discusses that loyalty points are The most popular cryptocurrency format. Second place is social game virtual currency, third place is social network virtual currency, and finally peer-to-peer network virtual currency.

(Jaideep & Jyoty, 2019) believe that a secure network of currency exchange is paramount to the proper implementation of cryptocurrency as a replacement for traditional currencies. Moreover, their paper states that specifically, the present study suggests that if crypto currencies are presented as part of Lakshmi Coins, then society can be motivated to adapt slowly to them, which in turn will pave the way for rapid progress in cryptocurrency use. In the future, bitcoin may provide benefits to Indians, but not everyone will benefit equally.

(Kumar, 2019) It is thought that the absence of legislation will lead to a negative impact on cryptocurrency systems. The RBI's silence regarding Bitcoin regulation can prove to be damaging. In India, a lot of activities are centred on Bitcoin, such as traders, exchanges, and merchants accepting cryptocurrency payments.

(Lloyd, 2021) talks about how the world has been focusing on just one cryptocurrency, i.e., bitcoin, and how there are over ten thousand cryptocurrencies in existence. New financial technologies, including Bitcoin, are challenging the law, lawyers, and regulators. Despite the overwhelming popularity of Bitcoin primers and publications that claim to provide concise, holistic descriptions of Bitcoin, the dissertation shows that they mislead communities and nations into ignoring critical Bitcoin network features (such as tax policy).In addition to demonstrating how flexible communities have been when accepting 'things' as mediums of exchange, examining the history of money, currency, and legal tender provides useful insights into how money changed over time. The dissertation outlines the consequences that can occur if one cryptocurrency's mechanics are misrepresented. It remains to be understood how the law applies to other cryptocurrencies with unique features.

(Read, 2022) states Bitcoin's rise in price and demand is greater than the rate of coin reward decay, so electricity consumption unquestionably and inevitably rises, limited only by the price of Bitcoin itself. Since Bitcoin is used in reward, it is almost impossible to track, and therefore impossible to tax, revenues and profits from this highly mobile industry. Seeing as how public policy to constrain this industry is frustrating, it is best if Bitcoin is less popular as new currencies emerge.

6. Conclusion and Recommendation

Specifically in relation to A brand-new, efficient, and appealing payment method

called bitcoin is available, and it has the potential to increase business and operator revenue. Along with accepting actual money, it also offers alternate payment options that make it possible for users to conduct simple financial transactions in the worldwide market, including buying, selling, transferring, and exchanging. The e-Business and e-payment sectors can benefit from cryptocurrency in numerous ways. Still, there isn't a lot of trust in cryptocurrencies yet. Numerous cryptocurrency platforms have numerous worries, difficulties, and problems. Users must exercise extra caution while utilizing virtual currencies like cryptocurrencies until they are properly regulated and managed. Consequently, the main issue with cryptocurrency systems is the absence of laws.

Cryptocurrencies are a new, effective, and attractive mode of payment that can increase company and operator revenues. They also provide additional payment methods beyond real money, enabling users to carry out financial activities such as buying and selling & transferring and exchanging easily. Despite the fact that cryptocurrency platforms offer new mechanisms and methods for digital financial transactions, they are not regulated and controlled as well as they should be. The lack of regulations was found to be the main concern in cryptocurrency systems. The results show a preliminary assessment of cryptocurrency's use, growth, trust, and future expectations. Considering the large number of currencies flowing through different systems all over the globe, the analysis suggests that cryptocurrency may become the next currency platform. In addition to these results, the confidence and trust rate of cryptocurrency users is high, as evident in several cases around the globe.

The average user, however, has not yet fully grasped the potential of crypto currencies. Several concerns, issues, and challenges exist on most cryptocurrency platforms. While cryptocurrency is still unregulated and uncontrolled, users are advised to take extra precautions when using it. Based on the results of the study conducted by Mr. Shailak Jani more than 58% of respondents agreed that virtual currencies of various types and forms will become the language of financial transactions in the future. In contrast, 22.58% of participants were indifferent, while 19.35% were against virtual currencies as the means of financial transactions in the future. Blockchain technology shows great potential to positively impact the e-Business and e-Payments sectors in the future, allowing for more opportunities to be realized. With the advance of technology, cryptocurrency will continue to advance. Many people are now aware of the potential and opportunities offered by cryptocurrency. Moreover, vendors are now accepting payment with different types of cryptocurrencies. The world has also witnessed the advent of new forms of virtual currency recently. Research opportunities in the cryptocurrency field are significant. Various different perspectives need to be analysed to determine if the real financial laws relate to the implementation of cryptocurrency platforms. Furthermore, the adoption and acceptance levels of cryptocurrencies also require further analysis with large samples. Trust and confidence are important factors that should be explored in terms of usage and trade of Cryptocurrency forms. In India, the research scope can further be extended to develop use-cases for applications of cryptocurrency across a range of industries.

As the number of platforms providing virtual currency grows and the volume of virtual currency trades increases, we can predict the future of virtual currencies. Furthermore, as our society becomes more cashless (we are already using credit cards, debit cards, and online banking to complete financial transactions) we will, sooner or later, accept and integrate virtual currencies. Virtual currency usage has increased dramatically over the past few years, and many issues need to be taken into consideration to control such a financial system. The absence of clear and strict regulations and policies leads to increased risks and problems which the virtual currency industry would face.

This upcoming aeon of virtual money, online transactions and digital lockers must be controlled and regulated by strict legislations and laws, in all countries.

Forbes' Ed Sperling (Editor in Chief of Semiconductor Engineering) reported that while cryptocurrency is not real money, it still deserves serious attention from lawmakers. His statement summarizes the need for legal and policy regulations regarding virtual money.

REFERENCE

1. Dr Mubarak, & Manjunath, H. (2021). A Study on Cryptocurrency in India. IJRAR Journal, 10.
2. Nandal et al., a B-schools service quality measure: scale development and validation, Academy of Strategic Management Journal, Volume 20, Special Issue 6, 2021
3. Jaideep, J., & Jyoty, K. P. (2019). A Study on Cryptocurrency in India – Boon or Bane. Journal of Emerging Technologies and Innovative Research (JETIR), 6.

4. Jani, S. (2018). The Growth of Cryptocurrency in India: Its Challenges & Potential Impacts on Legislation.
5. Kumar, A. (2019). A study on opportunities and challenges of bitcoin. IJRAR, 6.
6. Lloyd, D. (2021). Bitcoin: property, money, currency or legal tender? SSRN, 69.
7. Read, C. (2022). The Inevitability of Increasing Energy Consumption in the Cryptocurrency Mining Arms Race. SSRN, 34.
8. Rueckert, C. (2019). Cryptocurrencies and fundamental rights. Journal of Cyber Security, 12.
9. Shirakawa, J., & Korwatanasakul, U. (2019). cryptocurrency regulations: institutions and financial openness. ADBI Working Paper, 17.
10. Shovkhalov, S., & Idrisov, H. (2021). The Economic and Legal Analysis of Cryptocurrency. MDPI Journal, 17.

*Integrating Advancements in Education, and Society for Achieving
Sustainability – Dimitrios A. Karras et al. (eds)*
© 2024 Taylor & Francis Group, London, ISBN 978-1-032-70841-6

37

Analysis on the Online English Teaching System based on Integrated Artificial Intelligence and 5G Technologies

Prabhdeep Singh*

BBD University, Lucknow, UP, India

Abstract: We live in a technologically advanced culture where robots and systems are intelligent and resemble humans. This is due to artificial intelligence (AI) advances. The machines are utilised both directly and indirectly in a variety of industries, including healthcare, and sophisticated decision-making, as well as in educational institutions. Several educational advancements for both instructors and students have been made possible by the use of AI-based technologies and the Internet. With online learning systems built on AI principles, 5G has changed teaching and learning processes by enabling smoother, quicker access to educational information. This research suggests integrating AI and 5G Internet technology to implement the advanced energy-optimized low-energy adaptive clustering hierarchical protocol (A-LEACH) for online English teaching. First, the dataset was normalised. Artificial Neural Network (ANN) classifiers accurately identify features. Experiments demonstrate that our technique is more efficient than others.

Keywords: 5G, Artificial intelligence (AI), A-LEACH, Artificial Neural Network (ANN)

1. Introduction

The teaching of English has been significantly impacted by the advancement of technologies such as artificial intelligence and 5G wireless networks in the areas of language recognition and natural language processing. The traditional method of teaching English in colleges and universities will almost certainly need to be revised in light of the technical advances brought about by artificial intelligence and 5G wireless networks (Navaneetha et al., 2022). Because of the impersonal nature of online learning, some students may feel uncomfortable speaking out in class due to a lack of trust, while others may choose to sit on the sidelines completely. It's hard to have productive conversations, and lessons suffer as a result. Teachers are unable to

*Corresponding author: prabhdeepcs@gmail.com

DOI: 10.4324/9781032708461-37

really understand the students' mental states. When teaching spoken Chinese online, teachers encounter previously unheard-of difficulties (Rajesh, et al., 2021). The amount of incentives accessible to online second language learners has decreased as a consequence of the difficulty in adapting study findings on enthusiasm in traditional foreign language classes to the online English learning environment (Liu et al., 2021). The fusion of artificial intelligence and education has been supported by advancements in these technology and goods as well. The conventional teaching approach is also starting to be challenged by internet-based education. With the use of modern technology tools, many traditional classroom lectures may be made available online (Hu, Z. 2022). For the development of more sophisticated educational communication networks, 5G and AI are two intriguing advances. Accelerating the establishment of 5G test environments, 5G educational application solutions, study and improvement of 5G unified innovations (Sun, X. 2021). Thus, using AI educational 5g technologies into the online teaching of English may assist English majors in expanding their worldviews, improving their online English teaching abilities, and achieving even better teaching outcomes.

2. Related Works

Lei, X. 2020 proposed an optimization technique of English teaching mode that is dependent on 5G technology and AI technology. The approach is intended to be used in colleges and institutions. Gao, Y. 2021 examined how the current state of innovation might be used to the English major at one of China's educational institutions and recommended a 5G era. It is of special use in English instruction to

enable students to easily and unrestrictedly have access to material that is pertinent to the subject matter. This is made possible by the increased speed of data transmission and the proliferation of high-quality online tools. Yu et al., 2021 offered an improved method for overcoming the obstacles and difficulties associated with learning English at colleges and universities in China. English classrooms are now clever and efficient for learning English as a result of the arrival of AI and 5G technologies. Wang, M. 2022 devised a Genetic Algorithm (GA)-based method for translating, writing, and lecturing college English. The recommended technique may promote resource sharing, improve teacher-student interactions, and improve students' general English writing skill by looking at this cutting-edge type of system. Liu et al., 2022 examining the innovative kind of network, according to this study, resource sharing will be encouraged through network interaction innovations, and teacher-student interactions will be strengthened, as well as students' general ability in English writing. The efficiency of the proposed methodology is enhanced in this paper using English writing and interaction techniques.

3. Proposed Methodology

Figure 37.1 denotes the proposed methodology.

Dataset: We used a dataset that included 6,610,000 items and 3,250,000 positive and negative sentences written in English. The dataset was applied. Normalization is used as part of the preprocessing of the dataset in order to remove the impulsive noises. Before being classified as either favorable or unfavorable, each and every item that was included in our data set for the evaluation was first obtained and examined from online English teaching domains.

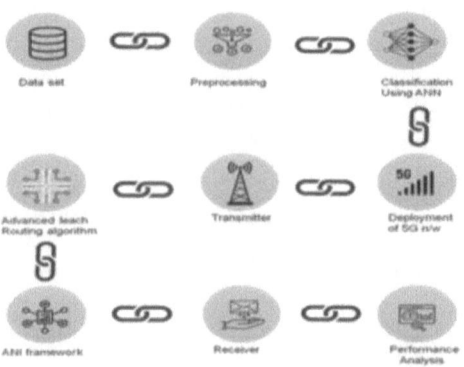

Fig. 37.1 Proposed methodology

Preprocessing using Min-max normalization: min-max normalization, the following equation is employed to normalize features with a range of [0, 1].

$$u' = \frac{u - \min_B}{\max_B - \min_B} \quad (1)$$

The minimum and maximum values of feature B are shown here by and , respectively. The attributes' original and normalized values are represented by the values u and u', accordingly. As can be seen from the formula above, the maximum and minimum feature values are converted to 1 and 0, respectively.

Classification using ANN: Data predictions are made using the back propagation ANN method. A neural network's training phase employs forward propagation. After the forward pass, a value is generated by the output layer nodes. Prior to utilizing the activation function to calculate the node's output during the forward pass, the node's whole input must first be known. This equation is employed to determine the total input that each neuron in a feed-forward neural network has received.

$$\text{Total Input} = w_1 * n_1 + w_2 * n_2 + ... + e_c * n_c + 1 * n_b \quad (2)$$

$w_1, w_2, ..., w_w$ - Input neurons. $n_1, n_2,......,$ n_c – Weights associated with input neurons. n_b – Weight associated with bias Output of neuron is calculated using the activation function. The activation function is employed in order to determine the neuron's output.

$$\text{Activation function} = 1/(1 + z^{(-Total\ Input)}) \quad (3)$$

Where: Total Input – The total input to the neuron. It is common practice to train artificial neural networks using back propagation in conjunction with an optimization strategy like gradient descent. The two stages of the method's two-step cycle are propagation and weight updates. Following the back-propagation phase's comparison of the forward pass output to the projected output, link weights are modified.

Advanced LEACH Protocol (A-LEACH): A hierarchical technique was first utilized in wired broadband to conserve power; however, it was subsequently implemented in wireless sensing networks (WSNs) to gain a longer lifetime and reduced consumption of power. This was accomplished by transferring the general framework from wired broadband to WSNs. A-LEACH is the first adaptive protocol that is utilized in WSN for the choosing of the cluster head.

The primary objective of the A-LEACH method is to decrease the amount of energy that is used. This objective may be achieved by choosing groups very precisely in two stages: the stage, and the steady state stage. A-LEACH organizes the nodes into groups according to the results of their random process. During the stage of setting up, the base station will transmit a report to all of the control nodes, also known as cluster heads (CHs). According to the formula in (4), these

CHs are selected based on the threshold value Th(n), which takes into account the probability of becoming CH pr, the current round r, and the proportion of non-CHs that were included in the cycles before the current one by 1/pr.

$$Th(n) = \begin{cases} \dfrac{pr}{1 - pr\left(randmod\left(\dfrac{1}{pr}\right)\right)}, & n \in s, \\ 0, & otherwise \end{cases} \quad (4)$$

The value shared by all of the nodes in the e network is arbitrarily generated as a combination of 0s and 1s. If the resulting value is less than Th-(n), the node in question will take on the role of control node for the cluster, with the other nodes acting in the capacity of noncluster heads. The cluster head sends a request to the noncluster heads during phase 2 of the steady state to join the group in order to transmit data. However, neither the node location nor the amount of remaining battery is taken into consideration in the CH selection process.

4. Result and discussion

Figure 37.2 depicts the results of analysing the prospective questions on learning interest from the students' questionnaire to acquire more about how student interest in learning has evolved over time. The proportion of voters who said they were "very much in line" raised from 19.865 to 24.85%, a rise of 5.01%.

The higher education proficiency categorization is shown in Figure 37.3. The information in Figure 37.3 indicates that pair 1 of the outcome of students' language ability pre- and post-tests could be displayed. This shows a significant improvement in learners' language comprehension abilities and

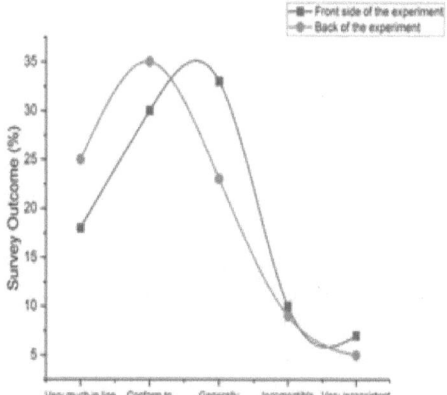

Fig. 37.2 Prediction Accuracy

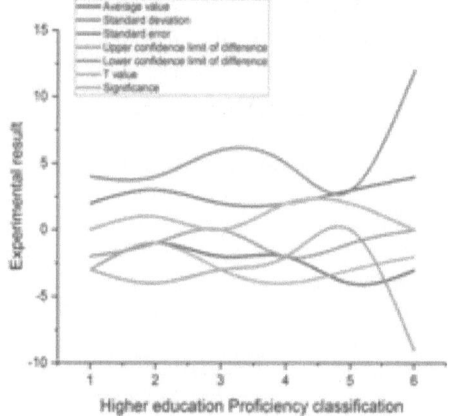

Fig. 37.3 Prediction Time

demonstrates positive learning variety. The outcome of the students' pre- and post-tests on their language comprehension abilities demonstrate that, within the experimental class, there is a large difference between before and after the experiment, demonstrating that the students' language comprehension ability has varied significantly. Data from the students' language acquisition in pair 3's pre- and post-tests shows that there has not been a significant development in their language assessment skills throughout the trial.

5. Conclusion

The use of AI-based technology and 5G has revolutionized online education for students and instructors alike. The use of online learning platforms powered by AI approaches has transformed English teaching and learning processes by enabling more attractive and rapid access to educational information. This work used AI and 5G Internet technology to develop a routing system for online English instruction advanced LEACH protocol. Findings show that, in comparison to previous protocols, the suggested A-Leach method and classification using ANN offers a greater transmission rate, nodes, and low energy usage.

REFERENCES

1. Rajagopal, N.K., Qureshi, N.I., Durga, S., Ramirez Asis, E.H., Huerta Soto, R.M., Gupta, S.K. and Deepak, S., 2022. Future of business culture: an artificial intelligence-driven digital framework for organization decision-making process. Complexity, 2022.

2. Rajesh, N. and Christodoss, P.R., 2021. Analysis of origin, risk factors influencing COVID-19 cases in India and its prediction using ensemble learning. International Journal of System Assurance Engineering and Management, pp. 1–8.

3. Liu, C., Wang, L. and Liu, H., 2021. 5G network education system based on multi-trip scheduling optimization model and artificial intelligence. Journal of Ambient Intelligence and Humanized Computing, pp. 1–14.

4. Hu, Z., 2022. Study of the Effectiveness of 5G Mobile Internet Technology to Promote the Reform of English Teaching in the Universities and Colleges. Computational Intelligence and Neuroscience, 2022.

5. Sun, X., 2021. 5G joint artificial intelligence technology in the innovation and reform of university english education. Wireless Communications and Mobile Computing, 2021.

6. Lei, X., 2020, July. The research on optimization strategy of college English teaching mode based on 5G technology and artificial intelligence technology. In 2020 International Conference on Virtual Reality and Intelligent Systems (ICVRIS) (pp. 869–872). IEEE.

7. Y. Gao, "A survey study on the application of modern educational technology in English major college teaching in the age of 5G communication," Ieory and Practice in Language Studies, vol. 11, no. 2, pp. 202–209, 2021

8. Yu, H. and Nazir, S., 2021. Role of 5G and artificial intelligence for research and transformation of English situational teaching in higher studies. Mobile Information Systems, 2021.

9. Wang, M., 2022. Design and Research of College English Reading, Writing, and Translation Teaching Classroom Based on 5G Technology. Mobile Information Systems, 2022.

10. Liu, X. and Huang, X., 2022. Design of Artificial Intelligence-Based English Network Teaching (AI-ENT) System. Mathematical Problems in Engineering, 2022.

Note: All the figures in this chapter were made by the Authors.

*Integrating Advancements in Education, and Society for Achieving
Sustainability – Dimitrios A. Karras et al. (eds)*
© 2024 Taylor & Francis Group, London, ISBN 978-1-032-70841-6

38

Comparative Legal Analysis of Adoption vs. Surrogacy in India

M. Garg[1], S. Jeet*, I. Kaur[2] and N. Singh[3]

K.R. Mangalam University, Gurugram, Haryana, India

Abstract: Owing to the change in the environment and the way of life in today's times, more and more couples are experiencing various medical difficulties during conception. Individuals who being unable to conceive a kid of their own due to certain causes, making use of various means like adopting a child or going through medical interventions for say Assisted Reproductive Techniques including surrogacy. Although many people find these techniques helpful, there are a number of implementation-related shortcomings. This study makes a modest effort to give a comparative analysis between the option of adoption and surrogacy in Indian framework and talk about the various factors in relation to cost, viability, etc. The paper will also discuss several legal rules, emphasising any gaps in distinct personal laws.

Keywords: Adoption, Assisted Reproductive Techniques, Personal laws

1. Introduction

It is the wish of every parent to have a kid to take their name further and to look after them in their old age. However, this is not at all times that their wish comes true due to various social and medical reasons. Medical science is offering many options for such kind of childless people like IVF, IUI, and ICSI, and several others, however, they don't always work out. So, when these options fail, a barren couple choose to either go for adoption which can be beneficial for childless people as well as the kid to be taken in adoption that may be a child without parents or any guardian or if they wish to have biological link with the child, they opt for surrogacy.

For the most part, both practises are idyllic for formation of families. In adoption, one can register for adoption through a central organisation called CARA (Central Adoption Resource Agency). It is a lengthy procedure that might take several years. Moreover, as

*Corresponding author: shobhna.jeet@krmangalam.edu.in
[1]meghaygarg.1987@gmail.com, [2]inderpreet.kaur@krmangalam.edu.in, [3]neha.singh@krmangalam.edu.in

India is a State of diversities, every sect has its own set of beliefs and specific adoption laws in absence of Uniform Civil Code. On the contrary, substitution is a process where a lady agrees to pregnant for another who is unable to have children owing to certain deformities. But it has been perceived that surrogacy involved many social, legal, moral, medical and ethical issues which make it far from reality.

Surrogacy and adoption are both wonderful ways for potential parents to have a child. Both has its own set of advantages and disadvantages. This paper is an attempt to discuss the exact position of these two methods to address the problem of childless couple.

2. Adoption

Adoption implies the procedure within which the child is isolated from his natural guardians and turns into the unpretentious offspring of his new parents with every right, advantages and consequences that come with the relationship. Thus, it can be stated that adoption is a transplantation of child with all the rights of naturally born child into the family where he was not born naturally.

As stated above, India is a nation of great diversity including many beliefs and each religion is governed by their own personal laws. As far as adoption is concerned, law of adoption is also not common for all religion. For example, Hindus can embrace under the Hindu Adoption and Maintenance Act, 1956, however, Muslim personal law is not having any concept of adoption. Provision of acknowledgment of paternity is there in Muslim law. Also, they can have child through adoption only through orphanage by getting authorization from the court in Guardians and Wards, Act, 1890. Likewise,

in Christian and Parsis, adoption is not a notion that exists, they have the ability to adopt a kid who has been placed in foster care as per the Guardians and Wards Act, 1890 which is not like adoption as if a child is taken in foster care, he is at liberty to break the ties with foster parents after attaining age of majority.

Indian legislations dealing with adoption

The Hindu Adoption and Maintenance Act, 1956

Main highlights of this Act are:

- A Hindu is allowed to adopt a kid under this Act.
- The age gap among parent who is adopting and an adopted child of opposite sex should be least 21 years.
- If an individual is already having a living natural or adopted child, the second child of same sex can't be adopted.
- A widowed, unmarried, or separated Hindu lady can accept a kid, but a married woman cannot adopt a child without her better half's consent, unless he has a poor brain and is incompetent of making choices or providing assent.
- A male who is having spouse living can have a kid through adoption with the approval of his wife except in case where she is not able to give a valid consent.
- If the embraced child is vested with some property that will remain his personal property and that cannot be taken away just because he is now a member of another family.
- Adoption in no way will be affecting sapinda relationship or prohibited relationship that adopted child was having in his natural family.

The Guardians and Wards Act, 1890

Main highlights of this Act are:

- Any person irrespective of his religion can take a child as his ward under this Act.
- Child can only be taken in guardianship and that too will be for 21 years that after attaining this age, child will be free to even break the ties from his guardian.
- There is no provision given regarding guardianship of orphan, abandoned and surrendered child.

Juvenile Justice Act, (Care and Protection of Children) Act

- Every citizen is entitled to adopt, and every kid has the potential to be adopted, regardless of their religious beliefs.
- Two children of the same sex can be adopted.
- It further provides that orphan, abandoned and surrendered children can be adopted for therapeutic reasons. These children can be given in adoption as per the direction of courts in acquiescence with the adoption guidelines published by the concerned authorities under this Act.

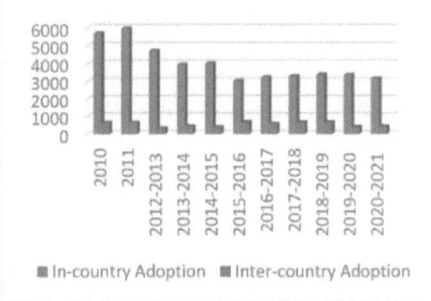

Fig. 38.1 Adoption Statistics available on the official website of Central Adoption Resource Authority Ministry of Women and Child Development, Government of India

Source: http://cara.nic.in/

Despite of having the provisions related to adoption and reforms made in the adoption laws and making it possible for them to get a child adoption irrespective of their personal laws, still the adoption rate is falling in last decade in India as per adoption statistics released by Central Adoption Resource Authority (CARA).

The reason for fall in adoption rate is that it involves complex and time taking procedure to adopt a child. Moreover, lack of genetic link is another factor driving people to look for alternatives like surrogacy to have child

3. Surrogacy

As exemplified by the Black's Law Dictionary, the procedure of bearing and carrying a child on behalf of another is known as surrogacy. The word surrogate means a substitute which means a person chosen to fill in favour of another. In this manner, a replacement mother is a female who transports a child for somebody else, either with her own egg or from the placement in her womb of a prepared egg of another woman, with the aim of carrying the kid to full term and then handing the child over to the prospective parents after delivery. The surrogacy procedure can be classified as:

- Conventional surrogacy and gestational surrogacy
- Commercial surrogacy and altruistic surrogacy

In customary surrogacy that has been in use since the dawn of time, when recent expertise was not there, a substitute mother's egg is fertilized with the prospective father's sperm, and the kid is then carried to full term. The natural mother of the baby is the substitute mother in this form of surrogacy.

In gestational surrogacy, the intended father's sperm is pollinated with the intended mother's egg. After this treatment, pollinated egg is then implanted in the surrogate mother's uterus that will to conceive a child. With this kind of surrogacy, the child is genetically related to the planned parents, not the substitute woman.

Another form of surrogacy is commercial surrogacy where surrogate women give birth to the child in lieu of financial assistance apart from the medical expenditure. In this process, parties enter into an agreement whereby the surrogate woman agrees with the intended parent to bear the child for them and in return intended parents agree to pay her for the same. On the contrary, in altruistic surrogacy, no financial assistance is given to surrogate women except medical expenditure for the service she provides.

In 2002, commercial surrogacy became permissible in India which gave rise in the emergence of many agencies that claimed themselves s as specialized in arranging surrogacy facilities. Moreover, it attracted foreign nationals also as in India they were able to get surrogate at vary cheap rate as India still have a large inhabitant who are there below poverty line and here, in India medicinal services are also not very expensive. But actual picture is very harsh. In the absence of any specific legislation, commercial surrogacy has given rise to many issues like selling of children, issue of custody of child, harassment of surrogate women, unhealthy living condition, mental trauma and may more.

In 2008, in the matter of Baby Manji Yamada v. Union of India and Another, the Indian judiciary principally dealt with the issue involved in surrogacy i.e., custody of child. In this case, Patel, who works at the Akanksha Infertility Clinic, agreed for Pritiben Mehta to have a surrogate baby for Japanese couple Ikufumi and Yuki Yamada. Yamada's sperm and an unknown Indian woman's egg were used to impregnate Pritiben. Yamada and his wife, meanwhile, applied for annulment of marriage after the pregnancy. None of the Indian laws addressed the main issue arose about the parentage of the child that who was the mother of the infant (Manji): Pritiben, Yuki Yamada, or the lady who contributed the egg. There was no law to address this issue. Dr. Patel was also accused of conducting a child trading ring by taking advantage of the want of surrogacy legislation, according to a subsequent court filing. The matter was concluded and Baby Manji was handed to her grandma Emiko but, this case forced to think for strict legislation do deal with issues in surrogacy business and in response the Surrogacy (Regulation) Bill, 2016, was passed further passed by the Lok Sabha in 2018, and then passed again as the Surrogacy (Regulation) Bill in 2019 which ultimately took the shape of law as The Surrogacy (Regulation) Act 2021 which came into force from 25th January, 2022. With this law commercial surrogacy was outlawed fully and only altruistic surrogacy remaining legal.

Legislation dealing with surrogacy in India

The Surrogacy (Regulation) Act, 2021

Main highlights of this Act are:

- Commercial substitution is outlawed, leaving only altruistic substitution as an option.
- Surrogacy procedure or actions by surrogacy clinic are prohibited and they are not permitted to serve anyone who fails to meet the Act's requirements.

- Clinics that provide surrogacy services must submit a tender for registration within sixty days from the day of the competent authority's appointment subject to renewal every 3 years.
- Only a legitimately wedded Indian man and woman can seek surrogacy where a male should be between the ages of twenty-six and fifty-five and a woman between the ages of twenty-five and fifty. Further, they must not previously have any genetic, adoptive, or substitute child whatsoever.
- In order to serve as surrogate, a woman must be an Indian lady between the ages of thirty-five and forty-five.
- A woman can be a surrogate only for a single time in her life.
- Intended couple must have a 'Certificate of Essentiality' before going for surrogacy.
- Surrogate mother must have 'Certificate of Eligibility' before going for surrogacy.
- Any known negative effects and risks of the operation must be disclosed to the surrogate mother. She must also provide written authorization in a language that she is comfortable with.
- The National Assisted Reproductive Technology and Surrogacy Registry records office shall be established for the purpose of registering substitute facilities under this Act.
- The Act mentions certain activities as offences such as commercial substitution, exploiting the substitute mother, ignoring, searching, or shunning a substitute kid and importing or exporting human embryos or gametes.
- Any couple found engaged in commercial surrogacy will face a maximum 5-year prison term and a fine up to Rs. 50,000 if they commit it for

the first time. It may enhance to 10 years in jail and a fine up to Rs 1,000,000 for repeated felony. And if any other person, institution, or a clinic that is discovered to have abused surrogate or proxy mothers or their offspring might receive up to ten years in prison and a fine up to Rs. ten lakhs.

But, it has been argued that despite the fact this law has been passed with a vision to ban commercial surrogacy in India to protect the surrogate woman from exploitation and to safeguard the rights of surrogate mother, intended parents and that of surrogate child but simultaneously, it has been asserted this Act is contravening Article 21 & Article 14 of Indian Constitution as it an attack to right to reproductive choice of a woman as well as it denies the right to have child to single parent or gay/homosexual couple even in an era where homosexuality is declared legal by the Apex Supreme Court itself.

4. Adoption v. Surrogacy

As far as comparison is concerned to learn about the best way to have a kid, we can analyses that both adoption and surrogacy existed in India a considerable time in one form or another. Adoption is the practice that not only supports the childless parents but it is beneficial for various children who are without parents. Adoption process is regulated in India through a centralized agency *i.e.*, CARA which keeps a data of all adoption agencies and adopted children but still, it involves a lot of issues and greatest of them is absence of genetic link and disparity in personal law that is reducing it to bare minimum. On the contrary gestational surrogacy is a modern concept. Primary justification for using surrogates is that it gives a biological connection with the baby and this process helps the parents to involve in their child's growth from the very beginning unlike adoption where child may be older one. Intended parents are able to get parental rights even before the child is born. This process is also not without issues which primarily includes exploitation of surrogate women and unfair practices done by ART clinics. To cope up with this problem The Surrogacy (Regulation) Act 2021 was passed.

5. Conclusion

Adoption as well as surrogacy both is the ways to help the childless couple. In adoption, where one can have a child who is biological child of someone else whereas in surrogacy a couple with proven infertility is capable of bearing its own genetic child with the support of gestational carrier. But it is quite difficult to state which among the two is better option for having a kid. Both have advantages and disadvantages, and the technique to be chosen is determined by the couple's needs. Surrogacy allows a couple who wants children of their own, on the contrary, if a person is willing to give home and parental care to the child of someone else, he or she may consider adoption. Monetary consideration is also a contributing aspect. Adoption process is far cheaper than surrogacy. In addition, the personal laws are also posing a hindrance when concerning adoption in the nonexistence of uniform adoption law. To conclude, it can only be argued that neither adoption nor surrogacy can be explicitly described as a benefit or a disadvantage. For surrogacy, it may be a curse if not properly controlled and handled, since it can lead to the exploitation of many children, & women so with the adoption which is extremely tedious and drawn-out process. However, if done correctly, these two practices may provide joy to a lot of childless couples.

REFERENCES

1. Bakshi, R (2018), Surrogacy and Adoption: An insight, The Times of India, Retrieved from, https://timesofindia.indiatimes.com/blogs/love-redux/surrogacy-and-adoption-an-insight/ visited on 15.04.2022
2. Retrieved from http://cara.nic.in/ last visited on 15.04.2022.
3. Satyan, Kanika (2015), Adoption v. Surrogacy - An Indian Comparative Legal Analysis, Retrieved from SSRN: https://ssrn.com/abstract=2625608 visited on 18.04.022
4. The Hindu Adoption and Maintenance Act, 1956
5. The Guardians and Wards Act, 1890
6. https://indiankanoon.org/doc/1874830/#:~:text=%E2%80%94When%20a%20guardian%20appointed%20or,Chapter%20II%2C%20may%2C%20if%20 last visited on 18.04.022
7. Bryan A. Garner, Black's Law Dictionary, 8th Edn. 2004, p. 4529
8. Retrieved from https://legalbots.in/blog/surrogacy-regulations-in-india-surrogacy-regulation-act-2021 last visited on 12.05.2022
9. https://egazette.nic.in/WriteReadData/2021/232118.pdf visited on 12.05.2022
10. Navtej Singh Johar v. Union of India, AIR 2018 SC 4321

Integrating Advancements in Education, and Society for Achieving Sustainability – Dimitrios A. Karras et al. (eds)
© 2024 Taylor & Francis Group, London, ISBN 978-1-032-70841-6

Effects of Teamwork, Financial Behaviours, and Human Resources on Cooperative Financial Management

39

Manoj Kumara N. V.*

Maharaja Institute of Technology, Mysore, India,

Manjunatha M. K.[1]

Visvesvaraya technological University,
Centre for Post Graduate Studies, Mysuru, Karnataka, India

Praveen Kumar M.[2]

Trinity Group of Institutions, Vijayanagar, Mysuru, India

Abstract: It is hoped that cooperation financial management may provide for community requirements in the areas of collaborative oversight and Human Resource Management (HRM). Human resources, financial attitudes, and coordinated effort were investigated to better understand their impact on cooperatives' financial management. We used a Likert-scale based study strategy to compile our data. The data was analysed using multiple regression methods, and samples were selected using a random sampling procedure. The study sample consists of 168 unions located in Buleleng Regency, Bali Province, and is determined using the Slovin formula. The results of the study show that cooperative financial management may be considerably enhanced by allocating partial human resources, considering financial attitude aspects, and coordinating various variables. The study's findings suggest that competent financial management will have a stronger impact on a cooperative's growth potential, leading to a need for more skilled cooperative managers.

Keywords: Human resources management (HRM), collaborative, cooperation, financial attitudes, multiple regressions

1. Introduction

Workplace transitions can have negative repercussions on employees and companies. HRM studies must now focus on worker happiness due to these changes. Pressure to be productive at work has intensified since the 2008 financial crisis, yet this has hurt worker satisfaction and output. As inequality and working conditions worsen,

*Corresponding author: drmanojkumara.2020mba@gmail.com
[1]drmkmanjunath11@gmail.com, [2]pkglobalreach@gmail.com

DOI: 10.4324/9781032708461-39

fairness and optimism fall. Human resources policies and procedures can increase morale and productivity in this scenario. We found certain difficulties in the HRM-wellness relationship (Salas-Vallina, et al., 2021). Economic growth has destroyed the planet's ecosystem and natural resources. Social activity is restricted. Studies show the human factor's impact on sustainable development and resource preservation. With a focus on social responsibility and sustainable performance, firms are also committed to social and environmental goals. 70% of 2800 worldwide companies surveyed include sustainability in their targets (Chams, et al., 2019). Traditional corporate wellness initiatives promote individual behaviour change (Jayesh, et al., 2022). As an indication of how far corporate wellness solutions have come beyond simply offering staff with health information and exercise recommendations, we may look to massive courses, identity systems, and specialised coach service with financial incentives (Rajagopal, et al., 2022). Wellness programmes have evolved from being seen as a major expense to reduce future care and support having spent to being viewed as a crucial investor for trying to shape organisation behavior, constructing brand recognition, and trying to recruit best talent in response to the younger generation demands for intent, society, and versatility in the worksite (Wieneke, et al., 2019).

2. Related Works

Cooper, et al., 2019 looked at how HRM strategies that prioritise workers' happiness might boost productivity. Anwar, et al., 2020 examined the impact of Green HRM practises on academic staff's environmental citizenship behaviour (OCBE) and environmental performance. Lee, H.W. (2019) combined collaboration, employee satisfaction, and work motivation to evaluate that sustainable HRM influences organisational performance. Kaufman, B.E. (2020) discussed psychologization, which lowers macro-level HRM data to individual psychological-behavioural aspects and variations. Data from 1666 Spanish manufacturing firms was analysed by Martnez-Sánchez et al. 2020 to test the hypothesis that a moderating role played by HR flexibility in R&D activities and knowledge absorption.

3. Research Methodology

This study endeavored to uncover and demonstrate the relationship between cooperative financial management and human resources, financial behaviors, and teamwork. This study employs a quantitative methodology. This technique for gathering data is a research technique. A set of questions is used in this research method to contact cooperatives in Bali's Buleleng Regency. The sample for this study was chosen based on calculations made with the Slovin method utilising straightforward formulae and computations. In this study, 2831 cooperatives in Buleleng Regency served as the population. 165 cooperatives in Buleleng Regency made up the research sample, which was determined using the Slovin formula. Simple random sampling is the foundation of the sampling process. The analytical approach makes use of multiple regressions, and the traditional assumption test is made up of "normality tests, multicollinearity tests, and heteroscedasticity tests". Data quality testing also includes validity and reliability tests.

4. Research Hypothesis

H1: Good human resource effects on teamwork financial management.

H2: Team - based financial management benefits from having a positive financial mindset.

H3: Cooperation financial management benefits from teamwork.

5. Result and Discussion

When the authenticity of the data was confirmed, it was found that each statement's significance value for each variable was less than 0.05, proving that the data were accurate and suitable for this study. The reliability analysis showed that all of the variables included in this study had a Cronbach's alpha value greater than 0.59, making them all appropriate for usage.

Exponential growth Kolmogorov-Smirnov Z test gives normally distributed data $0.427 > 0.05$. Tolerance levels > 0.10 and VIF values > 10 were achieved across the board for all research variables. Since all variables had significance levels > 0.05, none demonstrated heteroscedasticity. The summary model's adjusted R2 is 0.398, or 39.8%. Human resources, financial attitudes, and coordination explain 39.8% of cooperative financial management variables (100%-39%). "Human resources, financial behavior, and teamwork may all affect cooperative financial management, according to the "ANOVA or Simultaneous F test" regression table. Human resources are 0.002 as important as financial behaviour and collaboration in collaborative financial management.

Table 39.1 Determining measurement

Model	R	Adjusted R Square	Std. Error of the Estimate	R Square
1	0.638	0.397	2.08656	0.409

a. KO, SK, SDM, and (variable) as determinants
b. PKK is a response variable.

Table 39.2 Statistical F-test

Model		df	Sum of Squares	F	Mean Square	Sig
1	Regression	3	487.980	37.361	162.660	0.000b
	Residual	162	705.297		4.354	
	Total	165	1193.277			

a. PKK is a response variable
b. Determinants: (variable), Team work, Financialbehaviours, HRM

Table 39.3 Statistical t-test

Model		Unstandardized Coefficients		Standardized Coefficients	T	Sig.
1		**Std. Error**	**B**	**Beta**		
	(variable)	2.178	8.364		3.839	.000
	Co	.067	.208	.258	3.107	.002
	FA	.073	.238	.235	3.242	.001
	HRM	.031	.097	.267	3.079	.002

PKK is a response variable

According to the T-test, the value of the variable representing the degree to which employees work together was 3.107 > 1.653, the value of the variable representing the degree to which employees' financial attitudes were positive was 3.242 > 1.653, and the value of the variable representing employees' access to human resources was 3.079 > 1.653. Three separate factors substantially impact the dependent variable, according to the regression test. Table 3 shows the regression results and statistical relevance tests for each variable.

Because the t-value of 3.079 for the human resources variable is greater than the table value of 1.654, we accept H0 (hypothesis 1). Thehuman resource element positively affects cooperative financial management. Cooperatives' human resources are crucial to their future growth. Proper financial management of owned human resources can help to cooperative operations. Human capital that is equipped with the necessary skills, knowledge, abilities, and attitudes may manage cooperative funds effectively. Smart human resources, particularly in accounting, can assist collect richer and more trustworthy financial data. Better performance and more efficient use of funds may result from the availability of appropriate knowledge and skills in financial management, which could be put into practise by taking on responsibility. Human resource competences positively affect the quality of financial statements. Thus, skilled human resources improve cooperative financial management.

Multiple linear regression According to the regression test, financial behavior towards cooperative financial management has a significance value of 0.001 less than 0.05 and a t count of 3.242 greater than a t - statistic of 1.653, supporting hypothesis 2 that the variable has a statistically significant positive influence on cooperative financial management. A good financial behaviour may bring about improved fiscal responsibility and self-assurance. The way people think and act about money is what ultimately determines their financial success. The person or the office culture would suffer as a result of this mindset. A financially responsible person will think about their money in a methodical, well-planned manner. This study's financial attitude contributes to individual thinking, appraisal, and judgements about financial management methods for himself and his organisation or business. In this opinion, one's personal feelings about money have a role in how it's handled in the job

To have a better financial mentality is to manage money better. Conversely, as the cooperative is a corporation with several shareholders, the manager's commitment to fiscal responsibility is recognized. It found that financial attitudes influence financial management behaviour. Financial attitudes impact financial management behaviour. The ability to see the big picture financially has a significant impact on personal economic freedom. Multiple linear regression analysis on the teamwork variable supports the third null hypothesis, indicating that the coordination variable positively affects cooperative financial management (t = 3.107 > t = 1.653). Coordination is key to implementing activities and making decisions, especially in financial management. Coordination is done to discuss, direct, and synchronise financial arrangements for each cooperative unit or division. Without cooperation in financial management, abuse can affect entire management. Cooperative financial management is one indicator of cooperative coordination or collaboration. This research shows that the Foundation successfully organises its goals and operations by using a combination of horizontal and vertical management information systems. It found that coordination meetings are present in every academic activity and the incidental administration of the academic community, demonstrating the importance of cooperation and working relationships. Cooperation among team members is a key factor in the success of financial management.

6. Conclusion

Human resources, financial behavoiur, and teamwork have a favourable influence on cooperative financial management,

according to studies. Human resources with skills, knowledge, talents, and competent attitudes may handle cooperative funds successfully. A person's financial mindset can affect the surroundings, such as in financial management. Better financial mindset means better financial management. Teamwork is done to identify, manage, and synchronise financial arrangements for each cooperate unit or department. In this study, it is difficult to corroborate respondent replies using a questionnaire used to collect data on research hypotheses. Limiting this study's scope to the Buleleng regency helps keep things manageable. Internal controls, loci of control, local expertise, and similar concepts warrant more study. Since there are likely many more variables at play, as well as other purpose of the study, a replication of this study is possible.

REFERENCE

1. Salas-Vallina, A., Alegre, J. and López-Cabrales, Á., 2021. The challenge of increasing employees' well-being and performance: How human resource management practices and engaging leadership work together toward reaching this goal. Human Resource Management, 60(3), pp. 333–347.
2. Chams, N. and García-Blandón, N. and García-Blandón, J., 2019. On the importance of sustainable human resource management for the adoption of sustainable development goals. Resources, Conservation and Recycling, 141, pp. 109–122.
3. Wieneke, K.C., Egginton, J.S., Jenkins, S.M., Kruse, G.C., Lopez-Jimenez, F., Mungo, M.M., Riley, B.A. and Limburg, P.J., 2019. Well-being champion impact on employee engagement, staff satisfaction, and employee well-being. Mayo Clinic Proceedings: Innovations, Quality & Outcomes, 3(2), pp. 106–115.

4. Rajagopal, N.K., Saini, M., Huerta-Soto, R., Vílchez-Vásquez, R., Kumar, J.N.V.R., Gupta, S.K. and Perumal, S., 2022. Human resource demand prediction and configuration model based on grey wolf optimization and recurrent neural network. Computational Intelligence and Neuroscience, 2022.

5. Jayesh G. S. Dony Novaliendry, Shashi Kant Gupta, Amit Kumar Sharma and Bramah Hazela 2022 A Comprehensive Analysis of Technologies for Accounting and Finance in Manufacturing Firms. ECS Transvol. 107(1) pp: 2715.

6. Cooper, B., Wang, J., Bartram, T. and Cooke, F.L., 2019. Well-being-oriented human resource management practices and employee performance in the Chinese banking sector: The role of social climate and resilience. Human Resource Management, 58(1), pp. 85–97.

7. Anwar, N., Mahmood, N.H.N., Yusliza, M.Y., Ramayah, T., Faezah, J.N. and Khalid, W., 2020. Green Human Resource Management for organisational citizenship behaviour towards the environment and environmental performance on a university campus. Journal of Cleaner Production, 256, p.120401.

8. Lee, H.W., 2019. How does sustainability-oriented human resource management work? Examining mediators on organizational performance. International Journal of Public Administration, 42(11), pp. 974–984.

9. Kaufman, B.E., 2020. The real problem: The deadly combination of psychologisation, scientism, and normative promotionalism takes strategic human resource management down a 30-year dead end. Human Resource Management Journal, 30(1), pp. 49–72.

10. Martínez-Sánchez, A., Vicente-Oliva, S. and Pérez-Pérez, M., 2020. The relationship between R&D, the absorptive capacity of knowledge, human resource flexibility and innovation: Mediator effects on industrial firms. Journal of Business Research, 118, pp. 431–440.

Note: All the tables in this chapter were made by the authors

Integrating Advancements in Education, and Society for Achieving Sustainability – Dimitrios A. Karras et al. (eds)
© 2024 Taylor & Francis Group, London, ISBN 978-1-032-70841-6

Cultural Wars in School Mathematics from Philosophical Standpoint: Mathematics, Mathematicians, and Mathematics Education in India

40

Ashraf Alam*, Atasi Mohanty

Rekhi Centre of Excellence for the Science of Happiness,
Indian Institute of Technology Kharagpur, India

Abstract: Math is a major source of stress for students. Feelings of inadequacy are intrinsically tied to *arithmophobia*. Many kids in India give up on mathematics by the time they are in 3rd or 4th grade. Statistics show that the highest percentage of candidates who do not pass, fail in mathematics. Math education theorists have been at each other's throats for years about what constitutes 'good evidence' for their own beliefs. These debates occurred after World War II and included behaviourism and cognitivism. Standard cognitivist models of mathematics learning and radical constructivism have been at odds since the 1980s, mostly on epistemological issues like the claim that all knowledge is created by the learner. Raising epistemological problems in discussions about education has always been a catalyst. This article thus attempts to engage metaphysically in these nuanced philosophical debates on mathematics education.

Keywords: Mathematical thinking, Philosophy of mathematics education, Mathematics education in India, School mathematics, Indian education system, Mathematics pedagogy

1. Introduction

It was often believed that a school's curriculum is the set of courses students had to take in order to graduate. Although educational philosophers have been interested in the topic of curriculum since at least the 18th century, the discipline of curriculum studies did not emerge until the 1890s. Johann Friedrich Herbart (1776–1841), a German philosopher, is frequently cited as a pivotal thinker in the development of the modern curriculum. In his pedagogical ideas, Herbart emphasised the importance of selection and organisation of teaching-learning material. The first book dedicated to the topic of curriculum was written by Franklin Bobbitt in 1918 and titled 'The Curriculum'. In 1924, Franklin Bobbitt wrote the book 'How to Make a Curriculum'. Curriculum-related yearbook 'The Foundation and Technique of Curriculum - Construction' was published in 1926. Starting in the 1890s, the curriculum development movement gained traction and

*Corresponding author: ashraf_alam@kgpian.iitkgp.ac.in

DOI: 10.4324/9781032708461-40

spread throughout the world [1]. School mathematics curriculum in India has far-reaching consequences, and thus cannot be taken lightly while developing it. Many philosophical and theoretical questions shall be debated for effective learning and teaching of mathematics at schools in India [2]. Exactly which mathematical idea(s) depend on unstated assumptions, if any? In what way do we know that these presumptions are correct? Which ideas of knowledge and education are taken as givens? What evidence exists for radical constructivism and sociocultural perspectives on mathematical learning in India?

In what ways do learning theories, which are typically cognitive and often individualistic, take into account the social environment of learning mathematics in India? Will competing learning theories help or hinder mathematics education for Indian students? In what ways do these concepts diverge from typical pedagogical approaches in Indian classrooms? Is there evidence of discrepancies between common teaching methods in India and the most cited theories of education?

2. Feelings of Inadequacy are Intrinsically Tied to 'Arithmophobia'

What mathematical ideas are most useful in Indian arrangements? How exactly should these mathematical ideas be evaluated? How do the feedback loops established by various evaluation methods affect mathematical education in India? How similar are mathematical knowledge justification and mathematical learning assessment? What role do Indian students play in this scenario? What skills are

necessary for Indian students to succeed in mathematics? How does one's experience in math class shape who they become? Is there any research in the Indian context that shows whether learning math improves or harms a person? How often are these effects - positive or negative? What kinds of learning situations produce them, and what role does Indic culture play?

Is there a difference in how math instruction impacts children depending on their demographic characteristics? To what extent does mathematics education in India shape the lives of future scientists, engineers, and citizens? Exactly what part do subjective factors like feelings, outlooks, and values play in the field of mathematics? How can one develop mathematical ability? Does every Indian student have the potential to study mathematics? How closely should Indian students' mathematics education be modelled after research findings in mathematics education? How well do Indian students' mathematical abilities translate to their other real-life situations?

In all of academia, mathematics is the one subject that is certain to elicit a broad range of feelings from its students [3]. Anyone who grew up in a Tamil-speaking household would be hard-pressed to claim ignorance of the Tirukkural. Those who suffer from arithmophobia or arithmetic anxiety often appear in scientific literatures. In the Indian setting, this kind of dread takes on a very local flavour [4]. Statistics show that the highest percentage of candidates who do not pass, fail in mathematics. The factors that contribute to students' dislike of mathematics in the classroom have been the subject of much research and analysis [5]. The debate over whether knowledge and learning are primarily individual or societal has lately been bolstered by constructivism and socio-cultural perspectives. There is currently no clear-cut answer to this debate, despite several attempts to do so by past researchers, with novel theories, that blend aspects of individual cognition and communication. When compared to different learning theories, how effective is this one over the other? Do different learning models produce different results?

3. Self-Fulfilling Prophecies Produced by Pervasive Societal Attitudes

There are a lot of obstacles that need to get away in order to improve mathematics instruction in Indian schools. Indian students' unfavourable views regarding mathematics education are influenced by four factors that may serve as a framework for understanding these difficulties. A lack of confidence among the majority of students, a curriculum that disappoints both a talented minority and the non-participating majority, assessment methods that are overly simplistic and promote a view of mathematics as mechanical computation, and inadequate preparation and support for teachers. All of these factors affect the pedagogical foundation, which needs serious and immediate attention [6]. This would illuminate the many pathways through which a person grows as a moral agent. This refines the idea of autonomy such that it is linked to questioning one's own sense of the inevitable and natural in their sense of self. A classroom is a place where each student must take initiative to learn and internalise their own distinct meaning [7]. One of mathematics' defining features is that it builds up over time. If you have trouble with decimals, you

will have trouble with percentages, and if you have trouble with percentages, you will have trouble with algebra and other math topics. One other major factor is the prevalence of symbolic language. Some kids are bored and disinterested when they cannot understand the symbols being used around them [8].

Social cues may be used to predict mathematical failure. Education in India has structural obstacles that reflect unequal social structures based on socioeconomic status, gender, and caste, all of which contribute to failure (and the perception of failure) in mathematics education. This setback is compounded by the self-fulfilling prophecies produced by pervasive societal attitudes that portray women as incapable of mathematics or that, for ages, have linked formal mathematical proficiency with higher castes [9].

4. Mathematical Language: A Potent Source of Isolation

It is important to acknowledge the language employed in textbooks, particularly at the primary level, as a major source of challenge. The great majority of students in India struggle with maths because the language used is so different from their daily language. As so, it becomes a potent source of isolation all on its own. Anxiety is inevitable in a math class where rote learning of procedures and formulae takes precedence over developing a comprehension of the underlying concepts. It is common for students to give up on mathematics early on in school, either because they are afraid of failing or because they are only able to muster a smattering of success [10]. What ideas and principles guide the teaching of mathematics in India? Do advanced mathematical teaching theories exist? Do we have any other term that is very close to 'pedagogy'? Why do different approaches to teaching math to Indian school students have different implicit assumptions? In what way do we know that these assumptions are correct? Does each method of teaching mathematics by Indian math teachers have its own unique set of underlying pedagogical tenets?

Perhaps more so than in any other discipline, the success of a math teacher rests on her mastery of the subject matter, the nature of mathematics, and the methods at her disposal. Teachers lose interest in teaching math when they have to rely solely on textbooks. Math teachers have unique challenges at both ends of this spectrum. Most elementary school instructors incorrectly believe they know all there is to know about teaching mathematics and instead try to replicate the methods they were taught themselves while they were in school. Sometimes this works, but usually, it merely helps to postpone or conceal the real issue till later [11]. What results can we expect from various mathematical pedagogical approaches? Or, do other assumptions, like values, or the acceptance or promotion of particular philosophies or epistemologies, have to be made in order to significantly impact either classroom strategies or students' learning outcomes in Indian schools? Do the goals and methods work together? Can the different competing theories on how best to teach mathematics to Indian students be identified and analysed? What strategies, resources, and methods have been or are currently being utilised in the realm of mathematics instruction in India? Which of these has been useful, and in what cultural contexts has it been most useful?

5. Continuous Professional Development (CPD): Technology and Child Psychology

Some educators in secondary and higher education have a unique difficulty in the classroom. The lack of continuous professional development (CPD) programmes for educators and the major evolution of curricula since teachers' own schooling mean that many educators lack a solid foundational understanding of key concepts in their respective fields [12]. This leads them to depend on commercially provided 'notes' that offer inadequate background and explanation to pupils. Due to a lack of resources and training, elementary school mathematics teachers often fail to adapt their lessons to reflect the latest research in child psychology. When concepts are formally represented in algebra in upper elementary, a lack of prepared teachers shows itself in a failure to make connections between formal mathematics and practical learning. Inadvertently squelching students' interest and enthusiasm for math and science by ignoring their intrinsic connection has far-reaching consequences [13].

How do diverse technological strategies support different mathematical pedagogical tenets, and what ideas underpin them? In what ways, both expected and unexpected, do these technologies improve our lives? Is there a philosophical framework for understanding how technology connects us to the wider natural and social worlds? To what extent must one understand mathematics in order to achieve the goals of mathematics education? How can we gauge the success of our efforts to improve mathematics education in India? Where does the instructor fit in? What options does the teacher have as an intermediary in the relationship between the learner and the mathematics? It is troublesome that elementary, middle, and high school math instructors seldom talk to one another, let alone their colleagues at the university level [14]. Most teachers never talk to or consult with mathematicians who research in their field. Once again, those who work in teacher preparation are not found in places from where sophisticated mathematical content can be learned, like universities or laboratories. Your family's ancestors were probably forced to take an analytical geometry course in college. These topics are now included in standard curricula. Some areas of study, such as solid geometry and spherical geometry, have been eliminated so that more time may be spent on the new, faster methods. The prevalent belief that introducing calculus and differential equations to pupils at an early age is beneficial for those who choose careers in the sciences, technology, and engineering may be at least partially responsible for the narrowing of focus. Regardless of the motivation, mathematics education has become more vertical than horizontal.

6. Conclusion

Unfortunately, it is common knowledge that a very small number of young children show remarkable ability and enthusiasm in mathematics [15]. Gifted pupils get nothing from the mathematics programme except for a crushing sense of disappointment. The programme does not make sufficient use of the students' motivation since it lacks conceptual depth and is only moderately demanding. While some schoolwork might be a nuisance inflicting annoyance, exams tend to wreck more anxiety in students. The

methods used to evaluate students' work are largely to blame for the hegemony of rote memorising of procedures and formulas in India's K-12 mathematics classrooms, which has been related to a number of the aforementioned issues. Exam pressure might cause pupils to memorise procedures without fully understanding what they are doing or why. Kids who worry and fall short because they cannot find a suitable replacement, are no less [16].

The 'Math Wars' is the latest battlefield in a long-running debate about the relative merits of various approaches to teaching mathematics. Progressive educators have resisted those researchers who favour traditional teacher-centered instructional methods, while instead emphasising child-centered teaching methods like problem-solving and investigative approaches. But does the data favour one of these perspectives, and if so, under what conditions? The progressive/traditionalist pedagogies divide has only helped to intensify resistance and has deflected attention from the aims of teaching mathematics.

REFERENCES

1. Alam, A. (2022). Positive Psychology Goes to School: Conceptualizing Students' Happiness in 21st Century Schools While 'Minding the Mind!' Are We There Yet? Evidence-Backed, School-Based Positive Psychology Interventions. *ECS Transactions, 107*(1), 11199.

2. Doig, B., Williams, J., Swanson, D., Borromeo Ferri, R., & Drake, P. (2019). Interdisciplinary mathematics education: The state of the art and beyond.

3. Alam, A. (2022). Investigating Sustainable Education and Positive Psychology Interventions in Schools Towards

Achievement of Sustainable Happiness and Wellbeing for 21st Century Pedagogy and Curriculum. *ECS Transactions, 107*(1), 19481.

4. Chevallard, Y. (1990). On mathematics education and culture: Critical afterthoughts. *Educational Studies in Mathematics, 21*(1), 3–27.

5. Alam, A. (2022). Mapping a Sustainable Future Through Conceptualization of Transformative Learning Framework, Education for Sustainable Development, Critical Reflection, and Responsible Citizenship: An Exploration of Pedagogies for Twenty-First Century Learning. *ECS Transactions, 107*(1), 9827.

6. Zazkis, R., & Campbell, S. R. (Eds.). (2006). Number theory in mathematics education: Perspectives and prospects.

7. Alam, A. (2020). Test of Knowledge of Elementary Vectors Concepts (TKEVC) among First-Semester Bachelor of Engineering and Technology Students. *Alam, A. (2020). Test of Knowledge of Elementary Vectors Concepts (TKEVC) among First-Semester Bachelor of Engineering and Technology Students. Periódico Tchê Química, 17*(35), 477–494.

8. Nesher, P., & Kilpatrick, J. (Eds.). (1990). Mathematics and cognition: A research synthesis by the International Group for the Psychology of Mathematics Education.

9. Alam, A. (2020). Challenges and possibilities in teaching and learning of calculus: A case study of India. *Journal for the Education of Gifted Young Scientists, 8*(1), 407–433.

10. Darragh, L. (2016). Identity research in mathematics education. *Educational Studies in Mathematics, 93*(1), 19–33.

11. Alam, A. (2020). Possibilities and challenges of compounding artificial intelligence in India's educational landscape. *Alam, A. (2020). Possibilities and Challenges of Compounding Artificial Intelligence in India's Educational Landscape. International Journal of Advanced Science and Technology, 29*(5), 5077–5094.

12. Ernest, P. (1989). *Mathematics Teaching: The State of the Art*. The Falmer Press, Taylor & Francis, Inc., 1900 Frost Rd., Suite 101, Bristol, PA 19007.

13. Alam, A. (2021, November). Possibilities and Apprehensions in the Landscape of Artificial Intelligence in Education. In *2021 International Conference on Computational Intelligence and Computing Applications (ICCICA)* (pp. 1–8). IEEE.

14. Alam, A. (2022). Impact of University's Human Resources Practices on Professors' Occupational Performance: Empirical Evidence from India's Higher Education Sector. In *Inclusive Businesses in Developing Economies* (pp. 107–131). Palgrave Macmillan, Cham.

15. Alam, A. (2022, March). Educational Robotics and Computer Programming in Early Childhood Education: A Conceptual Framework for Assessing Elementary School Students' Computational Thinking for Designing Powerful Educational Scenarios. In *2022 International Conference on Smart Technologies and Systems for Next Generation Computing (ICSTSN)* (pp. 1–7). IEEE.

16. Alam, A. (2022). Social Robots in Education for Long-Term Human-Robot Interaction: Socially Supportive Behaviour of Robotic Tutor for Creating Robo-Tangible Learning Environment in a Guided Discovery Learning Interaction. *ECS Transactions, 107*(1), 123.

*Integrating Advancements in Education, and Society for Achieving
Sustainability – Dimitrios A. Karras et al. (eds)*
© 2024 Taylor & Francis Group, London, ISBN 978-1-032-70841-6

Can India's National Education Policy 2020 cut the Hidden Costs associated with 'Shadow Education'?

41

Ashraf Alam*, Atasi Mohanty

Rekhi Centre of Excellence for the Science of Happiness,
Indian Institute of Technology Kharagpur, India

Abstract: Shadow education refers to private, fee-charging educational services that supplement or substitute for formal public education systems. It includes test preparation coaching, private tutoring, and supplementary classes outside of regular schools. Despite its widespread popularity, shadow education has been criticized for exacerbating existing educational inequalities and diverting resources away from the public education system. This paper analyses the provisions of the new National Education Policy (NEP 2020) in relation to its potential to address the challenges posed by shadow education. This paper tries to answer whether shadow education provides equal opportunities or unequal outcomes, whether it is for the pursuit of excellence or is it the price of pressure. The findings suggest that the NEP 2020, through its focus on revamping the formal education system and promoting a culture of learning, has the potential to address some of the key issues arising from the emergence of shadow education in India.

Keywords: School education, Coaching institutions, Shadow education, Private tuition, Indian schools, NEP 2020

1. Introduction

Education is a critical aspect of human development, playing a pivotal role in shaping individuals' lives and determining their future opportunities. The formal education system, consisting of primary, secondary, and higher education, is the cornerstone of most societies. However, in many countries, a parallel system of education, known as shadow education, has emerged. Shadow education refers to private tutoring and test preparation services outside the formal education system. In recent years, the use of shadow education has been on the rise, particularly in developing countries, where the quality of formal education is often perceived to be inadequate. Shadow education is a complex and multifaceted phenomenon that has significant implications

*Corresponding author: ashraf_alam@kgpian.iitkgp.ac.in

DOI: 10.4324/9781032708461-41

for individuals, families, and societies. On the one hand, it can provide students with additional support and opportunities for academic advancement. On the other hand, it can create new challenges and exacerbate existing educational inequalities. Despite its growing importance, little is known about the nature and extent of shadow education, its impact on students and society, and how it can be regulated and improved.

In this paper, the researcher aims to provide a comprehensive overview of shadow education, its causes, and its consequences. By examining existing research and data, the researcher aims to shed light on this emerging trend and provide insights into its impact on the education system and wider society. This paper will also examine policy initiatives aimed at addressing the challenges posed by shadow education and exploring potential solutions for reducing its negative effects.

2. Viewing "Shadow Education" from a Philosophical Standpoint

From a philosophical standpoint, shadow education raises several important questions and issues regarding the nature and purpose of education.

The nature of education: Shadow education raises questions about the definition of education and what counts as valid learning. Is education limited to formal institutions and government-approved curriculums, or can non-formal educational activities such as coaching institutes and private tutoring be considered education as well?

The purpose of education: Shadow education highlights the tension between the pursuit of academic success and the development of the whole person. Is education primarily about getting good grades and preparing for exams, or should it also focus on fostering critical thinking, creativity, and personal growth?

The role of the state: Shadow education raises questions about the role of the state in education. Should the state have a monopoly on education and regulate all forms of learning, or should individuals have the freedom to choose the type of education that is best for them?

Equity and access: Shadow education highlights the issue of equity and access to education. Does the growth of shadow education create a two-tiered system, where students from higher-income families have access to superior education, while students from lower-income families receive a lower-quality education?

The commercialization of education: Shadow education raises concerns about the commercialization of education and the commodification of knowledge. Does the growth of the coaching culture perpetuate a market-based approach to education, where education is seen as a commodity to be bought and sold?

These philosophical questions and issues surrounding shadow education highlight the complex and multifaceted nature of education and its impact on society. They also raise important questions about the role of education in promoting equity, access, and personal growth, and the extent to which education should be regulated and controlled by the state. By exploring these questions, we can gain a deeper understanding of the challenges and opportunities presented by shadow education and work towards creating an education system that is equitable, accessible, and promotes the development of the whole person.

3. Viewing "Shadow Education" from a Psychological Standpoint

From a psychological standpoint, shadow education has a significant impact on students' academic and personal well-being.

Academic performance: Shadow education is often seen as a means of improving academic performance, but it can also lead to increased stress and pressure to perform. Students who engage in shadow education may experience higher levels of anxiety, stress, and burnout, which can negatively affect their academic performance.

Self-esteem and identity: The pressure to excel in both formal and shadow education can lead to an excessive workload for students, negatively impacting their self-esteem and sense of identity. This can lead to feelings of failure and inadequacy, which can have long-term effects on students' personal and academic development.

Socialization: Shadow education can also have an impact on students' socialization. By devoting a significant amount of time to coaching institutes and private tutoring, students may have less time for other activities and relationships that are important for their personal and social development.

Family relationships: The pressure to excel in shadow education can also strain family relationships. Parents may feel pressure to provide their children with additional educational support, which can lead to financial strain and conflicts within the family.

Education beliefs and values: Shadow education can also influence students' beliefs and values about education. By emphasizing rote memorization and test preparation, shadow education can reinforce the belief that education is only about getting good grades, rather than fostering critical thinking and creativity.

These psychological impacts of shadow education highlight the importance of considering the well-being of students in the education system. It is essential to create an education system that promotes academic excellence and personal growth, without putting undue stress and pressure on students.

4. Viewing "Shadow Education" from an Economical Standpoint

From an economic standpoint, shadow education has a significant impact on the allocation of resources and the distribution of educational opportunities and outcomes.

Resource allocation: Shadow education diverts resources away from the public education system, potentially reducing the quality of formal education for all students. The excessive expenditure on private tuition and coaching institutes can reduce the resources available for the improvement of public schools, resulting in lower-quality education for students who do not have access to shadow education.

Income disparities: Shadow education tends to benefit students from higher-income families, exacerbating the already existing income-based educational disparities. These students are more likely to have access to high-quality shadow education, which can give them an advantage in the formal education system and in the labor market.

Market growth: The growth of the coaching industry has led to the creation of a large market for test preparation and private tuition services. This has resulted in the commodification of education, where education is viewed as a marketable product rather than a public good.

Market inefficiencies: The market for shadow education is often characterized by inefficiencies and market failures, such as information asymmetry and quality issues. Students and parents may have difficulty evaluating the quality of shadow education providers, which can result in suboptimal investments in shadow education.

Labor market implications: The shadow education industry has significant implications for the labor market, particularly for educators. The growth of shadow education has created a large demand for educators, but it has also resulted in lower wages and poor working conditions for teachers in the formal education system.

By understanding the economic effects of shadow education, policymakers and educators can work towards creating an education system that is equitable, efficient, and sustainable. This can involve policies and initiatives that support the improvement of public education, reduce income-based disparities in educational opportunities and outcomes, and promote the development of a well-functioning market for shadow education.

5. Viewing "Shadow Education" from a Political Standpoint

From a political standpoint, shadow education has a significant impact on the allocation of educational opportunities and the distribution of power and resources in society.

State vs. market control: Shadow education challenges the state's monopoly on education and shifts control over educational opportunities from the state to the market. This can result in a fragmented education system, where different segments of the population have access to different types and quality of education.

Political accountability: Shadow education can make it more difficult for the state to monitor and regulate the quality of education and ensure that educational opportunities are equitably distributed. The lack of transparency and accountability in the shadow education sector can undermine public trust in the education system and reduce political support for educational reform.

Education policy: The emergence of shadow education has implications for education policy, particularly regarding the role of the state in providing educational opportunities. Policymakers must consider the potential trade-offs between supporting the growth of shadow education and maintaining the quality and equity of formal education.

Political ideology: Shadow education is often seen as a reflection of broader political and ideological debates about the role of the state in society, the balance between state and market control over education, and the distribution of educational opportunities. Some view shadow education as a threat to the public education system, while others view it as a complement to formal education that offers greater choice and competition.

In conclusion, shadow education has significant political implications, as it challenges the state's monopoly on education, reinforces existing inequalities in access to education, and raises questions about the role of the state in providing educational opportunities.

6. Viewing "Shadow Education" from a Sociological Standpoint

From a sociological perspective, shadow education has several impacts on society. It contributes to the stratification of society, creating a two-tiered education system where students from higher-income families have access to better resources and opportunities, while those from lower-income families have limited access. This perpetuates existing income-based educational disparities and exacerbates social inequalities. Additionally, the pressure to excel in both formal and shadow education can lead to an excessive workload for students, negatively impacting their well-being. This can result in students feeling overwhelmed and stressed, leading to burnout, and negatively affecting their mental health.

Furthermore, the emergence of the shadow education industry highlights the limitations and deficiencies of the formal education system. It reveals a mistrust in the ability of the formal education system to provide students with the skills and knowledge they need to succeed in life. This undermines the social value and credibility of the formal education system and questions its ability to meet the needs of students. Finally, shadow education also has broader implications for society. It diverts resources away from the public education system, potentially reducing the quality of formal education for all students. This creates a vicious cycle where students from lower-income families are unable to access quality education and are disadvantaged in the job market, further exacerbating social inequalities.

In conclusion, shadow education, from a sociological perspective, highlights the complex interplay between education and society. It sheds light on the societal factors that drive the demand for shadow education, the consequences of its emergence, and the challenges faced by policymakers in addressing this issue.

7. India's Coaching Industry

India's coaching industry is estimated to be worth over $15 billion, with an annual growth rate of 20–25%. The coaching industry in India has grown rapidly over the past decade, with an increasing number of students seeking additional support in preparation for competitive exams, such as the Joint Entrance Examination (JEE) for engineering and the National Eligibility cum Entrance Test (NEET) for medical studies. As of 2021, there are over 40,000 coaching institutes in India, with the majority located in Tier 1 cities such as Delhi, Mumbai, Bangalore, and Kolkata. It is estimated that over 20 million students are enrolled in these institutes, with an average expenditure of $500–$1,500 per year. The growth of the coaching industry in India can be attributed to a number of factors, including the increasing competition for limited seats in top colleges, the growing demand for professional courses, and the lack of quality education in the formal education system.

8. India's new National Education Policy 2020 to Combat the Growing Coaching Culture in India

India's new National Education Policy 2020 aims to address the issues arising from the growing coaching culture in India through several strategies, including strengthening the formal education system [1]. By improving the quality of formal education and reducing the reliance on shadow education, the NEP aims to make education more accessible and equitable for all students [2]. It further emphasizes the importance of a holistic education that focuses on the all-round development of the student, rather than just academic achievement. This will help to reduce the pressure on students to attend coaching institutes and private tutors and improve their overall well-being [3]. The NEP further aims to increase access to quality education for all students, regardless of their socio-economic background. This shall be achieved by improving the infrastructure and resources of public schools, providing financial assistance to students from low-income families, and expanding access to online learning resources.

9. Conclusion

Shadow education is a complex and rapidly growing phenomenon that has significant implications for individuals, families, and societies [4]. Despite its potential benefits, such as providing additional academic support, it also has several drawbacks, including exacerbating existing educational inequalities, increasing workloads, and diverting resources away from the formal education system. The growing popularity of shadow education is a reflection of the wider educational challenges facing many countries, including the quality of formal education and the increasing importance of academic performance in determining future opportunities. However, it is crucial that policymakers and educators work together to address these challenges in a comprehensive and holistic manner, rather than relying on shadow education as a short-term solution.

In light of these challenges, it is important to explore alternative solutions that can improve the quality of formal education, such as investing in teacher training, expanding access to technology, and developing more comprehensive and inclusive assessment systems [5]. By doing so, it will be possible to reduce the need for shadow education and ensure that all students have access to a high-quality education that equips them with the skills and knowledge they need to succeed in the 21st century.

REFERENCES

1. Alam, A. (2022). Positive Psychology Goes to School: Conceptualizing Students' Happiness in 21st Century Schools While 'Minding the Mind!' Are We There Yet? Evidence-Backed, School-Based Positive Psychology Interventions. *ECS Transactions, 107*(1), 11199.
2. Kumar, A. (2021). New education policy (NEP) 2020: A roadmap for India 2.0. *University of South Florida M3 Center Publishing, 3*(2021), 36.
3. Alam, A. (2022). Investigating Sustainable Education and Positive Psychology Interventions in Schools Towards Achievement of Sustainable Happiness and Wellbeing for 21st Century Pedagogy and Curriculum. *ECS Transactions, 107*(1), 19481.
4. Zhang, W., & Bray, M. (2020). Comparative research on shadow education: Achievements, challenges, and the agenda ahead. *European Journal of Education, 55*(3), 322–341.
5. Alam, A. (2022). Mapping a Sustainable Future Through Conceptualization of Transformative Learning Framework, Education for Sustainable Development, Critical Reflection, and Responsible Citizenship: An Exploration of Pedagogies for Twenty-First Century Learning. *ECS Transactions, 107*(1), 9827.

*Integrating Advancements in Education, and Society for Achieving
Sustainability – Dimitrios A. Karras et al. (eds)*
© 2024 Taylor & Francis Group, London, ISBN 978-1-032-70841-6

Breaking the Taboos by Navigating the Maze of Sex-Ed and Sextracurricular Revolution: Re-Imagining 'Sexuality Education'

Ashraf Alam*, Atasi Mohanty

Rekhi Centre of Excellence for the Science of Happiness,
Indian Institute of Technology Kharagpur, India

Abstract: Despite its importance, 'sexuality education' remains controversial and often taboo in many communities and schools. The lack of comprehensive and inclusive 'sexuality education' can have serious consequences, including increased rates of unintended pregnancy, sexually transmitted infections (STIs), and sexual violence. This scientific paper explores the importance of teaching 'sexuality education' to school students in India. The paper emphasizes teaching 'sexuality education' that shall promote healthy sexual behaviours and relationships. The researcher has designed and developed a comprehensive curriculum on "Sexuality Education" for Indian school students, providing appropriate teaching-learning materials (TLMs) and pedagogical tools. The researcher has also suggested several games and activities, thus making the course more engaging and interesting. Furthermore, several ways to use storytelling method to effectively engage school students in classroom discourse has been proposed. The developed curriculum is evidence-based, culturally sensitive, and responsive to the needs of students.

Keywords: Indian Schools, Sexuality Education, Masturbation, Consent in Relationships, Abortion, Sexual Orientation, Gender Identity, Curriculum, LGBTQ+

1. Introduction

Sexuality education is a critical component of comprehensive health education and plays a vital role in promoting positive sexual health outcomes and preventing negative behaviors [1, 2]. Sexuality education is typically delivered at various stages of development, with age-appropriate content and methods tailored to the developmental stage of the learner [3]. Effective sexuality education programs involve collaboration with families, communities, and healthcare professionals to ensure that students receive comprehensive and accurate information [4, 5]. "Sex education" refers to education about anatomy, physiology, and sexual intercourse, while "sexuality education" encompasses a

*Corresponding author: ashraf_alam@kgpian.iitkgp.ac.in

DOI: 10.4324/9781032708461-42

wider range of topics, including relationships, communication, consent, reproductive health, sexual orientation, and gender identity [6, 7]. In India, there is no school or university that teaches "sexuality education" as a standalone subject. However, some universities do offer modules on related subjects, such as reproductive health, gender studies, or human sexuality, as part of their public health, medical, or social sciences programs. Also, a few universities offer courses on family planning, sexually transmitted infections, and reproductive health services as part of their public health and medical programs. Other universities offer courses on gender and sexuality, including topics such as sexual orientation, gender identity, and sexual health, as part of their gender studies, sociology, and psychology programs. In India, while there is recognition of the importance of sexuality education in promoting sexual health and preventing negative behaviors, the implementation of comprehensive and inclusive sexuality education programs remains limited and inconsistent. There is no national curriculum for sexuality education, and the implementation of sexuality education programs varies widely across different states and school systems. Some schools unintentionally provide limited or stigmatizing information in this field. In certain cases, cultural and religious beliefs pose barriers to the implementation of comprehensive sexuality education programs. Nevertheless, there have been efforts in recent years to improve the quality and availability of sexuality education in India, including advocacy for comprehensive sexuality education programs, the development of evidence-based resources, and the strengthening of partnerships between schools, communities, and civil society organizations.

2. Abortion

The contemporary scientific discourse around abortion recognizes it as a safe and effective medical procedure when performed by trained medical professionals under appropriate conditions. Research has shown that abortion is a common practice, with approximately 1 in 4 women in the U.S. having an abortion at some point in their lifetime. On the political side, abortion remains one of the most contentious and divisive issues, with ongoing debates about the morality, legality, and accessibility of the procedure. Supporters of abortion rights argue that it is a matter of personal autonomy, privacy, and reproductive freedom, and that restrictions on abortion impede access to necessary medical care. Opponents of abortion argue that it is morally wrong and that it should be restricted or abolished, often on the basis of religious or cultural beliefs about the sanctity of life. The issue of abortion has been and continues to be litigated in courts, with different countries having varying laws regarding the legality and availability of the procedure. In some countries, abortion is widely available and legal, while in others it is restricted or illegal. In recent years, there have been efforts in some countries to further restrict or even criminalize abortion, leading to concerns about the impact on women's health and well-being. The scientific and political discourses around abortion are ongoing and highly polarized, with advocates on both sides presenting strongly held and deeply held beliefs. The issue continues to be a topic of political and social debate, with ongoing efforts to influence laws and policies around access to abortion.

3. Sexual Orientation and Gender Identity

The contemporary scientific discourse around sexual orientation and gender identity recognizes that sexual orientation and gender identity are complex and multi-faceted, shaped by a combination of biological, psychological, and social factors. Research has shown that sexual orientation and gender identity are normal aspects of human diversity and cannot be changed through therapy or other means. On the political side, there has been growing support for LGBTQ+ rights, including legal recognition of same-sex relationships and marriage, anti-discrimination protections in housing, employment, and public accommodations, and increased visibility and representation in media and public life. However, there is also ongoing opposition to these developments, particularly in areas where cultural norms remain hostile to non-heterosexual and non-cisgender individuals. There are ongoing debates about the appropriate balance between individual rights and religious freedom, with some religious groups opposing legal recognition of LGBTQ+ rights on moral or theological grounds. In recent years, there have also been discussions and controversies around the medical treatment and legal recognition of transgender individuals, including access to hormone therapy and gender-affirming surgeries, and the right to change one's legal gender marker. The issue of transgender rights continues to be a political flashpoint in many countries, with opponents claiming that it undermines traditional understandings of gender and sexuality, while proponents argue that it is a matter of basic human dignity and equality.

4. Masturbation

The political and religious discourses around masturbation have been varied and complex. Some key points include:

1. Historical views: Throughout history, masturbation has been stigmatized and condemned by various political and religious authorities, who viewed it as a sin or a moral transgression.

2. Contemporary views: In many contemporary societies, attitudes towards masturbation have become more accepting and tolerant, although there are still many political and religious groups that continue to view it as sinful or immoral.

3. Religious views: Different religious traditions have varying attitudes towards masturbation, with some viewing it as a serious sin and others taking a more permissive stance.

4. Political views: Political attitudes towards masturbation can vary widely depending on the country and region, with some governments actively promoting it as a healthy sexual behavior, while others may criminalize it or restrict access to information about it.

5. Feminism: The feminist movement has challenged traditional attitudes towards masturbation and has encouraged women to claim control over their own bodies and sexuality.

5. Consent in Relationships

Consent is viewed as a necessary component of healthy and respectful sexual relationships and a key aspect of sexual autonomy and agency. Research has shown that consent education can lead to a better understanding of communication, boundaries, and negotiation in sexual relationships, as well as a reduction in instances of sexual assault and coercion. The scientific discourse emphasizes the importance of respecting the autonomy and agency of all partners. Consent education often includes a discussion of factors that can impact the ability to give informed consent, such as power imbalances, alcohol or drug use, and past experiences of trauma or abuse. It also highlights the importance of being aware of nonverbal cues and body language, and recognizing that consent can be withdrawn at any time.

6. Course Curriculum and Course Objectives

In this research investigation, the researcher has developed a curriculum for a course on "Sexuality Education" to be taught to students in Indian schools. The detailed curriculum and pedagogical tools for its effective transaction are detailed in ensuing sub-sections. The course objectives of teaching sexuality education to students include:

1. To increase students' knowledge about sexual anatomy, physiology, and related health issues.
2. To foster an understanding of gender identity, sexual orientation, and diversity.
3. To educate students about healthy relationships, communication skills, and consent.
4. To provide students with accurate information about contraception, sexually transmitted infections (STIs), and HIV/AIDS prevention.
5. To empower students to make informed decisions about their sexual health and well-being.

6. To encourage students to respect the privacy and rights of others in matters related to sexuality.

7. Outline of the Curriculum

Unit 1: Introduction to Sexuality Education

    Chapter 1: Overview of Sexuality Education

    Chapter 2: The Importance of Sexuality Education

Unit 2: Anatomy and Physiology of the Reproductive System

    Chapter 1: Understanding the Reproductive System

    Chapter 2: The Menstrual Cycle

Unit 3: Puberty and Adolescent Development

    Chapter 1: Physical Changes During Puberty

    Chapter 2: Emotional and Psychological Changes During Puberty

    Chapter 3: Impact of Puberty on Sexual Health and Well-being

Unit 4: Sexual Health and Hygiene

    Chapter 1: Understanding Sexually Transmitted Infections (STIs)

    Chapter 2: Understanding HIV/AIDS

    Chapter 3: Sexual Health and Hygiene

    Chapter 4: Contraception and Safe Sex

Unit 5: Pregnancy and Childbirth

    Chapter 1: Pregnancy Prevention and Family Planning

    Chapter 2: Options for Pregnancy and Childbirth

    Chapter 3: Benefits and Risks of Different Types of Childbirth

Unit 6: Sexual Orientation and Gender Identity

    Chapter 1: Understanding Sexual Orientation and Gender Identity

Chapter 2: Experiences and Challenges of LGBTQ+ Individuals

Chapter 3: Health and Well-being of LGBTQ+ Individuals

Unit 7: Relationships and Communication Skills

Chapter 1: Understanding Healthy Relationships

Chapter 2: The Importance of Consent in Sexual Relationships

Chapter 3: Impact of Media and Technology on Relationships

Chapter 4: Communication Skills for Healthy Relationships

Unit 8: Abortion

Chapter 1: Medical and Ethical Aspects of Abortion

Chapter 2: Political Aspects of Abortion

Chapter 3: Access to Abortion Services

Chapter 4: Impact of Restrictive Abortion Laws on Women's Health

Unit 9: Sexual Pleasures

Chapter 1: Understanding Sexual Pleasure

Chapter 2: Safe and Responsible Sexual Behaviors

Chapter 3: Cultural and Social Norms Impacting Sexual Pleasure

Unit 10: Rape and Sexual Violence

Chapter 1: Definition of Sexual Violence

Chapter 2: Prevalence and Impact of Sexual Violence

Chapter 3: Supporting Survivors of Sexual Violence

Chapter 4: The Importance of Consent in Preventing Sexual Violence

Unit 11: Media Literacy and Sexuality Education

Chapter 1: The Role of Media in Sexuality Education

Chapter 2: Evaluating the Accuracy and Impact of Media on Sexuality and Relationships

Chapter 3: Media Literacy Skills for Sexuality Education

The curriculum shall be transacted such that it remains flexible and inclusive, and subject matter is taught in a manner that is developmentally appropriate, culturally sensitive, and respectful.

The following teaching-learning materials (TLMs) and pedagogical tools shall be used:

1. *Audio-Visual Materials:* Films, documentaries, videos, and podcasts can provide a visual and auditory experience for students and help them understand different perspectives on sexuality education.

2. *Online Platforms:* Online platforms such as discussion forums, chat rooms, and e-learning platforms can be used to encourage students to share their thoughts, opinions, and experiences with others.

3. *Case Studies:* Using case studies can increase their engagement in the course.

The teachers are encouraged to conduct open discussions and be aware of the impact that their teaching approach may have on students.

8. Storytelling for "Sexuality Education"

This method can:

1. *Engage students:* Storytelling can create an emotional connection with students.

2. *Foster empathy:* Hearing real-life stories can help students understand and empathize with the experiences of others.
3. *Provide a safe space:* Storytelling can create a supportive and non-judgmental environment where students can openly discuss their thoughts and experiences.
4. *Promote reflection:* Encouraging students to reflect on the stories they hear can help them internalize the issues and concepts.

9. Games and Activities

1. *Icebreakers:* Using icebreaker games can help students get to know each other and create a supportive and inclusive learning environment.
2. *Role-plays:* Role-plays can help students practice communication and decision-making skills in real-life situations.
3. *Group discussions and debates:* Group discussions and debates can be used to explore different perspectives and encourage critical thinking on sensitive topics such as abortion, consent, and gender identity.

The teacher should ensure that students have the opportunity to reflect on what they have learned.

REFERENCES

1. Alam, A. (2022). Positive Psychology Goes to School: Conceptualizing Students' Happiness in 21st Century Schools While 'Minding the Mind!' Are We There Yet? Evidence-Backed, School-Based Positive Psychology Interventions. *ECS Transactions, 107*(1), 11199.

2. Leung, H., Shek, D. T., Leung, E., & Shek, E. Y. (2019). Development of contextually-relevant sexuality education: Lessons from a comprehensive review of adolescent sexuality education across cultures. *International journal of environmental research and public health, 16*(4), 621.

3. Alam, A. (2022). Investigating Sustainable Education and Positive Psychology Interventions in Schools Towards Achievement of Sustainable Happiness and Wellbeing for 21st Century Pedagogy and Curriculum. *ECS Transactions, 107*(1), 19481.

4. O'Brien, H., Hendriks, J., & Burns, S. (2021). Teacher training organisations and their preparation of the pre-service teacher to deliver comprehensive sexuality education in the school setting: a systematic literature review. *Sex Education, 21*(3), 284–303.

5. Alam, A. (2022). Mapping a Sustainable Future Through Conceptualization of Transformative Learning Framework, Education for Sustainable Development, Critical Reflection, and Responsible Citizenship: An Exploration of Pedagogies for Twenty-First Century Learning. *ECS Transactions, 107*(1), 9827.

6. Miedema, E., Le Mat, M. L., & Hague, F. (2020). But is it comprehensive? Unpacking the 'comprehensive' in comprehensive sexuality education. *Health Education Journal, 79*(7), 747–762.

7. Alam, A. (2022). Impact of University's Human Resources Practices on Professors' Occupational Performance: Empirical Evidence from India's Higher Education Sector. In: Rajagopal, Behl, R. (eds) Inclusive Businesses in Developing Economies. Palgrave Studies in Democracy, Innovation, and Entrepreneurship for Growth. Palgrave Macmillan, Cham.

Integrating Advancements in Education, and Society for Achieving
Sustainability – Dimitrios A. Karras et al. (eds)
© 2024 Taylor & Francis Group, London, ISBN 978-1-032-70841-6

Ancestral Perception Reintroduced: Assessing Consumption Patterns and Perceptions of Superfood to Bridge the Knowledge Gap among the Youth of India

P. Manjunath[1]

Research Scholar - School of business and management,
Christ (Deemed to be University), Bangalore

L. N. Fukey[2]

Professor- Christ (Deemed to be University), Bangalore

Abstract: Foods rich in nutrients and essential to the human body to sustain and survive in good health are a growing concept amongst the youth. Superfood, what one believes to be essential to living healthy lives are being misinterpreted in the world as something exclusive and expensive. Little do we realize that if we follow exactly what our ancestors depended on, the entire concept of superfood would cease to exist. This paper aims at bridging the gap between misunderstood food patterns and how they are being portrayed as a superfood and reintroduce them back into their lives. India is a country rich in heritage and extremely healthy food patterns, but the question is why are people taking a major detour away from what we have and in search of something away from our habits? Bringing back conventional eating patterns and methods in which they can incorporate them into their daily routine. Our own superfood are excluded from our daily diets as they are told to be rich in fats and are termed unhealthy. A prevalent archetype would be the exclusion of rice from one's diet because it is rich in starch and sugars, but rice is the grain that helps sustain. Because of this knowledge gap between facts and fads, the promotion of "superfood" from an international perspective with modern labels, are taking place in people's mind as the only healthy option available. Breaking this pattern and educating Indians on what they are already familiar with and fail to acknowledge is reintroduced to shift the limelight from just a tag to a way of life.

Keywords: Superfood, Perception, Conventional consumption pattern, Health

1. Introduction

Superfoods as a concept have been floating around for quite some time now and possibly created an image in consumers' heads. But how often or how many of these images are true? There are perceptions where people think to get the maximum out of superfoods one has to spend a lot of money, another image can be a person

Corresponding author: [1]priyanka.m@bhm.christuniversity.in, [2]leena.n.fukey@christuniversity.in

DOI: 10.4324/9781032708461-43

consuming food in quantities which might have reverse effects on them and while some think that it is something very healthy and will transform their lives. We co-live in this vast environment and seasons are transitions that our bodies adapt to with the help of seasonal fruits and vegetables. Summer fruits like mango and watermelon are perfect examples to describe this. These fruits provide necessary vitamins and the hydration our bodies require during summer and when there is extensive water loss in the form of perspiration. These same fruits if genetically modified and made available during the winter season will make us fall sick as our bodies do not require that particular nutrient then. (Mansi Natutiyal, 2022)

The Indian diet for many generations has been very well balanced on its own. Because of the Western world's influence, many different and unnecessary alterations have been introduced to an already well-balanced food pattern. For example, in south India, people used to eat the previous day's cooked rice soaked overnight in water first thing in the morning, also known as kanji, which was rich in probiotics and helped maintain gut health. A superfood doesn't have to necessarily be eaten to incorporate necessary nutrients in our bodies, but also to prevent problems arising from common foods. As a snack, many Indians eat boiled peanuts, and since they are so easy to eat we tend to lose track of how much we eat. Back in the day, our grandparents and parents used to give us a small piece of jaggery to balance out the dominant "pitha" from the peanuts. It also helps in increasing hemoglobin levels. The popular chikki also revolves around the same concept but has now been replicated with sugar as it is inexpensive. (Diwekar, Rujuta, 2016)

An overview and background of Indian Superfood and the various aspects associated with it

Incorporating superfood in our daily diets

The present world thinks that in order to eat well, we must spend a lot of money on things that even our forefathers couldn't identify. We believe that eating healthfully must be expensive and bland, and that achieving one's goals requires hardship. Even if people work hard and seek to accomplish their objectives, once they do, they lose sight of their routines and revert to their initial state. (Diwekar, Rujuta, 2016) Indians have a long tradition of regularly eating foods that are good for the body in their diets. The use of the term "superfood" in the western economy has led people all over the world to believe that only western foods can provide our bodies with all the essential nutrients. People in India don't fully grasp that we do not need to alter our meals in order to obtain something that is already in front of us because our foods are already abundant in all the phytochemicals. (Mansi Natutiyal, 2022) Millets have long been the main component of the diet of Indians, but due to the rise in popularity of other cereals, we have diverged from our traditional diet and now eat a variety of western bowls of cereal, including oats, quinoa, and buckwheat. While their ancestors used to eat fermented foods like dosa and porridge in the morning to help improve gut health, companies like Quakers and Kellogs have been heavily marketing their products made up of foreign cereals, which has caused many Indians to change their diets to one that is more influenced by the west. Amaranth and millet, two Indian superfoods, now fail to stand out in the marketplace. (Erler, Mirka, et al, 2020)

Numerous US research examining Americans' health came to the conclusion that they are both overfed and undernourished. The health advantages are not taken into account, but taste, cost, and convenience are important considerations. Despite being readily available, ultra-processed foods fall short of meeting our systems' needs for vitamins and fiber. People who eat less food overall and less processed food are found to have better overall health. Foods with a high content of processed vegetable oils are extremely inflammatory. Superfood help to boost immunity and intestinal health. A superfood is typically unprocessed and high in nutrients. (Gerard Mullin, et al., 2019)

Consumption practices

People frequently confuse hunger for dehydration, according to a study that has been conducted on the subject. We frequently have the misconception that we always feel the want to eat when we aren't doing anything. These erroneous alarms are to blame for the rise in childhood obesity. When we fail to hydrate ourselves, we experience these "hunger pains," which are frequently caused by dehydration. On average, humans should drink 3 to 4 litres of water every day. Another study that was done The Holy Basil is termed as a superfood and has been used in small quantities in Indian households for traditional practices and customs. Even today, in temples, as theertham, water boiled with holy basil is given to all devotees. (Rodrigues, Rios, Daniela, et al, 2021)

Traditional superfood

One can get the most from a particular ingredient if they don't search for it and just consume what's available. This way people get into the habit of eating seasonally and are able to get the maximum out of that particular

food, the hard-to-digest fact is that people miss out on these superfoods because of common food misconceptions. For example, people avoid eating ghee and cashews as according to them they have cholesterol and some are not even aware of certain foods like Aaliv and Ambadi. (Diwekar, Rujuta, 2016)

Trends associated with superfood

With social media engagement rising globally, influencers who highlight certain aspects of the typical Indian kitchen have influenced so many of their followers. One such item that was brought to light was the Indian spice box. The Indian spice box consists of a mix of spices commonly used in the Indian kitchen such as mustard, turmeric, chili powder, cumin, coriander powder, etc. this blend and use of spices has led to the discovery of anticancer properties from this one item of the Indian kitchen alone. This age-old item has gone way ahead of what modern medicine can do. (Chaitanya, M, V, N, L, et al, 2020)

2. Methodology

The current chapter conducts a systematic review of research articles relating to four themes: (1) ways to incorporate superfood in our diet (2) consumption practices (3) traditional superfood and (4) trends associated with superfood. The referred articles were considered qualified for the present chapter based on their resonance; date back to the last twelve years; and published in cited journals such as Emerald, Elsevier, Springer and Taylor and Francis in Scopus Indexed. The scholarly search engines used to search for related articles are Google Scholar, Scopus, and CORE. Keywords such as Superfood, Perception, Conventional consumption pattern, Health were included in the filtration process.

50 articles were considered eligible after referring to 72 articles, while 22 articles were disqualified on the basis that they were ahead of print; some were not compatible with the intent of the current chapter, and others were not published in Scopus Indexed Journals. The articles were chronologically organized under each theme with Excel's assistance, which allowed organizing the data and re-tracing the articles.

Table 43.1 Table showing number of articles reviewed under each theme

| Year | Themes | | | | Total |
	Ways to incorporate Indian superfoods into our daily diet	Consumption practices	Traditional superfoods	Direction of modern superfoods	
2009-2010	1			1	2
2013-2014			1	1	2
2016-2017	3		2		5
2018-2019	2	3	3	6	14
2020-2021	5	10	4	5	24
2022	1	1	2		4
Total	12	14	12	13	**51**

Source: Authors

3. Conclusion

This paper aims at understanding the gap between traditional Indian superfood and the rising trends of superfood not native to the region of India. With the help of articles and facts available on how "superfood has an altering effect on the human body, this paper projects the idea of sticking to ethnic dietary requirements instead of expanding into a glamour-filled market of nonnative superfoods which is expensive and does not provide the body with what's required.

REFERENCES

1. Erler, M., Keck, M., & Dittrich, C. (2020). The changing meaning of millets: Organic shops and distinctive consumption practices in Bengaluru, India. Journal of Consumer Culture, 22(1), 124–142. https://doi.org/10.1177/1469540520902508 and distinctive consumption practices in Bengaluru, India. Journal of Consumer Culture, 146954052090250. doi:10.1177/1469540520902508

2. Wani, S. A., & Kumar, P. (2018). Fenugreek: A review on its nutraceutical properties and utilization in various food products. Journal of the Saudi Society of Agricultural Sciences, 17(2), 97–106. doi:10.1016/j.jssas.2016.01.007

3. Pal, A. deb. (2022). Phyllanthus emblica: The superfood with anti-ulcer potential. Journal. v3, pp. 84.

4. Dickinson, B. (2022). How to grow & eat your own superfoods. Google Books.

5. Ríos-Rodríguez, D., Sahi, V. P., & Nick, P. (2021). Authentication of holy basil using

markers relating to a toxicology-relevant compound - european food research and technology. https://doi.org/10.1007/s00217-021-03812-z

5. Forêt, R. d. l. (2017). Alchemy of Herbs: Transform Everyday Ingredients Into Foods and Remedies That Heal. United States: Hay House. Edition: 01, ISBN:9781401950064, 140195006X

6. Nautiyal, M. (2022). Superfood: Value and need., Volume: 04, Issue: 06

7. M. V. N. L., S., & Chaitanya, C. (2020). Anticancer drug discovery from the Indian spice box: 12: Drug Develo. Taylor & Francis. E-book ISBN-9780429330490

8. Diwekar, R. (n.d.). Indian superfoods. ISBN: 9788193237236, 8193237234

9. Arumugam, D. T., Sona, C. L., & Maheshwari, D. M. U. (2021). Fruits and vegetables as superfoods: Scope and demand. The Pharma Innovation Journal. TPI 2021; 10(3): 119–129.

*Integrating Advancements in Education, and Society for Achieving
Sustainability – Dimitrios A. Karras et al. (eds)*
© 2024 Taylor & Francis Group, London, ISBN 978-1-032-70841-6

44

Government's and Community's Role for Sustainable Coastal Ecosystem (Case Study: Ecomarine Muara Angke Mangrove Vegetation)

Adelina*
College Graduate of Student Environmental Management,
State University of Jakarta, Indonesia

Achmad Husen
Lecturer of Environmental Management Philosophy,
State University of Jakarta, Indonesia

Diana Vivanti Sigit
Lecturer of Environmental Management Concepts at Postgraduate
Environmental Management, Jakarta State University, Indonesia

Abstract: The mangrove woodland "Ecomarine Muara Angke" is found in North Jakarta, Indonesia. The causes of damage to the mangrove vegetation at Ecomarine Muara Angke are settlements, piles of garbage and wastes entering the site. The rationale behind this consideration is to analyze the role of government and society in monitoring mangrove vegetation in the coastal economic environment. A type of subjective research can be a case consultation plan. The facts proved that the Muara Angke Mangrove Ecomarine could be a mangrove forest managed by the Muara Angke Mangrove Community (KOMMA) Muara Angke Ecomarine planted mangroves: mangroves, bees, turis and nipa palms. The government in anticipated to move unlawful inhabitants who live around Muara Angke Ecomarine ought to be beneath private administration but beneath government supervision and made into a nature preservation range. Based on the dialog investigation, it can be concluded that around Ecomarine Muara Angke there are numerous illicit settlement.

Keywords: Eco-marine, Ecosystem, Mangrove, Coastal, Sustainable

1. Introduction

This chapter will discuss about the research background and the research objective.

Management according to Prajudi Atmosuryo (1982:282) is an activity of utilizing and managing resources that will be used in activities to achieve a

*Corresponding author: adelinapgbn@gmail.com

DOI: 10.4324/9781032708461-44

certain goal. In the regulation of the coordinating minister foe economiv affairs of the Republic of Indonesia number 4 of 2017 concerning Ppolicies, strategies, programs and performance indicators for management of the national mangrove ecosystem, the objectives and roles of the parties are explained as follows: Ecological important values, institutional important values, legislative important values.

According to the regulation of the governor of the province of the special capital region of Jakarta no. 149 of 2018 concerning the organization and work procedure of the forest service.

The forestry service has the task of managing forest areas, distribution of forest product, urban forests, parks, green belt and cemeteries.

2. The Research Background

Indonesia is known as an archipelagic country with a strategic geographical location. It is a coastal area, where land and sea meet. The territory of Indonesia stretches from Sabang to Merauke, and has 17,499 islands with a total area of Indonesia about 7.81 million km^2. The area of Indonesian mangroves is 3,364,076 Ha. The highest distribution of mangroves is in Papua (Ministry of Marine Affairs and Fisheries; 2021). There was a downward trend in mangrove area in Indonesia during the period from 2000–2012 with an average rate of 0.26%-0.66% per year (Hamilton & Casey, 2016). Meanwhile, the area of mangrove in DKI Jakarta Province is 259.93 ha, of which 220.84 are in good condition and 39.09 Ha are in moderate damage. (Cecep; 2009). The decline in the quality and quantity of mangrove vegetation

is an important issue in urban environmental problems. This is in line with population growth which is one of the causes of changes in the urban and coastal environment with the increase in the expansion of population settlements around the Muara Angke area, requiring the attention of the local government (Pluit Village).

The mangrove ecosystem on the coast of Muara Angke is partially damaged (IPB Study; 2011). One of the impacts that has caused the decline in the quality and quantity of the mangrove forest vegetation "Ecomarine Muara Angke" is the amount of garbage that has accumulated up to 1.5 meters thick and filled with mixed mud. Mangrove vegetation in the Ecomarine Muara Angke Mangrove forest is a problem that requires serious handling by the government and the community. Mangrove vegetation in the Ecomarine Muara Angke Mangrove forest is a problem that requires serious handling by the government and the community. The "Ecomarine Muara Angke" mangrove forest area is located at the end of the fishing village. Mangrove forests have ecological functions such as coastal protection, prevention of sea water intrusion, habitat (place to live), feeding ground, nursery and nursery. soil), a spawning ground for various aquatic biota, and a buffer against the tsunami disaster (Dadang; 2012). Mangrove forest management "Ecomarine Muara Angke" has limited funds, human resources, infrastructure, and the availability of information so that it becomes the concern of Pluit village (Said KOMMA; 2021)

3. The Research Objective

Ecomarine mangrove Muara Angke Penjaringan village, Nort Jakarta, Indonesia.

4. Literature Study

This chapter will discuss about the definition of research variables.

Definition of Role

A role is dynamic aspect of someone's position. If a person fulfills his rights and duties according to this duties, then he has fulfilled a certain role (Soeharto, 2002; Soekamto, 1984: 237).

Definition of Government

Government is a form of organization that cooperates and carries out the task of managing the system of government and making policies to achieve the goals of the state. The definition of government in a broad sense is a form of organization that functions to carry out the tasks of a system of government to achieve common goals. Meanwhile, in a narrow sense, the government is an association body that has its own policies in managing an organization, carrying out management, and regulating the running of a government system in an area.

Definition of Public

Horton et al. (1991) define a society as a group of relatively independent people who have lived together for a sufficiently long time, live in a certain area, share the same culture and carry out most of their activities as a group. Meanwhile, Ralp Linton (1956) defines a society as a group of people who have lived and worked together long enough to organize themselves and think of themselves as a social unit with well-defined boundaries.

Definition of Vegetation

Vegetation is several types of plant, usually consisting of several types and live together in one place.

Definition of Mangrove

As per Odum, mangrove is derived from the term 'mangal' which signifies a grouping of flora. Additionally, Supriyanto has demonstrated that mangrove has dual connotations-primarily as a congregation of vegetation or woodlands that endure high salinity levels (due to the tidal movement of seawater), and secondarily as distinct species.

Ecosystem

A biogeocoenosis, or ecosystem, refers to a harmonious community of plants, animals, organisms, and inanimate entities, along with the processes that interconnect them to establish equilibrium, steadiness, and efficiency. The biosphere or ecosphere encompasses all beings on the planet that engage with the physical milieu (Djohar, 2017). Furthermore, ecosystems comprise trophic levels, biotic variety, and substance cycles. Through these interconnections, an ecosystem can maintain its equilibrium. The hallmark of an ecosystem is its arrangements to ensure this balance. Failure to achieve this balance can instigate shifts in the ecosystem's dynamics, leading to a new balance. Ecosystems are part of a larger group of biomes that can typically be identified by the flora that characterizes them.

Definition of Coastal

The Littoral Region serves as a buffer zone between terrestrial and marine environments that undergo transformations due to fluctuations in both domains. Littoral waters encompass the seas that lie adjacent to the coast, stretching up to a distance of 12 nautical miles from the shoreline, as well as the waterbodies that interconnect the coast and islands, estuaries, bays, shallows, saline marshes, and lagoons (as stated in the UU No. 27, 2007).

Sustainable Development

Numerous nations across the globe are engaged in progress, including Indonesia, which is utilizing technological advancements to develop its regions. It is imperative to prioritize environmental protection during development to prevent damage and pollution issues.

The concept of sustainability entails maintaining a specific level or stage as intended. Sustainable business practices involve meeting present needs without jeopardizing future generations' ability to fulfill their own requirements.

All nations must prioritize sustainability to ensure the survival of future generations. While it is possible to achieve sustainability now, it must be maintained in the long term.

5. Research Method

The research methods using qualitative.

6. Results and Discussion

Ecomarine Muara Angke serves as a natural habitat for various bird and animal species. To preserve the mangrove forest area, the government and the community need to give special attention, such as planting mangrove seedlings in an optimal manner. In 2010, young organizations, nature enthusiasts, volunteers, and the community joined hands to manage the mangrove vegetation with utmost care and responsibility. The atmosphere around the mangroves used to be cool and was home to a rare type of fish bird in 1978. Active community participation can foster sustainable coastal ecosystems. Empowering the community is crucial in managing mangroves. The government plays a vital role in educating people about the significance of preserving mangrove forests. The plan is to plant mangrove trees throughout the area to serve as a habitat (Marine Service, 2022). Ecomarine Muara Angke began planting 500 mangrove seedlings in 2010, but some failed to survive due to plastic waste suffocating the roots. Mangrove plantation along the coast depends on sediment. Mangroves can grow on mud. Mangrove plantation restores and protects the ecosystem and serves as a research site for students and other researchers. DKI Jakarta customer service (UP3) continues to assist the Muara Angke Ecomarine area in cleaning up the location from garbage piles, which pose a threat to the mangrove habitat and other ecosystems. Waste management in Muara Angke is a top priority for the government. Overpopulation, household waste, and factory waste have caused damage to mangroves (Santoso, 2011). The government has urged the public to maintain cleanliness and preserve mangroves to prevent tidal flooding and seawater abrasion. The Muara Angke mangrove forest has dwindled due to abrasion and rising seawater levels, leading to the disappearance of some mangrove forests. Ecomarine was once a sea of garbage, but now it has a thriving population of mangrove trees, thanks to the planting efforts that began in 2009. The management of mangrove vegetation in the Muara Angke Ecomarine has been under the supervision of KOMMA, a government agency, and the Java-Bali Power Plant (PJB) since the beginning. Comprehensive planning that involves all stakeholders, including the government, the community, and the private sector, is necessary to preserve the Muara Angke Ecomarine area as a mangrove and marine biota

7. Conclusion

The result of this perception advises almost the community has played a part within the administration of the Muara Angke ecomarine but ought to be advance made strides but there are numerous illicit settlements around it. This has an affect on the mangrove biological system since of the transfer of family squander, in expansion to mechanical squander, and dispatches of trash that can aggravate the living space of the mangrove ecosystem.

The government still needs a share in the management of the Muara Angke ecomarine, such as the return of money for the development and support of mangrove forest, efforts to go to Muara by road.

The management of mangrove vegetation in the Muara Angke Ecomarine is still ineffective, seen from the lack of synergy between the government or related ministries, local goverments, communities, the business world, academia, and the international community. Mangrove vegetation management is constrained as there is still a lot of land that needs to be planted with mangrove trees.

8. Acknowledgmet

The authors wish to express their gratitude to the Supervisors of University of Jakarta Indonesia for their support.

REFERENCES

1. Achmad Sofian, Cecep Kusmana, Akhmad Fauzi dan Omo Rusdiana. Evaluating the conditions of Kapuk mangrove ecosystem Jakarta Bay and its consequences ecosystem services. (2019)
2. Adriana Renwarin, Octavianus A.H. Rogi, Rieneke L.E. Sela. Studi of the Identification of residential waste management systems in the coastal areas of the city of Manado. https://ejournal.unsrat.ac.id/index.php/spasial/article/view/9675. Vol 2. No. 3. (2015).
3. Amida Urfah Khoirun Nisa, Bambang Sulardiono, Djoko Suprapto. Ecotorism development strategy for mangrove conservation area of Kertomulyo beach, Trangkil, Pati. https://ejournal3.undip.ac.id/index.php/maquares/article/view/24252. Vol 8 No 3 (2019).
4. Aswin, Ario Damar, Gatot Yulianto. Vegetation conditions and the changing of mangrove ecosystem cover in Tanakeke island Takalar regency south Sulawesi province. https://journal.ipb.ac.id/index.php/jurnalikt/article/view/33636. Vol 13 No 2 (2021)
5. Bima Agung Saputra, Rudhi Pribadi, Chrisna Adi Suryono. Mangrove biology in terms of the important value of diversity, dominance, diversity in the coastal district of Bonang, Demak https://ejournal3.undip.ac.id/index.php/jmr/article/view/24887. Vol 8 No 4 (2019).
6. California Ocean Protection Council Resolution. (2008). *An Implementation Strategy for the California Ocean Protection Council Resolution to Reduce and Prevent Ocean Litter.*
7. Cecep Kusmana Integrated management of mangrove systems: silviculture department, Faculty of Forestry IPB Bogor.
8. Creswell, J. W. (2016). *Reasearch Design Qualitative, Quantitative and Mixed Methods Approach. Student Libraries.*
9. Dadang Setiawan. Economic Valuation of Mangrove Forest Areas of Muara Angke Jakarta Comparison of Research Results 2002 dan 2012. University of Indonesia. 2016
10. Desi Melinda Sari, Suryanti, Bambang Surlandiono. Management of mangrove ecosystem as ecotourism area in Maroon mangrove edu park (MMEP) Semarang, Central Java. https://ejournal3.undip.ac.id/index.php/maquares/article/view/24224. Vol 8 No 1 (2019).